ROUTLEDGE LIBRARY EDITIONS: HISTORICAL SECURITY

Volume 9

STRATEGY WITHOUT SLIDE-RULE

STRATEGY WITHOUT SLIDE-RULE
British Air Strategy 1914–1939

BARRY D. POWERS

LONDON AND NEW YORK

First published in 1976 by Croom Helm Ltd

This edition first published in 2021
by Routledge
2 Park Square, Milton Park, Abingdon, Oxon OX14 4RN

and by Routledge
52 Vanderbilt Avenue, New York, NY 10017

Routledge is an imprint of the Taylor & Francis Group, an informa business

© 1976 by Barry D. Powers

All rights reserved. No part of this book may be reprinted or reproduced or utilised in any form or by any electronic, mechanical, or other means, now known or hereafter invented, including photocopying and recording, or in any information storage or retrieval system, without permission in writing from the publishers.

Trademark notice: Product or corporate names may be trademarks or registered trademarks, and are used only for identification and explanation without intent to infringe.

British Library Cataloguing in Publication Data
A catalogue record for this book is available from the British Library

ISBN: 978-0-367-61963-3 (Set)
ISBN: 978-1-00-314390-1 (Set) (ebk)
ISBN: 978-0-367-64446-8 (Volume 9) (hbk)
ISBN: 978-1-00-312456-6 (Volume 9) (ebk)

ISBN: 978-0-367-64450-5 (pbk)

Publisher's Note
The publisher has gone to great lengths to ensure the quality of this reprint but points out that some imperfections in the original copies may be apparent.

Disclaimer
The publisher has made every effort to trace copyright holders and would welcome correspondence from those they have been unable to trace.

Strategy Without Slide-Rule

BRITISH AIR STRATEGY 1914-1939

BARRY D. POWERS

CROOM HELM LONDON

© 1976 by Barry D. Powers

Croom Helm Ltd
2-10 St John's Road, London SW11

ISBN 0 85664-219-3

Printed in Great Britain
REDWOOD BURN LIMITED
Trowbridge & Esher

CONTENTS

Preface

1 The Zeppelin Era: The Challenge	11
2 The Zeppelin Era: The Successful Response	28
3 Defence in the Gotha Era	52
4 The Birth and Early Infancy of the Royal Air Force	75
5 The Formative Years of Non-Military British Concepts of Aerial Warfare (to 1931)	107
6 The Fledgling Years of the Royal Air Force in Doctrine and Development	158
Epilogue	208
Notes	210
Bibliography	267
Index	289

PREFACE

> Then the LORD rained upon Sodom and upon Gomorrah,
> brimstone and fire from the LORD out of Heaven ...
> Genesis 19:24

Man has dreaded an attack from the air for millennia. He has also long been in fear of new weapons. Thus the topic of this study fits into a centuries old pattern. It is, indeed, something of an historical convenience to begin this account with events connected with World War I. There were events prior to that war which will have to be noted later.

Nevertheless, there is much to justify this particular historical convenience. The air raids suffered by England in the Great War and their immediate results had the greatest influence on the formation and character of British military air doctrines and practices until (and, in part, during) World War II. Moreover, these raids and their immediate aftermaths offer early evidence of thinking and behaviour which characterised the general British reaction to air power thereafter, both in official and popular circles. These seminal events, then, furnish many insights for an understanding of this history during the interwar years.

Among the enlightening precedents which World War I furnishes are two aspects of British aerial history most often neglected by the numerous monographs. Firstly, studies of military air policy often analyse this problem in far too restricted a framework. There is a tendency to see the development of aerial doctrine as a purely military concern. Less often, the political implications in its development are noted. Very rarely is it seen as a complicated interaction of the many factors involved — popular conceptions, press campaigns, political thinking *and* military concerns. Very often little distinction can be made between ideas held by all elements. Certainly the military thinking is never done in a vacuum, uninfluenced by outside forces. This interaction will be given a continuing emphasis in this study.

Secondly, there is always the problem of an ambiguity between defensive and offensive concepts in aerial doctrines. Far too often these matters are discussed as though they are separate entities which can be independently analysed. Again, the experiences of World War I will illustrate this problem and show the extremely close interconnections between defensive concerns and offensive planning.

For many reasons, then, this account will examine closely the World War I roots of British aerial history. Following this, the profoundly important early interwar years will be discussed. The trends in popular, political and military thinking during this era will be analysed. The key changes brought on by technological discoveries will be noted.

Obviously, the history of Britain's interwar aerial defence preparations has an unavoidable teleological framework implicit in it. The Battle of Britain awaits and all have the hindsight to see the history in that context. But one key point of this interwar account is that this history is not a straightforward logical progress in defence preparations culminating in the success of 1940. Rather, the history will reveal many years of effort which were quite misguided if the Battle of Britain is used as the yardstick. England did not adopt the defensive doctrines which predominated by the start of World War II until quite late in the interwar era. This study will attempt to analyse closely the reasons for Britain's early policies which are so often ignored in studies featuring the aerial successes of 1940.

The reader will find herein many echoes of current air power rhetoric and practice. The nuclear age has hardly diminished interest and fear with respect to strategic bombing. Various writers have noted the rather obvious point that the history of the military use of air power reflects a continuity rather than a decisive break with respect to the advent of the atomic bomb.[1] Hyperbolic rhetoric about the impact of air warfare remains strikingly similar.[2] We still show confusion about the distinction between offence and defence in air power when we speak of first strike capability and then of deterrence. The point is that this study is concerned with an ongoing history.

I am indebted to many friends and colleagues in connection with this effort. Professors Robert A. Smith of Emory University and Brian Bond of King's College, London, have both read the manuscript and both made invaluable suggestions. Mr E.B. Haslam, Head of the Air Historical Branch of the Royal Air Force, and his staff were most accommodating and helpful. Mr Charles Porter provided an essential impulse. Joyce and Robert Meadows gave needed sustenance during two research trips to London. My 'Wichita Falls-London connection' (Mrs Dorothy Wilson and Mr Leonard Lickorish, C.B.E.) was very effective. Various libraries were memorable in their efficiency and usefulness; above all the Royal United Service Institution and the Public Record Office, but also excellent were the Air University and West Point facilities and staff. The Midwestern State University library staff also were always fully cooperative. Research funds came from a National Defense Fellowship and grants by Murray State and Midwestern State Universities. And above all, my gratitude goes to my treasured Joyce.

1 THE ZEPPELIN ERA: THE CHALLENGE

The air threat to Great Britain began with the opening of World War I. The 1914 version of the Schlieffen plan included the project of capturing Calais to use it as a base for aerial attacks on England. This was the enthusiastic programme of Major Wilhelm Siegert and it was sanctioned by the German Army High Command (*Oberste Heeresleitung*).[1] Thirty-six planes were assigned to this mission (six squadrons of six planes each) and Major Siegert was put in command of the operation.

Neither Calais nor other suitable Channel ports were taken by the Germans in their war-opening sweep and the bombing plans thus had to be set aside because the short range possessed by the 1914 airplanes required bases that close to England. Indeed, the primitive planes owned by all the combatant powers at the start of the Great War precluded realistic application of any grandiose schemes of air power. These planes in Siegert's command, for example, were able to carry only some three or four twenty-pound bombs each.[2] In sum, then, early limits in technological development of heavier-than-air craft and failure in troop operations forced the shelving of these first plans to bomb England. But the danger to England had been present and was fated to reappear. The German military authorities saw that the lack of flight range had been the basic problem and that year (1914) they commissioned designers to produce airplanes better suited for long-range bombing missions.[3]

The first phase of World War I thus included a scheme involving the concept now termed 'strategic' aerial bombardment. To quote the major historian of this abortive operation, 'the avowed, primary purpose of the Ostend unit [Major Siegert's command] was the bombing of industrial targets in England'.[4] The common view that the military use of air in the Great War was at first for reconnaissance and aerial photography[5] must therefore be amended. Aerial bombardment started with the war. On 2 August 1914, a German pilot bombed Luneville (France) and on 14 August 1914, a French pilot reciprocated with a bombing raid on a Zeppelin hangar near Metz.[6] And on that same day, 14 August, a German pilot reached as far as the Paris suburbs where he dropped two bombs of four pounds each.[7] It was only the limitations in the early model war planes in their weight-carrying ability and range which precluded a greater stress on this type of offensive utilisation of aircraft.[8]

If 1914 model airplanes were not very suitable agents for bombing missions, the Germans, nevertheless, had another aerial development at hand which could be turned to this account — their famous dirigible-type airships developed by Count Ferdinand von Zeppelin and known by his name.[9]

Von Zeppelin had first considered the possibilities of lighter-than-air craft at least by 1874.[10] By 1900 the first successful launching of a Zeppelin occurred and by 1907 thousands of people had taken passenger flights on this 'finest airship in the world'.[11] But military rather than civilian usage of the airship slowly came to dominate the Count's thinking. In 1912 he produced his first model solely intended for military purposes.[12] Count von Zeppelin managed to convince many military authorities that his airships were of significant value and by the outbreak of war, many German naval authorities especially were proponents of using the Zeppelins for long-range bombing of England. The leading voice among this group was Rear Admiral Paul Behncke, the Deputy Chief of the German Naval Staff.[13]

He was not unopposed in these ideas. Other naval leaders would use the dirigibles only for sea patrol duty. Von Tirpitz had a 'low opinion' of their utility and had resisted the production of Zeppelins in large numbers.[14] Some German political leaders had reservations. The Chancellor, Theobold von Bethmann-Hollweg, was opposed to the very idea of civilian bombardment,[15] and even William II was hesitant. The Kaiser was especially solicitous about the safety of his English royal relations and of English historical monuments.[16]

But Admiral Behncke pushed hard for his proposed raids. He argued that German naval inequality might be readjusted by bombing the London docks and the Admiralty headquarters in Whitehall. He spoke of another advantage which might be expected from such Zeppelin raids in a letter dated 20 August 1914; that they 'may be expected, whether they involve London or the neighbourhood of London, to cause panic in the population which may possibly render it doubtful that the war can be continued'.[17]

It was this latter argument, it seems, which won over von Tirpitz's approval. He stressed that London should be the main target, not just for the possible damage that might be inflicted but also for 'the significant effect it will have in diminishing the enemy's determination to prosecute the war...'[18]

The bombing enthusiasts in fact did not have too much trouble gaining their point. Major Siegert's command after all had already established an approved theoretical precedent. Also, that common human rationalisation of retaliation for others' acts helped firm up German irresolution. The English blockade and a French air attack on Freiburg (4 December 1914) were especially useful in this regard.[19]

The official approval of Zeppelin raiding plans on England, then, was given in a secret order by the Kaiser on 12 February 1915. In it the following provisions are found:

(1) His Majesty, the Kaiser, has expressed great hopes that the air war against England will be carried out with the greatest energy.

(2) His Majesty has designated as attack targets: war material of every kind, military establishments, barracks, and also oil and petroleum tanks, and the London docks. No attack is to be made on the residential areas of London, or above all on royal palaces.[20]

This debate nevertheless brought up two factors of major importance that were almost inevitably present thereafter in discussions of air power: hope for decisive effects from strategic aerial bombardment and questioning about the morality of that policy.

The moral question concerning civilian bombardment was (and is) one that could not be avoided. It is hardly surprising that the Germans, including the Emperor, immediately found it a relevant problem. Of further importance is that the means by which William II personally resolved his moral doubts represent a repeated pattern throughout the years covered by this study. This point is seen in his order of 12 February 1915 quoted above. In that paper, he avoided the moral problem of bombing civilians by designating only military targets. As will be seen, this circumvention was used throughout the interwar years by many air power advocates. But neither the Zeppelins in 1915 nor the long-range bombers of World War II could so accurately distinguish between military targets and civilian deaths. Analyse the targets of the London docks, for example (as designated in the Kaiser's directive). The East End of London, the vast living area of the London poor and foreign immigrant population, backs on to the London docks. Houses are yards away from the dock facilities. Thus it is a wholly ambiguous position to designate the docks as a target and to ban residential areas and yet William II was able to do this in consecutive sentences. He was hardly unique in this ambiguous position, however. This paper will furnish much evidence of man's ability to take Janus-like positions on the moral questions of warfare. What seems to be true with respect to military theory and practice is what Professor E.H. Carr found to be true with respect to politics in his brilliant study of political hypocrisy during the interwar years: 'Every political [substitute "military"] situation contains mutually incompatible elements of utopia and reality, of morality and power.'[21] Furthermore, men find it no harder to ignore the incompatibilities in the military domain than Professor Carr describes them so doing in the realm of politics.

A second continuing theme in air policy formulation is evidenced in Behncke's wishful thinking that air raids on London could result in a panicked population, even possibly to the extent of forcing the British out of the war. This belief in a 'knockout blow' from the air[22] is an ever present idea among certain air power advocates thereafter. The background to this air power presumption is worth some analysis.

The wars of the twentieth century have been described as 'total wars' involving the entire populations of the participating nations.[23] The people waged war on the home front just as the armies fought on the battle lines. If one could break the peoples' 'will to war' (a shopworn phrase of our century), then victory might be obtained – speedily and cheaply. This reasoning was given an early sanctification by the writing of the great Karl von Clausewitz who saw the Napoleonic-revolutionary armies as heralding the entrance of 'absolute war'.[24] In answer to this new dimension of threat, Clausewitz saw the essential need was no longer to conquer provinces or seek other limited purposes but 'by seeking out constantly the heart of the hostile power, and staking everything in order to gain all, that we can effectually strike the enemy to the ground'. Thus he specifically stressed the 'capture of the enemy's capital city, if it is both the centre of the power of the State and the seat of political assemblies and factions'.[25]

The extension of this view to proposals of strategic aerial bombardments of enemy capitals was a natural continuation of the thought. But, as Professor Possony so well points out, this vague target of the popular 'will to war' was never really analysed. The idea of 'moral collapse' became a truism before it was ever tested for its truth. None of the theoretical justifications using this idea contained an 'analysis of the causes, forms and effects of moral collapse nor of its nature or consequences. They do not make clear how a breakdown of morale could lead to capitulation'.[26] This was certainly true for Behncke's 1914 appraisal and it surely remains true for like statements in recent times.

Before the secret order of February 1915 which approved raids on London, Zeppelin attacks on other English targets had already taken place. The first airship raid on English soil was on 19 January 1915 on Yarmouth and the Norfolk coast.[27] By then Admiral Tirpitz had

The Zeppelin Era: The Challenge

changed the tenor of his objections. Now – as seen in a letter of his dated 22 January 1915 – it was only the target and not the idea which was wrong. He had become one of those who were pushing for an all-out air attack on London.[28] The Admiral had gone through some superficial (at least) skirmishing with the moral problem involved before he had reached this decision. In mid-November 1914, he had written that he still disapproved of 'frightfulness' in war tactics – this was relative to Zeppelin attacks – and that raids which dropped just a few bombs did no good, they just killed people and were therefore 'odious'. *But* 'if one could set fire to London in thirty places, then what in a small way is odious would retire before something fine and powerful'.[29]

With Tirpitz and the Emperor converted, London was fated to be attacked and the city suffered its first raid on 31 May 1915 when 3,000 pounds of bombs were dropped by a single Zeppelin raider on its north-east suburbs, killing seven, injuring thirty-five, and destroying some houses.[30] The immediate – rather than the theoretical – need for a defence of London had arrived. And London's defence provisions could hardly have been worse.

By a peculiar turning in the early history of British aviation, the Navy had been assigned the air defence responsibility for London (and England generally). The army had specific plans for its air branch and these mostly involved tactical support for its troops in the field. Because the number of army aircraft was limited and also because British leaders envisioned a short war making an immediate full commitment of resources seem reasonable, almost all the Royal Flying Corps (RFC) planes had accompanied the British Expeditionary Force to France at the outbreak of the war.[31]

The aerial branch of the navy, the Royal Naval Air Service (RNAS), had no such definite and immediate task for its aircraft.[32] For these reasons, on 3 September 1914, the Secretary of State for War, Lord Kitchener, asked the First Lord of the Admiralty, Winston Churchill, to assume the responsibility of the aerial defence of the homeland, even though the War Office had jealously preserved that 'privilege' up to then.[33]

It was not an easy assignment for the Admiralty. In 1914 there were at best only thirty-three guns in England that could be used against aerial targets and the planes at hand did not have the ceiling or armament to be effective against airship attack.[34] As Churchill has admitted:

Taking over responsibility for the air defence of Great Britain when resources were practically non-existent and formidable air attacks imminent was from a personal point of view 'some love but little policy' . . . I could with perfect propriety, indeed with unanswerable reasons, have . . . left the burden to others.[35]

Nevertheless, considering Churchill's naturally pugnacious character, surely Lady Violet Bonham Carter is correct in saying that Churchill 'eagerly' accepted this burden and, in fact, could not have done otherwise.[36]

It was in this manner, then, that Winston Churchill became involved with air defences at the very outset of England's need in that regard. His life was to be strongly connected with that problem for three decades thereafter.[37] And the immediate answering response that he and the Admiralty staff gave to that problem set out precedent for decades thereafter also. As Churchill realised, there was no adequate defence available against Zeppelin raiders by the time they reached England so he decided the solution would best be sought in bombing the bases of these raiders. Evidence of this 'forward air defence'[38] policy is available as early as 1 September 1914 in a note from Churchill to his Director of the Air Division (then Captain Murray Sueter).[39] Further notes of 3 and 5 September 1914 show that Churchill was also thinking in terms of gun and searchlight defence but that 'after all the great defence against aerial menace is to attack the enemy's aircraft as near as possible to their point of departure.'[40]

In these directives another continuous pattern of British air doctrine is foreshadowed. To paraphrase a later statement in this pattern, there was no doubt that the Zeppelin would get through.[41] Churchill thus turned to offensive strikes as an extension of defence. From that time on, there is no clear line of demarcation between air offence and air defence in British aerial doctrine.[42] As will be noted later, during most of the interwar years, English home aerial defence was entrusted to the Air Defence of Great Britain Command of the Royal Air Force and this 'defensive' command was composed primarily of long-range bombers. The history of British air defence is filled with echoes and re-echoes.

In medicine it may be that 'an ounce of prevention is worth a pound of cure' and it is certainly true that throughout most of the interwar years most air power advocates stressed offensive 'prevention' rather than defensive 'cure', but nevertheless offensive air strikes did not answer England's air defence problems in either World War and active and passive defensive measures[43] did finally provide the answers to airship and bomber raids in World War I and to the Battle of Britain in

World War II. Churchill's offensive emphasis thus started Britain down a wrong path in air defence policy.

To sum up this point with respect to the era of the Zeppelin threat, the 'forward defence' policy did destroy some airships in or about their sheds (six such successes were recorded by RNAS pilots in the first twelve months of the war[44]) but this did not stop Germany from pursuing this form of strategic air warfare. However, defence measures in and over England did finally achieve this result. But it should also be stressed that it took a long time to develop the home defences to that level of effectiveness.[45]

At the start, anti-aircraft guns were almost useless except, possibly, to hearten the populace that some counteraction was being taken. As noted there were only thirty-three guns in all of England that could be considered at all useful for anti-aircraft fire. Most of these were not specially designed anti-aircraft weaponry but were existing guns converted to this use. As such, they were hardly very promising.[46] For just one example, a twelve-pounder Horse Artillery gun had been so positioned that its barrel pointed vertically into the air. Needless to say, such a makeshift gunnery device could not traverse in pursuit of the moving target.[47]

Churchill realised immediately that the number of guns on hand was far too few to cover an area so vast as London. So, in a directive of 5 September 1914, he instructed that the priority for gun placement should be given only to the most vulnerable military targets. However, he had already (on 3 September) called for increased production of anti-aircraft weapons so the priority list of 5 September represented only his current evaluation of gunnery usage.[48]

The production increase that Churchill called for hardly resulted in a sufficient supply on hand by the time of the first Zeppelin raids in 1915. By September 1915, according to the man then assigned the responsibility for the gunnery defence of London, there were twelve true anti-aircraft guns in London plus an assortment of the improvised makeshift weapons already mentioned.[49] It was in that month that another innovation was added to London's gunnery defences — mobile guns which were designed to race to threatened areas in a manner somewhat akin to fire engine methods. The original models were of French design, brought back from France in September 1915. These 'autocanons' were later added to by English weapons mounted in open-bed trucks.[50]

Modern gunnery experts would surely sneer at the idea of mobile guns rushing through the streets of London in the direction of a reported airship sighting. In fact, an early postwar report of their activity deprecates their effectiveness. These guns usually arrived at their destination long after the airship had left the reported area.[51] But the commander of this motorised gunnery brigade had few doubts about the activities and effectiveness of his unit — to read Lieutenant Colonel Rawlinson's account is a lesson in blind smugness. For example, he states that his brigade's efforts were making the Zeppelin personnel very nervous (this by later summer 1916) even though his unit had not yet been credited with downing one single airship.[52] Rawlinson also emphasised the effect on home morale contributed by his unit's efforts. In one passage he describes the reactions of the bystanders as his guns raced by:

> It was . . . impossible to mistake the sentiments of the people, and I have often seen poor women with streaming eyes, holding up their children to see the guns as we passed, making it easy for us to realise the relief which it must have been to these poor defenceless people to see that at any rate some sort of defence was being provided for them . . .[53]

England's anti-aircraft gunnery defence, then, was most primitive and insufficient by the time the airship threat arrived. Moreover, the ammunition used in these guns was also hardly up to required levels of efficiency. As the outspoken Admiral Scott expressed it, the early unimproved ammunition 'was more dangerous to the people in London than to the Zeppelins above'.[54] The worst problem in this regard was that the shells had a very small bursting charge and so they returned to earth in the form of heavy missiles. Only by late 1915 and early 1916 was high-fragmentation ammunition available. Of course the danger from spent shells could never be wholly eliminated but with these new shells, the danger was greatly lessened.[55] Until then, surely the Londoners would have been better off had no guns been fired. According to one authority, the casualties from spent anti-aircraft fire amounted at times to one third of those caused by enemy bombs.[56]

Other aspects of ground defence were just as primitive and unavailing. Effective defence depends, among other things, upon advanced warning based upon accurate observation techniques. And England's ground observation system was at first a combination of muddle and trial and error. The Admiralty first assigned the responsibility to watch for enemy air incursions to the greater London police force. They were to telephone the Admiralty whenever an enemy aircraft was seen or heard. The Admiralty was then supposed to spread the alarm to Scotland Yard,

The Zeppelin Era: The Challenge

to transportation officials, and to the Speaker of the House of Commons, and (finally) to the gun and searchlight crews.[57] The task of observation was later (in 1915) assigned to the Chief Constables in the counties. However, their task was not only to phone the Admiralty but also to keep each other informed. 'The result was overlapping and a fearful congestion of the telephone system'.[58] What did *not* result was adequate forewarning. The information thus gathered came either too late or with insufficient accuracy or both.[59]

Another need in aerial defence is lighting for adequate night-time defence and here again inefficiency and insufficiency were in early evidence. Churchill had seen the need for many more searchlights immediately and he requested thirty to forty of these for home defence on 3 September 1914.[60] At that time, only sixty-centimetre projectors were in production although one experimental ninety-centimetre version had been produced.[61] The maximum height possible for the sixty-centimetre models was some 2,200 yards.[62] Under optimum conditions — and with sufficient quantity — these models could illuminate the Zeppelins.[63] However, optimum conditions were hardly common. Even by late 1916, in a report which attempted to present a very optimistic view, it was admitted that even moonlight lessened the range of the lights, that damp and rain severely diminished their efficiency, and that fog and smoke could virtually block off their light completely.[64] It is certainly relevant to add that damp, rain, fog and smoke are frequent problems in the environs of London.

All in all, then, considering the poor state of development in gunnery, ammunition, observation techniques and lighting, it is not surprising that 'it soon became current talk among the ribald that, of the three most useless things in the world, one was the anti-aircraft gun'.[65]

Defence by airplane seemed as useless as ground defence in the early stages of the war. The RNAS in 1914 had very few planes 'capable of flight even in favourable conditions'.[66] Even worse, the only armament available to the defence pilots and their accompanying observers at that time were hand-carried rifles.[67] It is wholly understandable therefore that the naval pilots assigned to air defence stations around London were more tempted by the attractions of the city than by their duties in defence.[68] Besides their normal inclinations, they must have been highly aware of the weak threat that they represented.

The lack of early defence capability against the airships certainly helps explain the tremendous impact that the early 'Zepp' raiders had on the English. Moreover, the very fact that attack from the air was a wholly new threat and dimension in warfare must be taken into account. Europeans would react to such raids now in a radically different manner. The 'thousand-bomber' raids of World War II make these early attempts seem very puny indeed. The single largest Zeppelin force to attack England consisted of nine ships in an attempted raid on Liverpool on 31 January 1916.[69] It thus requires an effort of will for the modern mind to appreciate how the very novelty of this form of war tremendously magnified the extent of its threat at that time.[70] And this tendency towards magnifying the threat does not just apply to English recipients of that threat.

By July 1916, for example, the Commander of the German Naval Airship Division, Peter Strasser, felt so confident in this form of war that he wrote his superiors a request for more Zeppelins with the following justification:

> The performance of the big airships has reinforced my conviction that England can be overcome by means of airships, inasmuch as the country will be deprived of the means of existence through increasingly extensive destruction of cities, factory complexes, dockyards, harbor works with war and merchant ships lying therein, railroads, etc. . . .

And to achieve this result of 'overcoming' England, Strasser asked specifically for only four more airships beyond the eighteen he then had in his command, although he also spoke vaguely of the need to produce as many as possible.[71]

German confidence was matched by English concern. On 1 January 1915, Winston Churchill memoed the Cabinet that the Admiralty had reports from Germany about plans for forthcoming large-scale airship attacks on London and that the home defences were wholly unable to prevent such attacks. ('. . . There is no known means of preventing the airships coming, and not much chance of punishing them on their return. The unavenged destruction of non-combatant life may therefore be very considerable.')[72]

The Admiralty's inability to defend against airship raiders especially concerned the First Sea Lord, Baron Fisher of Kilverstone. As early as December 1914 he had written to Admiral Sir John Jellicoe telling of his fears of such raids, that there was 'nothing whatever to prevent it . . .', and that Jellicoe should write Fisher what decisions he (Jellicoe) will make when air damage destroys their communications ('the Admiralty buildings would be razed to the ground, *so no orders at all!*').[73]

Lloyd George has written about a similar warning by Lord Fisher who circulated a paper which predicted Zeppelins dropping a ton of bombs on Horse Guards Parade. Fisher said the result would be a gigantic pile of ruins covering the bodies of leading statesmen and government and service personnel; that all would be consumed 'in one red burial blent'.[74]

Lord Fisher's great concern about the airship threat and the lack of defence against it prompted him to propose the radical policy of killing a captured German civilian for every British civilian killed by air raids. The First Lord, Churchill, refused to endorse this proposal although he did agree to place it before the Cabinet. When his reprisal policy was not accepted, Lord Fisher sent Churchill a letter of resignation (dated 4 January 1915). Lord Fisher later explained this letter as follows: 'I submitted my resignation ... because of the supineness manifested by the High Authorities as regards Aircraft.'[75] The First Sea Lord was finally persuaded to continue in office but Churchill claims that this controversy was the beginning of worsening relations between Fisher and himself.

Lord Fisher was not unique in his scheme of bloody retaliation. In fact the desire for revenge is another continuous thread in the air warfare story. Although usually this desire is expressed in terms of reprisal bombing raids, a letter in the *Daily Mail* some months after the Zeppelin raids had started shows a striking similarity to Fisher's proposal.

> Sir,
> Do we want to stop these Zeppelin raids? If we do, here is the remedy, let the government announce that for every English civilian killed one German officer in our custody shall be taken to the Tower and shot; for every woman killed, two shot; for every child killed, three shot — not tortured, just humanely shot ...[76]

Great concern about Zeppelins was evidenced throughout all levels of British society.[77] One official in RFC administrative circles remembers the airship period as a two-year span in which 'the whole country lived and suffered under a Zeppelin psychosis'.[78] There is abundant evidence to substantiate this judgement. B.H. Liddell Hart, for example, has recalled the reaction in Hull to Zeppelin raids during the time he was posted there on convalescent leave. He had witnessed an attack by two dirigibles:

> As there was no defence, the two airships hovered low over the city, and one could see the gleam of light each time a trapdoor opened to

drop a bomb. The moral effect of that undisturbed attack was so great that every time the sirens sounded, in the weeks that followed, thousands of the population streamed out into the surrounding countryside...[79]

Hull was also the scene of rioting after one of the earliest raids (in June 1915) when stores reputedly owned by Germans were sacked. Troops were needed on this occasion to restore order.[80] By 1916, Hull had recruited forty women as 'Voluntary Patrollers' among whose 'most trying chores was the controlling of crowds during the air raids'.[81]

Fear of the Zeppelin raiders can be illustrated by yet more specific data. Naturally, if such fear existed, it would be manifested, among other ways, by work stoppages. And these certainly occurred in great number. In 1916, one area's pig iron production dropped by five-sixths due to alarm over threatened airship raids – even though the locality was never actually bombed![82] Many statements can be found as to the effect on industrial production by both true *and* false air raid alarms, such as by the military authority E.J. Kingston-McCloughry[83] and by the 'official historian' of civil defence.[84] Perhaps a contemporary personal note expresses the point as clearly as is needed – this by John Evelyn Wrench on 14 October 1915: 'the whole staff is demoralised by Zeppelins and I can't get any work down.'[85]

In sum, terror, industrial absenteeism, community riots and anguish over the lack of defence[86] were all present (and mutually interacting) during the Zeppelin era in Great Britain. But possibly the best proof of the general concern over the Zeppelin threat is the story of the political success of Noel Pemberton Billing.[87]

Almost without exception, historians have neglected Pemberton Billing.[88] Therefore, A.J.P. Taylor deserves high marks for not only noting him but for giving a splendid capsule account of his significance. This account, in fact, will serve well here as an introduction to the subject.

Pemberton Billing, the first man to challenge the party truce successfully (at East Hertfordshire in March 1916) owed his success mainly to being the 'member of air'. He can claim credit for initiating the modern doctrine that war should be directed indiscriminately against civilians, not against the armed forces of the enemy. The Royal Air Force was created before the war ended, specifically to practise what Pemberton Billing preached.[89]

Noel Pemberton Billing had long been associated with aerial interests. In 1908 he had started a monthly publication about aerial concerns (*Aerocraft*).[90] In 1912 he had founded the Supermarine Aviation Works which, among other things, helped pioneer the flying boats.[91] During the early stages of the war, he served as Temporary Flight Lieutenant in the Royal Naval Air Service, in which service he helped in the planning of the forward defence policy against Zeppelin factories.[92] He resigned his commission at the start of 1916 in order to run for Parliament as an independent candidate on the single platform of changing the government's impotence with respect to its home air defences.[93] To run as an independent meant that he would have to challenge the newly created unity of the two major parties now working within a coalition government. And it must be assumed that for such an independent to *win* against such unified opposition, his platform would have to attract tremendous popular support.[94]

Certainly his appearance, style and popular, if demagogic, oratory helped. He was a 'lanky', 'sharp-featured' man who conveyed a dashing, modern, vital image. (The fact that he wore a wrist watch and monocle helped contribute to that image.) His auditorium appearances would be preceded by movies showing him in flight action. His street speeches were made from the cockpit of an airplane transported about by motor car. Wherever he campaigned he attracted great crowds and cheering enthusiasm.[95]

This unique candidate failed in his first attempt to win a parliamentary by-election (25 January 1916 for the Mile End seat) but he did show surprising strength. He received strong support from Lord Northcliffe's *Daily Mail*; sidewalk throngs 'screeched in wild acclaim' and his losing margin was only 376 votes out of a total of 3,606 cast. In this campaign he also gave good evidence of the disrespectful, flashy tone of his oratory, a characteristic which certainly never diminished thereafter. ('I don't care two whoops for being an MP, but I do care about the people of London being murdered by Zeps.')[96]

His second — and successful — attempt against the coalition government came shortly afterwards in the by-election for East Hertfordshire. The holder of the seat, Sir John Rolleston, announced his plans to retire as of the end of February 1916 and on 18 February, Pemberton Billing (that 'ever-exuberant ... crank' according to one prejudiced commentator[97]) announced his candidacy for the seat.[98]

This time his campaign included giant enlargements of Zeppelin damage shown at many of the 116 speeches he gave at announced meetings, a book published during the campaign (*Air War: How to*

Wage It) and the issuing of a phonograph recording of some of his speeches (proceeds from both book and record to go into a supervised account to be spent for propagandising the need for better air defence), and – quite possibly the most helpful of all – a Zeppelin raid on London four days before the election, on Sunday, 5 March. *The Times* at least was convinced that this raid greatly enhanced Billing's chances.[99]

That raid also convinced *The Times'* editors that Billing's arguments had validity. In a lead editorial 'Air Raids and Air Policy' (7 March 1916) *The Times* agreed with the candidate that the government was showing an 'incapacity . . . to grasp the real question . . . that question is nothing less than the future mastery of the air'. The editorial continued that this question could be as vital to England's security as the centuries old problem of the mastery of the seas – and that the masses recognised this too. 'They want an air policy filled with the spirit of our sea policy – a spirit of offensive and not of passive defence.'[100]

The voters obviously did agree with the independent candidate's platform. In this constituency, a London suburb which had voted Conservative 'for a generation', Noel Pemberton Billing received 4,590 votes while his 'coalition' opponent (running as a Unionist) received 3,559 votes, a winning margin of 1,031. Popular concern was also reflected in the size of the total vote which was over 2,000 more than expected.[101]

The new and self-proclaimed 'First Air Member'[102] of the House of Commons lost no time in delivering his maiden speech. On 14 March 1916, he found this opportunity during the Army Supply Estimates debate. He told the House that he had resigned his commission so that he could bring his aerial expertise to that body and thus gain the needed changes. If those changes were not made, 'the Air Service would continue to be a byword among its members and a subject of almost tragic mirth in its efforts to defend this country'. He stressed that such defence primarily depended upon offensive strikes and that these should therefore be pursued with all effort and that no attention should be paid to any 'so-called religious scruples' against their use. He concluded his remarks with his expressed expectation that the war could be 'determined in the air . . .'[103] Among those who were impressed by this speech was the Director of Air Organisation in the War Office, Brigadier General Sir Sefton Brancker: 'I heard him [Pemberton Billing] make a most creditable maiden speech in the House, and I went away rather pleased, feeling that he might be able to help the cause more than any of our champions in the House had done up to date.'[104]

General Brancker's mention of other air champions in the House indicates that Pemberton Billing was not being wholly exact when he called himself the 'First Air Member'. As has already been noted, Winston Churchill was an air power proponent in his official position as First Lord of the Admiralty. In his private life too, Churchill showed great interest in the new, exhilarating experience of flying. He took his first airplane flight in 1912 and was an enthusiast from this exposure on — even though he experienced several narrow escapes and one crash (from which he was able to walk away).[105] Another Member of Parliament who was connected with a strong advocacy of air power was William Joynson-Hicks. Like Churchill his interest in aerial matters predated the war (evidenced at least by 1912) and when the war heightened the House of Commons' interest in air power, he was called upon to chair its newly formed Parliamentary Air Committee.[106] And like Pemberton Billing, early in the war Joynson-Hicks published a book establishing his positions regarding air power. This book (primarily a reprint of his speeches plus a few other additions) shows, among other things, that Joynson-Hicks was a believer in the forward defence methods against the Zeppelins, that he deprecated home defence ability, especially with respect to anti-aircraft gunnery, that he endorsed a reprisal raid policy, and that he believed 'with an adequate Air Service you could very largely bring this war to a speedy and victorious conclusion'.[107]

Pemberton Billing was not alone, then, among parliamentary members in being an outspoken advocate of greater air power. But certainly he was the most extreme in this regard; certainly he was the most persistent and irritating in these views. Leaders of the House of Commons found him almost a daily pest. He introduced question after question during the House Question Time until he became the subject of intense exasperation among the backbenchers as well as the Government spokesmen. Many verbal exchanges in *Hansard* give testimony to the fact that he alienated all but a few of his brethren.[108] Nevertheless, day after day and with increasing sarcasm and vituperation, the 'Air Member' continued his attack upon the governmental air policy. He kept stressing that air defence really must mean offensive air strikes, that morality was not a consideration but that the need to retaliate was, and that air power could bring victory if the air forces were adequately financed and given sufficient independence from control by the older services. The House membership could not fail to be affected, no matter how exasperated at the same time. Moreover, the continuing presence of the Zeppelin raiders would add its counterpoint to heighten the effect and impact of Billing's words. These raids also meant a growing support for his ideas by the press as has been seen in the case of *The Times*.[109]

After his election, furthermore, Pemberton Billing did not cease his efforts to attract popular support. He was the chief speaker at meeting after meeting, month after month. To just cite one typical example among innumerable choices, he was the principal speaker at a large meeting concerning the aerial defence of London sponsored by the United Wards Club of the City of London. This was shortly after his election, on 27 March 1916. In his speech, Billing reiterated his normal themes of the gross inadequacies of the Government's air programme and of his conviction that the war's duration would be greatly shortened if Britain invested in a large aerial fleet of bombers. He also made the following significant statement:

> It has been suggested that I am a man of one idea. I glory in that accusation. Before many years have passed that one idea will occupy the minds of many men in this country and women too. Every inland town lies on the coast of the ocean of the air, liable to instant and violent attack. When you think that in about ten years' time countries will possess not 1,000 but 100,000 aeroplanes at the cost of a few battleships, it is a terrible thought. These aeroplanes will fly at a speed of 100 or 120 miles an hour. Their powers of mobilisation will be alarming. It means that if our relationship with another country 100 or 150 miles distant is strained at six o'clock in the evening, before we arise in the morning it will be possible for our principal towns and cities to be laid waste.[110]

This vision of an overnight air attack at the outbreak of war — an attack which would result in wholesale devastation of large urban areas — would later be associated with the Italian air theorist, General Giulio Douhet. Indeed, this vision would be generally discussed under the term 'Douhetism'.[111] However, as this quotation obviously indicates, it was not necessary to wait upon Douhet for the appearance of this aerial doctrine. This study will frequently show how this vision was a *natural* prophecy with respect to the new air weapon within the European theatre of war. (America, isolated by its oceans, was not so tempted to see air warfare within this extremely destructive setting. But even in America, this vision had its outspoken advocates, most notably General 'Billy' Mitchell.[112])

As usual, at this meeting Pemberton Billing's views received popular support. Those in attendance unanimously passed the following

The Zeppelin Era: The Challenge

resolution:

> That this public meeting of citizens and inhabitants of the City of London respectfully urges upon the Government the pressing necessity for prompt measures being taken for the adequate protection of life and property in the City against air raids by the enemy, and considers the most effective means of affording such protection would be by the creation and maintenance of an efficient air fleet, in addition to, and independent of, the existing naval and military requirements, to enable this country to carry out a vigorous offensive in the enemy's territory.[113]

One must remember that the various comments cited above were from an era when England had not yet received any concerted attacks by heavier-than-air craft. The emotions shown were roused by the relatively light damage and casualties caused by airship attacks. Naturally, not all the English were so troubled by the airships. Princess Victoria, for example, did not share Queen Alexandra's anxiety.[114] Ralph David Blumenfeld attempted to downgrade the Zeppelin threat in a series of contemporary articles.[115] Viscount Norwich (the former Alfred Duff Cooper) in his memoirs minimised the general reaction – but the title he chose for that work seems all too true with respect to this particular remembrance,[116] for grave concern was the more typical response and in its expression can be seen basic patterns in future British air policy, above all in the stated need for an independent air arm to strike at the enemy without being hampered by army and navy requirements. Also reflecting much future air policy was the expressed belief that defence measures as such were not effective; that the best means of defence was the pursuance of a strong offence.

Some historians state that history is best understood within the framework of a 'challenge-response' continuity. Certainly early British aerial history can usefully be viewed in this light. This chapter has been concerned with the first major aerial challenge to England. It has shown the emotions and fears generated by this challenge, and the first and generally impotent reactions to the challenge. The next chapter will show that England gradually achieved an effective response and thereby formed its first practical air defence.

2 THE ZEPPELIN ERA: THE SUCCESSFUL RESPONSE

The appearance of the Zeppelins, the first serious aerial threat in English history, created grave concern within all sectors of British society, from royalty to governmental and military leaders to the masses in general. The earliest available home defences against aerial attack were hardly such as to bolster confidence; they were extremely primitive and hardly up to the challenge. Therefore, the tendency was to rely mostly upon a 'forward air defence',[1] to bomb the Zeppelins in their sheds or to bomb their assembly plants — or in the words of one of the most colourful aerial spokesmen in England, Noel Pemberton Billing, 'to trap them in their nests'.[2]

This offensive form of defence was the approach taken by the Admiralty and its Royal Naval Air Service (RNAS), the Admiralty having the responsibility, as has been seen, of home aerial defence. The two key personalities involved were in complete agreement on this policy, Winston Churchill then First Lord of the Admiralty, and Captain (soon to be Commodore) Murray Sueter, then Director of Air Division, i.e. director of the RNAS.[3] Results of their planning were soon evidenced; an attack by an RNAS plane on a Zeppelin hangar in Düsseldorf on 8 October 1914, successfully destroyed an airship with the use of only two twenty-pound bombs, and was soon followed by an attack on a Zeppelin factory in Friedrichshafen (by Lake Constance) in late November 1914.[4]

This policy of a forward defence naturally demanded continental bases for the naval air squadrons assigned to the task. At first these planes used makeshift landing strips along the beach by Dunkirk but in February 1915, they changed from such temporary facilities to an established airfield near Dunkirk.[5] By then the first Zeppelin raid on England had already occurred (on 19 January 1915, on the eastern coast)[6] and the Admiralty was thus naturally giving more emphasis to this policy. This sense of urgency is quite clearly seen in a telegram sent by the Admiralty to the French Ministry of Marine to arrange for European bases:

> The Admiralty considers it extremely important to deny the use of territory within a hundred miles of Dunkirk to German Zeppelins,

and to attack by aeroplanes all airships found replenishing there. With your permission the Admiralty wish [*sic*] to take all necessary measures to maintain aerial command of this region. The Admiralty proposes therefore to place thirty or forty naval aeroplanes at Dunkirk or other convenient coast points. In order that these may have a good radius of action they must be able to establish temporary bases forty to fifty miles inland. The Admiralty desires to reinforce officer commanding aeroplanes with fifty to sixty motor-cars and two hundred to three hundred men. This small force will operate in conformity with the wishes of the French military authorities, but we hope it may be accorded a free initiative. The immunity of Portsmouth, Chatham, and London from dangerous aerial attack is clearly involved.[7]

Commander C.R. Samson was given the command of this Dunkirk air defence operation and his activities there involved not just air defence but also the development of the tank. Winston Churchill has succinctly stated the connection: 'The air was the first cause that took us to Dunkirk. The armoured car was the child of the air; and the tank its grandchild.'[8]

Commander Samson had to concern himself with his own unit's defence as well as with England's aerial home defence. The Germans were continually pushing in on the perimeter of his position. One of the methods Samson used in his ground defence was to bolt sections of boiler plating to standard touring cars thus creating, basically, a prototype model of an armoured car.[9] These cars were so successful in repulsing the enemy troops that Churchill ordered the navy to form several armoured car squadrons. And once research had begun in mobile armoured equipment, fertile minds could jump to tank-like visions.[10] Winston Churchill is not wholly accurate in placing the early history of tank development exclusively on these incidents just described – there were others who had like ideas and there was a confluence of diverse factors involved – but it is true that this naval/air history did play an important role in that development.[11]

In the event, the forward defence policy was the only really effective aerial defence that England adopted in the early stages of the Zeppelin era. The forward defence policy attained the destruction of six Zeppelins in the first year of war[12] while, in this same period, aerial defence in England achieved no credited 'kills' at all.

Among the six airships destroyed, surely the most satisfying was that of the first Zeppelin to raid London. This was the LZ.38 commanded by the audacious Hauptmann Karl Linnarz. On his return run from this raid (31 May 1915) Linnarz, on a whim, wrote a defiant message on his engraved calling card and dropped it, weighted, over the Dunkirk area where it was found the next day. It read 'You English! We have come and we will come again soon to kill or cure! Linnarz.' It was thus particularly gratifying to Samson's unit when it destroyed the LZ.38 at its base in Evère (just north of Brussels) on the night of 6 June 1915.[13]

Clearly the forward defence policy had some significant successes. It even forced the Germans to cease using Belgium as an area for Zeppelin bases by the close of 1915.[14] But it did not stop the airship raids on England. This policy did not provide the needed response to the challenge. What was needed were much better techniques in all forms of *home* defence. It proved not to be true that the best defence was a strong offence with respect to the airship threat, even though this was the position taken by the most outspoken air power advocates.[15]

Changes in technique were needed throughout the *whole* defence system, in technological expertise, in the handling and use of manpower, and in organisation. Certainly, for example, improvements in anti-aircraft gunnery were obviously and sorely needed. The very poor quality of England's anti-aircraft gunnery in the opening stages of the war — to the point where it was a much greater threat to the citizenry than to the Zeppelins — has already been described.[16] Winston Churchill, First Lord of the Admiralty, had immediately seen, at war's outbreak, the need for improvements in techniques and production of anti-aircraft weaponry.[17] The army started out, perhaps, with more optimistic delusions if comments from a 1913 speech by Col. J.E.B. Seely, Secretary of State for War, are indicative: 'Everybody concerned has been surprised beyond measure at the apparent ease and remarkable accuracy that can be attained in firing at aerial targets.'[18] But experiences on the Western front soon convinced army authorities that improvement in ground aerial defences was needed, and they requested action be taken in this regard.[19]

One result of this urging by army and navy authorities was the first systematised operational research activity associated with war technology in Great Britain. This was the establishment of the Anti-Aircraft Experimental Section of the Ministry of Munitions whose primary task was to achieve greater accuracy of fire and which was first directed by the young physiologist, Captain A.V. Hill, who was later to play an important role as a 'boffin' for the Royal Air Force.[20] Professor Hill later recalled this work in a number of sprightly speeches. He noted that this Anti-Aircraft Experimental Section was the first operational

The Zeppelin Era: The Successful Response 31

research establishment ever, that when it was formed, the anti-aircraft gunsights then in use 'were hopelessly wrong', that the time-fuses 'exploded most of the shells – if at all – in the wrong place', and, in sum, that anti-aircraft gunnery was an 'almost unknown subject'.[21]

The experimental station had some notable results. By its efforts fuse-timing devices were improved, a height calculator was developed (the Hartree height-finder), a 1,100 page textbook on AA gunnery was produced, and various other technical papers also resulted.[22] This early success story in operational research for purposes of aerial defence is noteworthy because scientific knowledge would seem to be an obvious factor in air warfare and yet too often its importance was discounted or ignored. One major British problem in the interwar era with respect to aerial doctrine was that too little of it was based upon scientifically proven data rather than upon 'gusts of emotion'.[23] To sum up this point then, aerial defence is a military concern which critically needs operational research by scientifically trained personnel and this continuous fact was first indicated in World War I. Moreover, the military mind has not been easily persuaded to change on the grounds of this scientific research, and this too was evidenced in World War I.[24]

Improvements in gunnery also came from other sources, especially from personnel directly involved with London's AA defences. Better fragmentation of the shells was achieved so that the populace were no longer pelted with heavy chunks of spent ammunition. Another safety precaution was added when the shells were prevented from exploding when they hit the earth if the fuses did not work on high. The range of fire was extended.[25] But no technical changes achieved would make AA gunnery anything but a seldom-hit-largely-miss affair until the advent of radar directional control which only appeared in operational use during World War II.[26] It is true that a barrage fire technique was introduced later in the Great War and that this did prove to be somewhat effective (as will be noted later) but this came after the Zeppelin era. In the first two years of war there was neither the number of guns available nor the overall planning which later produced the barrage fire defence.

In the Zeppelin era, then, the AA gunnery commanders had little to boast of with respect to actual number of successes. The first known achievement by AA batteries was in mid-August 1915, when guns situated at Dover hit the Zeppelin L.12 which was thereby forced to ditch in the water off Zeebrugge. This Zeppelin was thereafter towed to Ostend for repairs but then the Dunkirk based naval flyers were able to get to it, and by their bombing efforts the L.12 was effectively destroyed.

This victory must therefore be shared by both home and forward defence efforts.[27] The next success achieved by home defence gunnery was not until 31 March 1916, when a Zeppelin was hit near Woolwich and went down in the Channel off Essex.[28] Thus, in almost two years of war, English home aerial gunnery had accomplished little to justify much confidence.[29]

This first wholly successful 'kill' by home defence gunnery was in part made possible by an improved searchlight defence system.[30] There had been requests for stronger searchlights as part of the demands for technological improvements in defence. In 1915, production had been started on a 90cm searchlight with a range of 3,300 yards. This was better by a third over the older 60cm models.[31] By May 1915, there were only a dozen searchlights available to the London defences but less than a year later, there was a sufficient number on hand to provide a protective belt of lights situated some twenty-five miles inland in an arch extending from Northumberland to Sussex. (It may be noted here that the first organised postwar defence planning turned away from this perimeter searchlight defence operation. They were deemed to be too much like signposts for rather than threats to the invading aircraft.)[32]

Naturally, war production increases also augmented the number of guns allocated to London's air defences, and this would certainly help increase their potential effectiveness. About a month and a half before the 31 March 1916 success, the London defences had 148 guns, either at hand or in process of being delivered.[33] Nevertheless, ground defence was something of a *faute de mieux* measure as their number of successes indicates in this early stage of the war. Perhaps, as some argue, the guns were useful in keeping the Zeppelins too high to allow for accurate sighting of targets. But considering the primitive development in bomb sighting and control devices, this was not as advantageous as it may first appear.[34] It is at least true that the Zeppelins did fly higher due, in part, to anti-aircraft fire. This is confirmed by Commanding General of the German air forces, Ernest Wilhelm von Hoeppner.[35]

It was the airplane which had the greater success against the Zeppelins and here too, of course, greater numbers and technological improvements were of crucial importance. Naturally, war necessity hurried these developments.[36] They were essential. The British aircraft at the start of the war did not have the altitude capability or climbing speed ability to seriously endanger the Zeppelins.[37] However, by February 1916, twenty-five of the fighter-type B.E.2C models had been delivered to home defence forces.[38] This plane was destined to be the most effective airship destroyer in the war. (B.E.2Cs destroyed nine of

The Zeppelin Era: The Successful Response   33

the twenty-one total airships brought down by British airplanes. The next highest total was seven credited to the later model Sopwith 'Camels'.)[39]

The first of the B.E.2 models appeared August 1912, and had been designed by Geoffrey de Havilland at the state-owned Royal Aircraft Factory at Farnborough.[40] The expansion of this factory complex at Farnborough is one excellent indication of the acceleration given to the aircraft industry by wartime pressure. In 1909, it employed about one hundred workers in a plant operation containing one small machine shop, one shed for balloon construction and one shed for other aircraft. By 1916, there were some 4,600 workers there in a large multi-building factory system.[41]

The 1916 B.E.2C model was a biplane powered with a 90 h.p. engine and a ceiling of at least 11,000 feet. Of special moment for its role against airships was that it was especially suited for night flying, being designed to maximise stable flight.[42] In fact, these planes were made so stable that they could not be manoeuvred sufficiently for effective defence against bombers, the 1917 threat to Britain. But manoeuverability was obviously not a major consideration with respect to fighting the unwieldy airships.[43] Also of great importance was that the B.E.2C had special armament suited to its home defence tasks. It was one of the earliest British planes equipped with a machine gun (the .303 Lewis gun) and it also had a special missile-rocket firing attachment, the Le Prieur rocket device. This had been originally designed by the French for use against balloons but the usefulness of a fiery rocket fired at the Zeppelins was easily seen.[44]

Two other points in connection with the development of the B.E.2C are worth noting. When the plane was first designed, the idea was still present that planes could fulfil various disparate functions. Thus they were originally used as bombers (late in 1915) but they were quite inferior for this task. It was found that their load capability was wholly inadequate unless flown with only a pilot and no accompanying gunner. And this made them very vulnerable to German defending aircraft. The result was that the RFC sent along other fighter aircraft to protect the bombing B.E.2Cs. Thus occurred the first fighter escorted bombing missions in history. And lastly, the Royal Aircraft Factory subcontracted part of the production of the B.E.2Cs, and one company so commissioned was G. and T. Weir Ltd, of Glasgow. Thus Mr William Weir received his first introduction to military aeronautics. His contributions thereafter to the development of the British military air service were many, and profoundly important. He is, of course, better know by his later title of Lord Weir of Eastwood.[45]

The armament on the B.E.2Cs deserves separate mention. Defence planes without guns are almost useless — and that was their condition at war's start.[46] It was all very well for the First Lord of the Admiralty, Winston Churchill, to say of England's air defence strength (in a speech in the House of Commons on 17 March 1914) that:

> With the Military Wing, with which great progress had been made, we are already in a position of effective strength, and any hostile aircraft, airships or aeroplanes, which reached our coast during the coming year would be promptly attacked in superior force by a swarm of very formidable hornets.[47]

But hornets would be little feared without their stingers.[48]

Nevertheless, it was not by machine guns or rockets that the first Zeppelin in flight was brought down by a British pilot. On the night of 6 June 1915, Reginald Alexander John Warneford downed the LZ.37 over Ghent by dropping fire bombs upon her from above. For this much publicised achievement, Warneford received the Victoria Cross and the Cross of Chevalier of the Legion of Honour. Warneford's feat gave the British new hope — the Zeppelins could be attacked successfully from the air.[49] But chance hits on the top surfaces of the airships by the overside throwing of bombs was hardly a reliable way to defeat the dirigible threat. On 2 October 1914, Churchill had asked for the development of anti-Zeppelin grenades to be dropped on them from overhead,[50] but by 1916 the authorities decided against all such attempts because the misses would only add to the ground dangers.[51]

In that same letter of 2 October Churchill had also called for the production of incendiary bullets, and this was a far more practical suggestion. Naturally, the combustible gas-filled airships were highly vulnerable to incendiary fire. But it took much longer than necessary for this means to be adopted. Incendiary bullets were first successfully tested in 1908 by their New Zealander inventor, a Mr Pomeroy. He submitted this invention to the British War Office in 1914, but it was not accepted until 1916, and only put to extensive use in the autumn of that year.[52] By that time, other varieties of incendiary bullets had also been tried and put into use: the phosphorus-based Buckingham bullet first used by the RNAS in December 1915, and then improved in 1916 and the Brock bullet, invented in 1915 and thereupon adopted with great hopes (one half-million of these were ordered and produced) but which proved to be the least effective of all the varieties.[53] According to one knowledgeable article, it was the incendiary bullet which proved to be the most effective answer to the Zeppelins.[54] It was, of course, ideally suited to take advantage of their Achilles' heel of high combustibility. And, mercifully for England, they were on hand when the Germans (in the summer of 1916) produced a new and much larger Zeppelin which had a carrying capacity of 9,000 + pounds of bombs for the distances involved in their English bombing missions.[55]

As has been noted, successful aerial defence is very dependent upon applied technology. Improvements in gunnery methods, planes and aerial weaponry directly affected England's ability to defend herself. But this is not the whole story. Improved means were necessary throughout *all* of her defence facilities. Air defence is a complicated task involving both active measures to prevent bombardment and passive measures to ameliorate the effects of bombardment. Active defence involves four different sub-phases, each of which must be effectively accomplished in order to achieve an overall effective active defence. These four phases held true for World War I conditions, and they still hold true in the atomic age. They are, in order: detection of the threat, identification of its nature, interception, and, finally, destruction.[56] Naturally, authorities differ as to the order of priority in importance they give to these different defence needs. But three highly knowledgeable voices indicate how important the first two of the above categories are, the categories which involve the information gained by a ground observation system. Major-General E.B. Ashmore who was in charge of London's aerial defences in the later stages of the first World War has stated that 'the whole system of defence should depend on this knowledge'.[57] General Sir Frederick Pile who was Commander-in-Chief of the home Anti-Aircraft Command in World War II has said that it is the communication of that intelligence from the observation team upon which 'the whole success of the defences depended . . .'[58] And for one last example, the man who commanded almost all phases of Germany's air effort from 1916 on, Wilhelm von Hoeppner, also stressed the crucial importance of ground observation and especially the need for its factual accuracy.[59]

In the interwar years, until the development of radar, the English air defences were deemed to be alarmingly vulnerable, in large part because of their insufficiencies with respect to this matter of intelligence. For many reasons, then, the story of England's early efforts in this regard is significant.

The very rudimentary steps in the development of England's aerial observation and warning system have already been described. Briefly, police constables were delegated the responsibility to inform the Admiralty whenever they had information about incoming enemy raiders. The Admiralty then alerted a set list of important functionaries. City policemen and country constables proved to be unreliable sources for this information. Their reports were not sufficiently accurate, detailed, or, in many instances, sufficiently prompt.[60]

Early in 1916, for reasons that will soon be discussed, responsibility for home air defence was transferred from the Admiralty to the War Office. Among the many changes that resulted was a reorganisation of the observation and warning system. Army personnel were assigned to this task and they were placed in cordons some thirty miles in advance of vulnerable target areas. London was given a double cordon protection. In addition, a coastal observation line was established. However, this arrangement also soon proved faulty. Soldiers were all too prone to be prey to distractions when based in their own country with too little supervision and not enough to do. So police were once again resorted to, but now specifically assigned to these newly established cordon placements. Two companies of soldiers which had proved to be the most trustworthy were retained as observers in the sites needing the most constant attention.[61]

Many more innovations in the observation system followed. Indeed one way to judge the importance that the authorities attributed to this area is to note the frequency with which they changed it. For example, the reports were at first all phoned in to one headquarters at the War Office, but this overburdened the telephone communications there. To ease this telephone load, all of England was divided into seven subsidiary Warning Control Areas whose boundaries were determined by telephone trunk line centres where feasible. The seven divisions were each under a Warning Controller who was the anti-aircraft commanding officer of the same division. These officers' headquarters were maintained at the divisions' telephone centres so that they could keep their own districts informed as well as the War Office central headquarters.[62]

Each of these subsidiary headquarters featured a large operations table consisting of a large map of its region divided into some few hundred major squares, each of these squares subdivided into four squares within which were four still smaller squares. Each of the smallest squares, then, had identifying coordinates of three symbols. Each observation post had its own local map with corresponding coordinates. Thus any sighting of enemy aircraft could be reported with fairly accurate intelligence. The observers were responsible for as full a report as possible including estimations of the speed, altitude and direction of the

The Zeppelin Era: The Successful Response

enemy craft. To sharpen their ability in these skills, frequent practice raids were staged. These tests also helped officials in the control centres to judge the relative reliability of their informants.[63]

The above description carries the history of the evolving observation system into the early months of 1917. It thus depicts the system in its various phases during the Zeppelin era. Further changes were made in 1917 and 1918 during the period when the threat involved German heavier-than-air bombers, and these later changes will be described in their appropriate place. However, the account already presented indicates the general pattern — blundering confusion to relative efficiency — and the first steps toward the later developments have been taken. In fact, the famous war operations rooms of the Battle of Britain period of World War II can be seen in prototype in these early Warning Control Areas' operations tables.[64]

The least progress made by Britain in aerial defence during the Zeppelin phase of the Great War was in that general area today known as passive or civil (civilian) defence. However, here too the story is one of slowly improving effort and increasing knowledge.

As the term civil defence implies, this aspect of aerial defence directly concerns activities of the civilian population. It often involves regulations which would not be accepted during peacetime and includes restrictions which would normally be considered infringements upon the rights and privileges of the citizenry. The legal catchall which allowed these wartime civilian restrictions was the famous DORA, the Defence of the Realm Act originally issued on 8 August 1914, and variously amended and revised in the war years thereafter. The legal authority given in the original Act to such measures as were necessary for civil defence was clear, if vaguely all-encompassing:

> His Majesty in Council has power during the continuance of the present war to issue regulations as to the powers and duties of the Admiralty and Army Council, and of the members of His Majesty's forces, and other persons acting on His Behalf, for securing the public safety and the defence of the realm . . .[65]

An interwar study of civilian air defence in the important military journal issued by the Royal United Service Institution detailed various activities that could be taken under this legal provision. The Government was thereby allowed to confiscate or control land and roads for air defence purposes,[66] to control public behaviour prior to and during air raid conditions, to convert private buildings into public shelters, to control light and sound which might lend aid to the enemy raiders, to erect hoardings in front of damaged buildings and to place dangerous areas off limits, to establish special police and fire brigade areas, etc. And eventually, as if the Government powers were not already broad enough in these matters, one amendment (Regulation 25b of DORA) allowed the execution of any plan regarding air raids.[67]

The first use of such powers with respect to passive defence was an Order in Council of 12 August 1914, which enabled military authorities to restrict lights in harbour areas. Often the local military authorities passed on this task to local officials which certainly did not lead to uniform results. 'Seaside resorts were afraid that if their own authorities were too zealous, their visitors would be driven to less conscientious and better lighted rivals.'[68] This seeming indifference by the public to the Zeppelin danger can be explained by the fact that this was during the earliest months of the war when that danger had not yet materialised.

The lights of London were first dimmed on 11 September 1914. Also, about that time, the lake in St James Park was drained so that its reflection would not guide enemy flyers to such choice nearby targets as Buckingham Palace and Whitehall. But efforts to efface London from overhead eyes were – and were always to remain – unavailing. The Thames is always there, an infallible (when visible) directional arrow pointing to the heart of London. Only a heavy fog or cloud cover could protect London from the enemy's night vision, and this, of course, only in the period before navigation by technology.[69]

When the Zeppelin attacks started, bringing the realisation of immediate danger, lighting restrictions were imposed upon ever wider areas until finally, by October 1915, they included all the Home Counties. There was no attempt to achieve a total blackout but just a general reduction in brightness. Main streets were darkened to compare with side streets. Outdoor advertising lights were turned off. High-standing clock faces were darkened, most notably that of Big Ben. Car lights were dimmed. In some cases, false street lights were set up in dark areas to help confuse identification.[70] Naturally, such a widespread dim-out operation brought problems. Automobile accidents increased.[71] Public behaviour was adversely affected as the followng official report indicates:

> ... May 1916. An outstanding feature of this month has been the effect of the new order for darkening the streets at 8.45 p.m. Men of unpleasant character have appeared in the crowded better streets and speak to women freely. Large numbers of giddy girls find the novelty

exciting, and gather in dark corners giggling and shrieking . . .[72]

Another special problem was confusion due to overlapping authorities. For some time, lighting restrictions were issued by the Home Office, the War Office, the Admiralty and various local military and civil authorities down to and including individual decisions by local constables. Eventually, sole responsibility was given to the Home Office which thereupon issued blanket instructions superseding all previous statements from whatever source.[73] One current military authority, Air Vice-Marshal E.J. Kingston-McCloughry, has written, with respect to this decision which placed civil control over this military-connected problem of passive defence, that 'the first lesson in home defence problems had been learned'.[74]

An air raid warning alert system is another common civil defence method. In this regard also, the English story in this earliest period of need was one of trial and error and frequent change. At first it was deemed necessary to alert only a few key bodies and individuals – military authorities, Scotland Yard, transportation officials, the Speaker of the House of Commons, the 'ack-ack' crews[75] – but with the advent of the raids, the public demanded some general notice. The police were then ordered to circulate among the threatened population areas with placards reading 'TAKE COVER' and factory sites were also notified. However, false alarms brought on much unnecessary work stoppages, and by 1916 the government had again reverted to a policy of alerting only a selective list of contacts. The public demand for a general warning system continued, however, and it would later become strong enough to be decisive during the era of the Gotha bomber raids which will be discussed in the following chapter.[76]

Other than public warnings, official instructions were also given to the British people as to how to act during air raids. Scotland Yard gave the sensible advice to stay inside at such times, to keep protected, and the like. Then, on 17 June 1915,[77] Scotland Yard issued further instructions as to how to cope with dangers connected with the raids. Interestingly, the main dangers stressed were fire and *gas* and not blast effects. The public were advised to store water and sand for use in combating fires. To lessen the dangers of gas attack, they were advised to keep windows and doors closed and to keep homemade respirators handy. One such device was described – a breathing container of cotton waste saturated with washing soda. But the instructions also admitted that there was really no way to know the best formula to use in these gas masks ahead of time 'for until the poison used is known no antidote can be indicated'.[78]

It is certainly remarkable that the British Government issued detailed warnings to its civilian population about aerial gas attack as early as June 1915. While it is true that the Germans had used gas earlier than the well-known instance of the battle of Ypres (22 April 1915), neither of the two earlier attempts had received much publicity or impressive results. (A gas which irritated the eyes and nasal passages was used on 27 October 1914, at Neuve-Chapelle and an improved lachrymatory gas was used against the Russians on the Polish battlefront on 31 January 1915.)[79] Further, there had been no aerial gas attacks at all, nor would there be throughout the war.[80]

One sees here, with respect to gas fears, another pattern being established which will keep affecting the history of British air defence policy thereafter. This 1915 warning was only the first public notice of English official concern about gas raids on England. The interwar years would see an immense increase in this concern. Eventually, by April 1939, the Government would approve an expenditure for fifty-seven and a half million gas masks for the civilian population in preparation for the expected gas raids of World War II — which did not occur either.[81] Also indicative of the future was the confusion expressed in these 1915 instructions concerning methods of prevention. English officialdom was to stay confused in this matter throughout the interwar years with an ever increasing sense of anguished impotence.

The instructions given by Scotland Yard for controlling fire were far more to the immediate point. The majority of the bombs dropped by Zeppelins upon London in 1915 were incendiaries.[82] One such raid, on 8 September 1915, resulted in the greatest material loss to England of all the air raids the country underwent during the war.[83] On the other hand, loss of life was very light from fire raids. During the whole war only eight Londoners died (and seven were injured) because of incendiary attack.[84] Certainly England was fortunate that the incendiary bombs of that day were relatively inefficient as were the bombing devices with respect to accuracy.[85] Later in the war, fire service officials admitted to the Home Office that they were not prepared for really large-scale fire attacks,[86] but nevertheless, the efforts of the London Fire Brigade must also be listed as being instrumental in restricting the fire damage and casualties.[87]

Naturally, there were attempts to improve the fire services, and one such attempt involved another early instance of operational research for war-associated purposes. A committee was appointed to investigate which of various fire extinguishing devices available was best for air raid caused fires. Rooms were set on fire by means of incendiary bombs (and by other means for comparative tests) and then the fires were treated by the various means being tested. The tests uniformly showed that liquid chemical extinguishers were the most effective.[88]

The Government also took a step which foreshadowed later actions in World War II by issuing insurance against air raid damage (especially fire damage) that had formerly been left uncovered by private insurance. For some years, going back long before the war, fire insurers had limited their responsibility to fires from various *specified* causes 'in order to avoid the possibility (however remote) of being involved in a loss too great for their resources'.[89] Among the causes usually omitted by special clause were fires started by any act of war. War-caused blast damage was also usually uncovered. The whole problem of uninsured loss thus became of great public concern after the Zeppelin raids started. A committee was appointed by the Government to enquire into this need and the means required to meet it. The committee's report was issued on 9 July 1915 (Cmd. 7997), and its recommendations were adopted in full. Under its plan, His Majesty's Government insured war damage by either aircraft alone or by aircraft and other sources of bombardment (the rate increased if the second choice was desired). All potential property to be insured was divided into five rating classifications, roughly into divisions of private, agricultural and varied commercial specifications. Obviously, such protection eased public concern about war damage and this precedent was again adopted by Great Britain during World War II.[90]

The last major aspect of the defence by Great Britain against the Zeppelin threat which must be described is an analysis of the many re-organisations of the various controlling agencies of that defence. These reorganisations included both military and political changes. Another proof of the Government concern about this defence is the frequency of changes that were made to help find the right answers to the problem. This concern was politically pragmatic among other reasons. Naturally, His Majesty's Government could not turn a deaf ear to the 'ominous rumblings' from the press, the public and Members of Parliament.[91] As was noted in Chapter 1, these 'rumblings' were coming from all sectors of the populace, even including the Queen Dowager.[92]

The Admiralty made the first major organisational change in September 1915, when it put the redoubtable, if controversial, Admiral Sir Percy Scott in charge of the ground defences of London.[93] By then Arthur James Balfour had superseded Churchill as First Lord of the Admiralty[94] so it was he who made this appointment.[95] Scott was an obvious choice because of his outstanding work in naval gunnery before the war.[96] And much of the improvement in anti-aircraft guns and ammunition was initiated during his short tenure (five months and six days) in this position.[97] Nevertheless, the navy was acting as only a stop-gap authority over home air defence because of the army's early inability to assume this role. It was understood that the task would revert to the War Office when they were ready to assume it.[98] Thus the next organisational shift was the very considerable one of returning air defence responsibility to the army. This occurred, with respect to London's air defences, on 16 February 1916, and in April 1916, the army's responsibility was extended to all of England.[99] The army immediately showed the importance it attached to this charge by making Field Marshal Lord French, then Commander-in-Chief, Home Forces, responsible for the air defences.[100] Viscount French (formerly Sir John French) had been, of course, Commander-in-Chief of the British Expeditionary Force until recently replaced in that post by Sir Douglas Haig.[101]

Such major shifts in authority are not made without fault-finding and obloquy. Sir Percy Scott was certainly bitter. He was later to write, 'as my scheme of defence was not complete, it seemed a pity that new people with new ideas should take it over, but we did many peculiar things during the war'.[102] The choice of Lord French as Scott's replacement was severely criticised by William Joynson-Hicks in Parliament on the day it was announced. Jix (as Joynson-Hicks is commonly known) stated that French had far too much to do as Commander-in-Chief of the Home Forces to be sufficiently effective as air defence chief as well. He stressed that England really needed an Air Minister to handle all phases of aerial concerns. (Thus Jix is another prominent voice of authority who stressed the unity of aerial concerns.)[103]

Jix's speech was the first of a number of speeches on air policy that day (16 February 1916). All of the speeches from non-ministerial MPs were highly critical of the air defence operations. A recent raid (on 31 January 1916) by several Zeppelins upon the Midlands especially provoked harsh criticism. Furthermore, all these speakers supported Joynson-Hicks in his idea of a separate Air Minister. Some would have this official be over the air services as presently divided, others would have the air services merged into a separate air branch under the new air minister. All expressed great concern that England was shamefully inadequate in its preparations for air warfare. (Mr Warwick Brookes in a maiden speech said that the proposed Air Minister should receive Cabinet rank, be appointed to the War Committee, and be given similar powers to the civilian chiefs of the two older services because 'it is quite possible to see in the near future that they [aircraft] may become a deciding factor in this war'.[104] This general attack on the government was answered by various government spokesman – H.J. Tennant for the War Office, Herbert Samuel for the Home Office, A.J. Balfour for the Admiralty and Andrew Bonar Law for the Conservatives within the coalition Government – and it is equally noteworthy that all of these gentlemen saw fit to *apologise* for the present state of air defence. Various excuses were offered, such as the new nature of the threat and of air warfare in general, the extent of country that must be protected,[105] and the problem of *matériel* especially when one considers competing needs from other vital phases of the overall war effort.[106]

This debate shows many of the tensions and resulting pressures in England to effect the reorganisation of air defence and the air services in general. It was the January 1916 raid on the Midlands which prompted the new governmental decisions. The raid had been very successful and the defences had been almost nil. (In Birmingham there was only one plane with sufficient motor power available and an RFC pilot flew it with a passenger armed only with a rifle.) The Cabinet was forced to listen and react to the 'profound effect' that this raid caused throughout England.[107]

The Government also had to listen and react to that other strong trend evidenced in the parliamentary debate summarised above, the questioning of the overall aerial policy in England. The various suggestions for an Air Minister that were voiced in that debate reflected this widespread questioning of a totally divided control of air power between the army and navy. There was much justification for this criticism. The two services maintained different standards for supply. They contracted for their planes, motors, etc. at varying prices so that each branch tended to bid up costs, and such competition resulted in delays in delivery.[108]

The call for an Air Minister did not necessarily reflect the wish for a merged air service. Again, the Parliamentary debate reflects the general opinion. Some did want the merger while others believed such a radical change would be impossible in the midst of war. So the latter group called for one man with overall authority in aerial concerns to be placed over both services to prevent their wasteful competition. Many also believed such overall authority would result in a much stronger air offensive. This thinking reflected the felt need for reprisal raids on Germany, and the belief that military and naval control over air power resulted merely in tactical support by air to operations on land and sea.[109]

Lloyd George has written a very clear expression of that argument:

> Concentration on the use of aeroplanes as handmaids to the operations of fleets and armies meant a failure to develop their possibilities as an independent arm, and even left it uncertain which of the two branches should properly undertake such tasks as the defence of London against attacks by Zeppelins and German aeroplanes, or the carrying out of reprisal raids on German towns.[110]

Names were already being bruited about for the job of Air Minister, Lord Northcliffe[111] and Winston Churchill[112] being two examples.

It was, then, in answer to this pressure for an aerial authority separated from strictly navy and army concerns that the government announced the creation of a Joint War Air Committee headed by the Earl of Derby who was to be aided by the fervent air power advocate in the House of Lords, Lord Montagu of Beaulieu.[113] This was the first step in governmental organisational policy which was finally to result in the creation of the first independent air service in history, the Royal Air Force. The impulses behind this first step continued to pressure England toward that final result: public, press and parliamentary agitation; inter-service rivalry and its resulting wastage and inefficiency; and the wish for an air force which could carry out strategic raids on Germany. The major phase of this history will be an important concern in Chapter 4, but the beginning steps occur in the Zeppelin era and will be discussed briefly here.

The Derby Committee (as this first committee was commonly called) was hardly a success. Typical of beginning governmental attempts to answer new problems, it was given too few powers over too narrow a purview. Its terms of reference concerned solely questions of supply, such as standardising models, allocating equipment and attempting to increase production. And even in this restricted scope of interest, it had only advisory powers.[114]

What this committee did bring conclusively into the open was the bitter inter-service rivalry over aerial prerogatives. The Admiralty was especially intransigent about compromising in any area of its independent action.[115] Both Derby and Montagu resigned their 'impossible task' after some six weeks,[116] and in his letter of resignation to Prime Minister Asquith, Derby claimed that it would be 'quite impossible to bring the two wings closer together' unless they were merged into a single air service.[117]

Certainly, the Derby Committee failed to silence the public clamour over air policy. By March 1916, the demagogic Noel Pemberton Billing was in the Commons, and he was particularly scornful with respect to Derby and his committee. He claimed that the Government was just tossing a sop to the popular demand for air defence action by way of one of England's long-established titles.[118] He railed against the limited scope of the committee and at its lack of power and responsibility. He charged that the Earl had no knowledge of aerial matters. (Lord Derby later admitted to this ignorance in a speech to the House of Lords.[119]) Billing concluded by saying the whole approach was wrong in that this type of answer (the Derby Committee) maintained the divided control of the air service. He said that the air service would be stunted in its proper development by such divided authority.

> Somewhat of a Chinese principle has been adopted. The Army has bound one of its feet, and the Navy has bound the other foot, until it will never be able to walk by itself unless some definite change in the existing policy is adopted.[120]

Neville Chamberlain, a politician whose career was to be profoundly affected by air power concerns, was another who voiced early thoughts about air policy at this time. He was then Lord Mayor of Birmingham and on 12 April 1916, he chaired a meeting in Birmingham in which the key address was made by Lord Montagu of Beaulieu. This meeting gave Montagu a public forum in which to explain his and Lord Derby's resignations announced two days before. *The Times'* report of this meeting described Neville Chamberlain as a man 'who has taken a firm stand upon the defence of the country against Zeppelin attacks . . .' *The Times* also reported that Chamberlain stressed at this meeting the need to achieve air supremacy comparable to the old need for naval supremacy and that he said a separate air ministry 'had become absolutely essential for our national security'.[121]

The failure of the Derby Committee brought on the next governmental attempt to solve the puzzle of air policy and divided air authority, this being the creation of an Air Board with the eminent Earl Curzon of Kedleston as President. Lord Curzon's appointment indicated the importance that the Government placed in this Air Board. This former Viceroy of India was then Lord Privy Seal and a member of the Cabinet. Later, in 1923, he was even a strong candidate for Prime Minister.[122]

The Air Board's charge was a much greater one than that of the previous Derby Committee. General Brancker, one of the members of the Board, has furnished a brief description of its responsibilities:

> To consider and report on air policy; to coordinate the supply of aircraft material; and to arrange for the interchange of ideas between the Navy and the Army upon all air problems. The Board was advisory to the President and did not put debatable questions to the vote. It was described as 'free to discuss policy' and recommend course of action to the Admiralty and the War Office, together with definite proposals for the distribution of material. If the Admiralty and the War Office refused to take these recommendations, the President was to refer the matter to the War Cabinet.[123]

In his speech to the House of Lords announcing the formation of this Air Board, Lord Curzon showed that he considered his scope of purview to be very broad indeed. He freely admitted that the Government considered the Board to be a halfway measure between the impotent Derby Committee and an independent air ministry. He said that disagreements between the Admiralty and the War Office prohibited going further in the direction of a merger of the air services — *at that time*. Nevertheless, he prophesied, an independent air ministry 'is destined to come'. Moreover, 'the Board which has been appointed will undoubtedly hold this consummation in view. It is one of our duties to explore the ground and to examine the possibilities of such a solution.' He also stated that another task of the Air Board would be to examine possibilities of a long-range bombing offensive against Germany and to determine how England must organise its air power for that strategic task. He also mentioned air raid defence as another of their assigned areas to examine.[124]

Besides Curzon, the Board included among its members Admirals Tudor and Vaughan-Lee, respectively the Third Sea Lord and Director of the RNAS; General Sir David Henderson, Director-General of Aeronautics for the War Office (and a man who would play a most important role in the formative period of the RAF) and General Sir Sefton Brancker of the RFC; Lord Sydenham, the Deputy President, a man with much experience in the Committee of Imperial Defence; and Major John Baird who was the Board's spokesman in the House of Commons.[125] In sum, it was a powerful grouping of worthies and thus a strong proof of the Government's serious intent.

Nevertheless, the Curzon Air Board was basically a failure. It was never able to achieve the desired goal of cooperation between the two services. Since the Board was given no power to enforce its decisions, its function primarily had to be mediatory. This is a task which demands utmost tact, and Lord Curzon was never known for that quality. A.J.P. Taylor speaks of his 'pompous arrogance',[126] and Sir Sefton Brancker, although generally complimentary about Curzon, remarks on his 'complete disregard for other people's feeling'.[127]

Thus the task of prompting leaders of the army and navy into greater cooperation and eventual merger of their air services was certainly beyond Lord Curzon's ability — indeed it would have strained the wisdom of Solomon. The army's leading aerial field commander, Hugh Montagu Trenchard, held Lord Curzon in some personal contempt and argued against the broad scope and vague powers of his Board.[128] It was, however, the Admiralty who most resisted the Air Board's work.[129]

In his first report to the Cabinet (23 October 1916), Lord Curzon made no secret of where he believed the major fault lay in England's aerial development:

> The addition to the Navy of responsibilities for the air — not in itself necessarily impracticable — has in the manner in which it has been carried out, been attended with results that have been equally unfortunate to the Navy and the Air Service, and, if persisted in, will be incompatible not merely with the existence of an Air Board, but with the immense and almost incalculable development that ought to lie before a properly coordinated and conducted Air Service in the future.[130]

Then, in November 1916, Curzon again reported to the Cabinet about the Admiralty's lack of cooperation.[131] The Cabinet spokesman for the Navy, A.J. Balfour, First Lord of the Admiralty, was not one to leave such criticism unanswered. He had opposed the establishment of the Air Board[132] and he was certainly no friendlier to it after Curzon's attacks. He answered with a Memorandum to the Cabinet which was particularly bitter. Among its statements, for example, is the following:

> To do the Air Board justice, however, they are far more interested in abusing the Admiralty than in praising themselves. I do not suppose that in the whole history of the country any Government Department has ever indulged so recklessly in the luxury of inter-Departmental criticism.[133]

Struggles within the Air Board,[134] therefore, were reflected by comparable struggles within the Cabinet. And Lloyd George claimed that these latter squabbles so disrupted Cabinet business that they were a major cause of the impotence of Asquith's final months in office so far as Cabinet level affairs were concerned.[135] Since these *were* the final days of Asquith's tenure as Prime Minister, others have concluded that this weakness at Cabinet level, due to these tensions over air policy, helped push forward the fall of the Asquith coalition. Thus the excellent biographer of Lord Trenchard, Andrew Boyle, states that Curzon's disenchantment with Asquith over his vacillation on these matters 'easily persuaded [him] to throw in his lot with the major conspirators then engineering the overthrow of Asquith'.[136]

There was some contemporary thinking along parallel lines. For example, Brancker wrote Trenchard a letter on 3 December 1916 in which he said, 'Curzon is quite strong upon being determined to control supply and will resign unless Government agree. If he resigns it is possible that Government will go altogether'.[137]

However, while it is certainly true that questions about aerial policy have often affected British policies and statecraft, it is also true that writers on aerial matters are all too prone to inflate their claims. In this instance, the evidence is quite conclusive that the Cabinet dissolution of December 1916 was caused by factors mostly unconnected with these Air Board conflicts. Roy Jenkins' recent study of Asquith includes a long and very close analysis of the events of this dissolution, and Curzon's role (and the service problems in general) was relatively minor.[138] However, there *was* one stage in that history where the combined opinion of the so-called 'three C's' (Curzon, Robert Cecil, and Austen Chamberlain) did certainly affect Asquith's thinking,[139] and to that extent Boyle's argument has some validity. Lastly, in this regard, Lloyd George's contemporary opinion about Asquith and air policy supports Curzon's chagrin. Lord Riddell's diary entry of 10 June 1917 reports a dinner conversation with Lloyd George about a shortage of airplanes. Lloyd George stated in this regard, 'the shortage is due to Asquith's vacillation last year, when he delayed the settlement of the disputes between the Navy, the Army, and the Air Board'.[140]

Under the new ministry of David Lloyd George, the traditional game of ministerial musical chairs meant changes in the Air Board. Lord Curzon left its leadership for the major post of Lord President of the Council.[141] The Air Board itself was reconstituted under parliamentary legislation with, finally, some executive authority, this being over all aircraft design and production.[142] Lord Cowdray (W.D. Pearson, First Viscount Cowdray), highly reputed financier and construction engineer, was made President of the new board.[143]

The Zeppelin Era: The Successful Response 49

The era of the Cowdray Board extends beyond the period covered by this chapter, and so further discussion of this overall development will be continued in the next chapter. An early summary of this particular history can be made, however, from the evidence already presented. The attempt by the British Government to obtain some unity in the control and efforts of its air services was not an easy undertaking. The passions were great and the entrenched interests were strongly divided. One may conclude that a felt need for unity had to be adopted as an absolutely essential priority before success in that attempt could be attained. That felt need appeared in the Zeppelin era and so the attempt was started. Following chapters covering the 'Gotha' period will describe how that need became seen as indispensable, and thus will show how success — the creation of the Royal Air Force — was achieved.

It is time to recapitulate the measures taken which, in combination, produced the answer for England to the airship threat. Technical improvements and increased production were achieved in anti-aircraft weapons and shells, in searchlights and in defence planes and their gunnery (most especially in the development of incendiary bullets); increased efficiency was gained in ground observation and communications, in various phases of passive defence, and in overall direction and control, both militarily and politically. Naturally, in part these developments reflected the normal improvements that would come with experience; in part they also represented the fruits of the zealous efforts made to find answers to what was believed to be a threat of extreme danger. Naturally, also, these improvements were interacting. Changes in organisation brought new thinking into play. War Office control meant, for example, a turn to squadron-sized concentrations of the defence pursuit planes rather than the much smaller groupings used by the navy. The War Office also set up the belts of lights, gun and plane defences already described.[144] By the end of 1916, twelve defence squadrons were operational from the Thames estuary to Scotland, employing some 2,200 RFC personnel.[145] And, finally, all these ardent efforts started to produce results.

The first success against the airships that occurred in the skies over England itself happened on the night of 2 September 1916, and it was a particularly spectacular sight as the airship (in this case, a Shütte-Lanz type rather than a Zeppelin proper[146]) came down in flames in sight of the London populace. It crashed in Cuffley, Middlesex, and immediately the roads to Cuffley became crowded with the jubilant observers. The British pilot 'who gave London its most dramatic war spectacle . . .'[147] was awarded the Victoria Cross within hours of the event. His own report clearly shows that his plane (a B.E.2C), the new incendiary bullets and the effectiveness of the searchlights were all instrumental in his success.[148]

Further successes occurred with encouraging frequency. On 23 September 1916, an airship was forced down by Little Wigborough and its crew survived and were captured. Within a week three more victories were achieved. Two more airships were shot down over England in November 1916. This listing only includes those shot down over England; other victories were being scored over the Channel.[149]

Naturally, these achievements heartened the English. There is ample contemporary evidence of this. For example, the influential journalist, John Evelyn Wrench, wrote on 1 October 1916, 'we certainly are getting our air service much better'. (This with respect to a double victory over the Zeppelins on the day before.)[150] Lord Grey, for another example, had been one who doubted that England could survive the three-fold threat of the airship, the mine and the U-boat but, by November 1916, he wrote in a Cabinet paper, 'the Zeppelin, I believe, for offensive purposes, to be a complete failure, and I have ceased to be anxious about it'.[151] Of vital importance was the fact that the growing English confidence was met by growing German disillusionment in the airship's effectiveness. According to the German aerial chief, von Hoeppner, 'the practical use of dirigibles was very much limited towards the end of 1916 through the perfection of the hostile defences'.[152] By January 1916, Germany turned to only a very limited use of airships and this mostly against Russia and the Balkans.[153] England had won her first air defence challenge. There were, however, some intermittent Zeppelin raids on England until the final one occurred on 5 August 1918.[154] One of these later raids had been a very strong effort by the Germans. In October 1917, eleven Zeppelins were sent out for targets in the Midlands but in this case heavy winds caused the loss of four of these ships.[155] In January 1915, Winston Churchill cited weather conditions as the only real English defence against the airships[156] – in October 1917 weather still proved to be the *most* effective defence, if the wind was at gale force strength.

Wartime statistics are notoriously inexact. In January 1919, the British Government published its official figures for the air raid losses in England and they are still being cited without question.[157] However inexact, these figures nevertheless indicate how small the Zeppelin threat really turned out to be, notwithstanding the gravity with which it was viewed by contemporaries. In a total of fity-one raids, German airships dropped close to 6,000 bombs on England, the total weight of

The Zeppelin Era: The Successful Response 51

these bombs being under 500,000 pounds. Five hundred and fifty-six people were stated as killed and 1,357 wounded by these raids.[158] Certainly these figures would have been much greater but for the defence efforts described. If the threat was small (and, compared to most concentrated aerial campaigns, it was so), it nevertheless could have been much greater.

Unfortunately for England — and for later military trends in general, as shall be shown — her aerial fears were hardly over with the defeat of the Zeppelins. The writer W.O. Horsnaill gave a very prescient view of England's immediate aerial future in an article published in June 1916. He correctly saw that the Zeppelins would be a passing danger (this even though none had been destroyed over England up to that time). He realised that they would prove far too vulnerable. But he also forecast that Germany would soon replace the airships with long-range bombers which would prove to be a far more dangerous threat. He claimed that Germany was developing a fleet of bombers and, typically, he called for England to start doing likewise.[159] All that Mr Horsnaill forecast occurred and *that* history will be this study's next immediate concern.

3 DEFENCE IN THE GOTHA ERA

During the Zeppelin era, British aerial theory and doctrine slowly formed. In the following period when England experienced raids by heavier-than-air craft,[1] these rudimentary lines of development retained their basic patterns while they became more firmly set and supported. The first major airplane raid on London in history is a fruitful example of this process of development.

This historic raid occurred at midday on 13 June 1917. Fourteen German bombers appeared over London although twenty had started on the mission, the other six having been forced back because of varied mechanical troubles. The raiders dropped over one hundred bombs in the greater London area, killing 162 and injuring 432 more. Almost £130,000 material damage resulted, most of this within a mile of Liverpool Street Station, i.e. in the heart of the City of London itself. The East End of London experienced, however, the horrifying incident wherein one bomb pierced the roof of the Upper North Street School and continued on through three floors until it exploded in the basement where a kindergarten class was in session. The casualty count from that single bomb was eighteen children killed, forty-five more injured, plus two adults who were wounded. Another bomb also hit a London school building but fortunately did not explode. All told, this raid caused the highest casualty count of any experienced by England during the war. No German planes were lost.[2]

The bombing force which carried out this raid was a direct descendent of that original German air unit given the mission of the strategic bombardment of England, the Ostend unit under the command of Major Wilhelm Siegert.[3] Many of Siegert's pilots were assigned to this new unit (called *Kampfgeschwader* 1) to constitute a nucleus of experienced personnel. The new unit was assigned for training to the same airfield near Ostend as Siegert had used. The planes they used were the result of the attempt by German authorities to develop bombers which did not have the limitations of Siegert's primitive machines.[4]

These were the famous Gothas, specifically, the Gotha G.IV model, named after their producing factory, the Gothaer Waggonfabrik. The G.IVs were greatly improved models of former G-type bombers ('G' standing for *Grosskampfflugzeug*) which had seen war front service since 1916. The fuselage of the G.IV extended over forty feet, the wing span

Defence In The Gotha Era   53

was nearly seventy-eight feet (longer than any German bomber over England in World War II), two 260 h.p. Mercedes engines powered it and, depending upon the bomb load, it had a cruising speed of seventy to eighty miles per hour. The bomb load varied according to the ceiling desired — for daylight raids at heights approaching 18,000 feet, the G.IV could carry close to 700 lbs. of bombs and for lower flying night raids at about 10,000 feet, it could carry about 1,100 lbs. payload. The G.IVs were also equipped with protective armament of three machine guns for the three man crew. By World War I standards, these Gothas well deserved the sobriquet given them by one Parisian observer, 'les grands oiseaux de mort'.[5]

The era of the Gothas did not represent the first airplane raids on England. There had been some hit and run raids before by a few adventurous pilots; by the end of 1916 these had added up to twenty-five raids of which six did not involve any bombs dropped. The total casualty and damage results were very low, twenty killed (including fourteen in one raid alone) and under £12,000 in material damage.[6] Because these raids were few and produced light results, the public reaction was more temperate than it was to the Zeppelins. One airplane flew over and bombed London, for example, on 28 November 1916 (the first airplane raid on London ever), and the diary note of Lord Bertie (the English ambassador in Paris) offers a typical contemporary note: 'C. Hardinge writes that the German seaplane which passed over London dropped half a dozen bombs near Harrods' Stores without doing much damage'.[7] Sir Basil Thomson's diary note was equally dispassionate: 'An aeroplane dropped bombs in Knightsbridge, Eaton Place and the Victoria Palace. No doubt that they were aiming at Buckingham Palace. The plane was not seen'.[8] But the English were soon to lose whatever insouciance they held regarding German airplanes.[9]

Deliveries of the Gotha G.IV models to the Ostend-based unit started in March 1917. After the necessary test flights and other preparations, the bombing force was ready for its mission by mid-May. Field Marshal Paul von Hindenburg visited the unit to wish it well and to show the government's interest. Weather conditions delayed their first attack until 25 May when twenty-three planes took off for London. However London was saved in this instance by a cloud cover and so the first raid of the Gothas was on the alternate target area of Folkestone. Even so, the casualty rate was a third higher than the worst of the previous Zeppelin raids, 95 dead and 195 injured. A second raid on the English coast occurred on 5 June because London was again protected by a weather cover, and so Sheerness was the next town to suffer this new phase of the air war, at a cost of forty-five casualties. The third attempt by the Gotha raiders was the 13 June raid on London.[10]

Another continuing aspect of the German strategic bombing plans against England was political division within the state over these plans.[11] Chancellor Bethmann-Hollweg was, again, a strong opponent to the idea. At the time of the June raid on London, he was stoutly supporting the need to enter into peace negotiations with the English. On 25 June, he wrote von Hindenburg about this necessity and, in this context, he argued against the recent raid on London:

> We could help from outside [sic] by avoiding irritating the chauvinistic and fanatical instincts of the English nation without cause. There is no doubt that the last aerial attack on London has had a disastrous effect in this respect. According to reliable reports the anger of the English public has reached such a pitch that English statesmen who were not averse to making peace have declared that no English government which was willing to treat with Germany after such an occurrence would be able to withstand the indignation of the nation for a day. As I am unable to believe that such aerial attacks are absolutely necessary from the military point of view, may I be allowed to suggest that they be given up, in view of their disastrous political effects.[12]

The Field Marshal answered this letter on 7 July; perhaps not wholly coincidental is the fact that the second Gotha raid on London occurred that day. Hindenburg told the Chancellor that England could only be *forced* to negotiate, that a softening of the German effort would only mean a hardening in the English determination. He therefore concluded:

> We must, therefore, prosecute the war with all our resources and the greatest intensity. Your Excellency deprecates the aerial attacks on London. I do not think that the English nature is such that anything can be done with them by conciliation or revealing a desire to spare them. The military advantages are great. They keep a large amount of war material away from the French front and destroy important enemy establishments of various kinds. It is regrettable, but inevitable, that they cause the loss of innocent lives as well.[13]

It is also regrettable and perhaps inevitable that the purposes of strategic bombing are expressed so equivocally. The instructions given to the

Commander of the bombing squadron, Captain Ernst Brandenburg, were more explicit about the purposes of his assignment; that it was expected that such raids might help bring about peace by attacking 'the morale of the English people' and thus undermining their 'will to fight'. Moreover, such military targets were mentioned as war industries, docks and transport facilities and the like, and, as von Hindenburg did mention in his letter, there was also the expectation of a large withdrawal of defence *matériel* from the Western front.[14]

Bethmann-Hollweg lost his office on 12 July 1917, largely due to his position on peace negotiations and the opposition to him from the Army leadership.[15] Thus his opposition to the air raids was no impediment to their continuance. Fritz Fischer's statement in this regard has some interest. 'The bombing attacks on London were continued and the conflict between Britain and Germany sharpened into real hatred between the peoples'.[16] It would be fallacious to argue that 'real hatred' did not exist in many English minds before the Gotha raids[17] but certainly, passionate emotions were incited by these raids. This, in fact, is another of the continuous trends in this aerial history from the Zeppelin era through the Gotha period. Chapter 1 often noted the passions generated by Zeppelin raids but the response by the British to this newer danger was yet more fervid.

One *Times* reporter immediately recognised this point of increased hatred. In his covering story the following morning after the 13 June raid, he wrote, 'if it were possible at this time of day for the enemy to increase the utter and almost universal detestation in which he is held by the people of this country, he did it yesterday'.[18] Probably the most passionate single reaction was to the killing and maiming of the children in the Upper North Street School. Great masses of people crowded in attendance of the public funeral processions for the victims.[19] The *Daily Mail* used its pictorial back page for pictures of the faces of these casualties.[20] In the first of these pictorial presentations (on 15 June) this paper put in apposition a particularly significant dual message. Under the children's pictures was presented a map of the German towns within 150 miles of the Western front. The caption headline made the point: 'Unwarned Child Victims of the War — A Reprisal Map of Germany'. Both of these points, the need for public warnings and the call for reprisal raids, were repeated again and again in the post-raid furor.

As already noted, the public clamour for an air raid warning system was sufficiently strong during the Zeppelin era to affect governmental policy. The Government had originally followed a policy of alerting only a few designated people and services but under public pressure, it had changed, for a time, to general public alerts. However, the work stoppages brought about by false alarms caused the Government to return to the former system.[21] After the 13 June raid, pressure from public demands was given renewed justification as witnessed by a statement (made before the London raid) by the officer in charge of all home defences of England. After the Folkestone raid, Lord French (Commander-in-Chief, Home Forces) told the War Cabinet that the warning system then in use was practically worthless.[22]

The popular demand for a public warning system had good representation within Parliament. In the House of Commons, there were demands made for such a system in the Question Time period during every session from 14 June to 29 June with only one exception.[23] Among these demands, only one was answered by the Government with full confidence and firmness. Sir Frederick Banbury (representing the City of London) suggested that the bell of St Paul's or some such device should be used in the future to give the City early warning and thus allow time for the financiers' stocks of money to be transferred into protected safes. Chancellor of the Exchequer Andrew Bonar Law had no problem answering that statement when he stated that His Majesty's Government was more concerned about people's lives than their money.[24]

The difficulty was, however, that there was great disagreement about what was the best method to protect lives. Government spokesmen kept insisting that if an early warning system was established the result would be that *more* people would be out in the streets. The Earl of Derby (Secretary of State for War) emphasised this point in the debate on the day following the raid. 'I believe that if one did give notice it would only bring more people into the streets than were in the streets yesterday'. He remarked upon the fact that people rushed into the streets after the raid started. 'Of course, it shows that there has been no terrorising about the raids, but it equally shows that there is more danger in sounding a warning than in not doing so'. But Derby admitted that the question was still open and being discussed further.[25] (The Earl's comments about the lack of terror is indicative of the fact that the public morale factor was as much a consideration in official circles in England as it was in Germany. While the German military believed they were attacking the English will to war, the British leaders worried about whether that will was weakening.)

Lord Sydenham, who had been an able member of the Air Board under Lord Curzon,[26] gave a well reasoned rebuttal to Derby's argument. He admitted that many citizens had gone into the streets or on to roofs to witness the raid but he also noted that many sought shelter and went into basements or other cover. He thus concluded that a public warning system would at least give the citizenry time to seek refuge if they wished it and that the Government owed them the opportunity to choose.[27]

Nevertheless, reasoned debate on this point was somewhat irrelevant. Although the governmental position in Parliament most often followed the Derby approach cited above, what the Cabinet were primarily concerned about was the problem of work stoppages — the same problem that led to the decision to stop advanced public warnings in the Zeppelin era. The government leaders naturally did not want to jeopardise England's industrial ability to meet the heavy war demands.[28] And their concern was soundly placed. One military authority later estimated that 75 per cent of the labour force in a given area did not work when that area received an air raid warning.[29] Only once, however, was this argument used by official spokesmen in the debates at this time. The Home Secretary, Sir George Cave, did mention the danger of production losses due to false alarms but, still, his primary stress was that warnings could add to the casualty totals by bringing people into the streets.[30]

The next air raid on London occurred on 7 July 1917, a Saturday, and when Parliament met the following Monday, the attack on the Government's air defence programme was fiercely renewed.[31] The very first outburst in the House (from Sir H.H. Dalziel) emphasised the public pressure for early warning. 'There is a strong feeling in favour of sirens, or something of that kind, being used in regard to air raids'.[32] Sir Frederick Banbury's anxiety had changed from worry over exposed money to concern about exposed people. He argued that the Government should quit worrying about the fools who rushed outside during a raid and realise that it was their duty to give sensible people time to protect themselves.[33] Only two members spoke against the idea of a public warning system and, significantly, their concern was that the populace would get too panic-stricken. Sir William Pearce prophesied that just two weeks of such warnings would see 'London in such an extreme state of excitement that it would have disastrous effects'.[34]

The Home Secretary's answer indicated the Government's confused position on the issue. Sir George Cave said that the Cabinet was still against the proposal; that early warnings would tempt workers to return home to see to their families and that once there, they would be lost for the whole day. But he also admitted that people were seemingly more cautious, more prone to seek shelter, and that 'the matter must be reconsidered, and that we must have regard to the new facts'.[35] The reconsideration did not take very long. On 12 July, Sir George told the House that the Government had decided to establish an early warning system for London.[36]

The arrangements made for the public alert system, 'which to those accustomed to the sirens of 1939-1945 may appear somewhat primitive',[37] nevertheless established the pattern of public warnings which remained part of England's air defence doctrine thereafter up to and through World War II. Maroons — sound bombs fired from small brass mortars — were used as well as policemen (on foot, bicycle or car) carrying alerting posters. The populace were supposed to be notified at least five minutes before an expected raid. 'All Clear' signals were given after the raids, in part by boy scouts blowing on bugles as they were driven about London by car. When the Germans turned to night raids (by late summer, 1917), there was some hesitation about disturbing sleep by alarms sounded at any hour of the night but this too became established practice by spring, 1918.[38]

The governmental fears about people rushing outside when alerted proved unrealistic. Lady Cynthia Asquith's diary note of 1 October 1917 illustrates the actual results. She was out in the early evening hours when an alert was sounded; 'A policeman flew past with the cry: "Air-raid warning: TAKE COVER" ... The sudden silence that falls on London is wonderful: all the traffic ceases, and the streets magically empty — the whole population swallowed up into the houses'.[39] The problem, indeed, was not people rushing into the streets but, rather, people rushing for shelter. In one instance, a stampede towards two underground subway shelters was started when the bursting sounds of the alerting maroons were mistaken for bomb explosions and the result was fourteen dead and an equal number injured, mostly women and children.[40] (The government account of the tragedy given to Parliament the following day seems a bit inadequate as the casualties were described as being 'very few'.[41]

It was thus in the First World War that Londoners started that troglodyte activity of seeking refuge in the 'Tube'. The two summer daytime raids of 1917 (13 June and 7 July) caused sufficient numbers to do this to force the Commissioner of Police to send constables to all the underground stations to maintain order. When the Germans turned to night raids by the Gothas starting in September 1917, the practice became habit for upwards of three hundred thousand people on all moonlit nights, whether alerted or not. They came with bedding, food, pets and other personal possessions, and they came to stay until all possible chance of attack was over for the night.[42]

The poor and the alien peoples of the East End constituted the great majority of these Tube dwellers. The boroughs of the East End were the doorsteps of London' for the incoming bombers. They were the first to be attacked and they received whatever bombs were left over when the raiders turned back for home. Thus the areas of the flimsiest housing and where basements were a rarity were the hardest hit.[43]

The conditions in these underground refuges were described in fine English understatement by the biographer of Lloyd George when he wrote that 'the scenes were not pleasant'.[44] A parliamentary discussion about these conditions, in February 1918, was more outspoken. As usual, in debates about aerial concerns, Noel Pemberton Billing participated.[45] He noted in detail how unsanitary the conditions were in the Tube and warned about the possibility of epidemics.[46] Another member spoke of the lack of bathroom facilities while another spoke of the many children in the shelters with too little clothing for good health. The Government spokesman, Sir George Cave, felt obliged to agree with these criticisms while stating that attempts were being made to improve what was obviously a very difficult situation; the stations were, for example, being thoroughly cleansed and disinfected after each night's use as shelters. But what Cave would not agree with was Billing's suggestion to have the police see that British women and children had first chance to enter the Tubes before the alien population filled them up.[47]

Crowded conditions in the underground stations were naturally a problem; the crowds were just beyond the ability of station officials to manage. Masses of people blocked stairs and platforms. In many stations the crush was so bad that passengers could neither get on nor off the trains.[48] A diary note of 17 February 1918 by Colonel Repington (by then the military correspondent of the *Morning Post* but more famous for his former association with *The Times*) indicates, however, that the British sense of courtesy was not all lost in the Tubes. 'Got home after a crush in the Tube. The platforms five deep with women and children taking refuge, and some five more rows standing, largely men'.[49]

Of course, Londoners took shelter in a great variety of places. They trekked to tunnels under the Thames, to open fields beyond the city, to police stations and other public buildings which offered more sturdy protection than their own homes, etc.[50] The poor were not the only ones who journeyed forth seeking safer quarters – but the wealthier were able to go further. Bath and Bournemouth were popular refuges[51] and the exodus forced rents up in Maidenhead and Brighton.[52] The most notable evacuee from the air raids was the Prime Minister himself. 'Lloyd George lacked physical courage. The air raids of the first war, and still more those of the second, terrified him, and he rarely spent a night in London.'[53]

There had been demands for a massive shelter programme to be initiated by the Government. One proposal, which was to be repeated much more vehemently in the 1930s, was 'that large underground shelters should be provided, in all thickly-populated areas, where people could seek protection when an air raid was in progress'.[54] After Winston Churchill became Minister of Munitions (in July 1917) he soon indicated his agreement that the Government should do more about providing air raid protection. His 'Memorandum of Protection from Air Raids' of 5 October 1917 is an important statement in that connection:

1. I have given directions that dug-outs and shelters are to be immediately provided under approved schemes in the whole of the munition factories in the bombing areas. Many private firms have already taken these measures with great advantage. The labour will be found from the people employed in the factories. The work should not take long. The loss on output must be accepted; it will certainly be much less than the loss caused by the people scattering to their homes whenever an air-raid alarm is given. There will also be a great gain in the feeling of confidence imparted to the workers. I hope these arrangements will be complete within ten days.

2. I consider that generally speaking people are entitled to a safe shelter within reasonable distance of their homes or their work. I consider that in or near each street a house or houses should be prepared affording reasonable security to the residents, and that in the vicinity of all large works, whether munitions or other, an adequate provision should be made for everyone ... I am impressed by the rapidity by which shelters have been provided in some of the munitions areas already, and I do not believe the task will be found a very formidable one. I expect the Germans are already hard at work providing proper shelters in the cities likely to be attacked.[55] It is especially important to the confidence of the population that in working-class areas consisting almost entirely of frail two-storey dwellings there should be sufficient shelters prepared. Where there are larger houses an issue of sand-bags and of leaflets containing clear printed directions as to the parts of the house which are safe, the

dangers to be avoided, etc., should meet the case . . . As long as people have a safe place to go to when firing begins and are compensated for the damage done to their houses, they will stand a great deal of hammering and get back to their work promptly when it stops.[56]

As can be clearly seen from this memorandum, what greatly prompted the Government to action was the fear of possible panic among the populace. There was no lack of specific instances to prompt this fear. This study has already noted some examples; Hull during the Zeppelin era[57] and the tragic results of stampeding mobs rushing to the Tubes.[58] Another tragic example occurred during the 13 June raid when a factory employing some seven hundred workers, mostly women, was struck not by bombs but by panic and over seventy hospital cases resulted.[59] Naturally, political leaders must concern themselves about such signs of weakening home front morale — as they viewed these signs. Lloyd George reflected this political sense when he wrote of the 'grave and growing panic' in the East End after the Gotha raids started.[60] The Cabinet quite properly considered the problem of public panic in its deliberations.[61] But nevertheless, this aspect of panic was gravely overemphasised as a lesson to be taken from World War I, especially during the interwar years. To cite just one example, the Air Raid Precautions Committee in 1930 suggested that three battalions of troops should be prepared to maintain order during an attack; if not, the Chairman warned, 'he had little doubt that such a state of panic would be produced as might bring about a collapse, certainly of the community in London, if not of the whole country.' And the report referred to World War I English experiences.[62]

The actual reactions by the British people to aerial attack in the Great War does not, however, support such simplified conclusions. As usual, history is a study of complexities. Outbreaks of panic among the citizenry do not automatically signify a loss of the 'will to war' even though the mental jump to this conclusion is made by so many air power authorities.[63] The word 'panic' is one which has deeply disturbing overtones to statesmen as well as to militarists. One can sense H.H. Asquith's sensitivity to the word in a letter he wrote on 6 October 1917, when he stated that 'the raiding season . . . has left behind it an afterswell of nerve-shock, if not exactly of panic, which is very discreditable'.[64] Winston Churchill's concern about popular panic has been noted.[65] The result is the natural tendency — seen in Asquith's comment — towards wishful thinking, towards the denial of the occurrence of panic. Under such impulses, official wishful thinking and the rhetoric of the popular press can follow like lines. The former Chief of the Imperial General Staff and later Commander-in-Chief of the home forces, Sir William Robertson, has written that the raids did not bring panic but, rather, a greater determination within the British people to defeat the enemy;[66] which can easily be compared to the *Daily Mail's* editorial rhetoric: 'They [the Germans] imagine that by a wicked, purposeless act of murder, such as the bombardment of a poor cottage at defenceless Broadstairs, they can frighten the British people ... The only result in England is to remind us that the German is a Hun.'[67]

Nevertheless, the facts are that there was panic and that there was also a stiffening of morale.[68] Fear is not an equivalent word for cowardice. Nor is capitulation the only alternative open to the coward. One noticeable indication of the stiffening of morale brought on by air raids was that recruiting centres processed more volunteers than normal on days following the raids.[69] (An early propaganda attempt to capitalise on this tendency was a recruiting poster created during the Zeppelin era. It showed a Zeppelin flying over St Paul's, spotlighted in a searchlight cone, and it read in part: 'It is far better to face the bullets than to be killed at home by a bomb. Join the Army at once and help to stop an air raid.'[70]

The raids also brought forth increased popular pressure for more defence protection (and, as will be noted in detail in Chapter 4, for retaliation in kind upon the Germans). In sum, then, signs of panic occurred at the same time as the 'will to war' increased, at least by the yardstick of increased mass agitation for more action. And that agitation was heeded. The attitude taken by Sir W. Pearce in a speech in the House of Commons (on 21 February 1918) really characterised the Government's activity throughout this period. Pearce prophesied that the defence of London might become a crucial factor in the war, and he expressed his fears that general panic was an immediate danger. He thus warned that the London poor must be given a reason to feel confident that all possible was being done on their behalf or else the result could be disastrous.[71] Just so did the Government act to appease the populace and reduce the potential for panic.

The pressure for added defence did increase immediately after the 13 June raid. *The Times* editorialised this need on the following morning, although it also stressed the point that the Western front should not be weakened for home defence needs.[72] William Joynson-Hicks made similar comments in the House of Commons that same day.[73] And in the War Cabinet, decisions were made that day which reflected these calls for additional defence. These decisions were reported in a letter from the CIGS, Field Marshal Sir William Robertson, to Sir Douglas Haig, Commander-in-Chief of the British forces on the Western front. Robertson said that the Government was not letting itself be intimidated into rash actions but that it had decided it was necessary to give the German raiders some opposition. Therefore, in that the planes then serving for home defence were not up to the challenge of the Gothas, they had to have some of the newer models that were in action in Haig's command area. Thus, Robertson informed Haig that he was to send one or two defence squadrons back to England. It was assumed that they would not be kept away from the front lines for long; that an expensive lesson furnished the Germans would be enough to end that threat.[74]

The War Cabinet knew that borrowing from the Western front was not the most desirable approach — the advice of *The Times* and of Joynson-Hicks was hardly necessary in that regard. Thus, in that same meeting, the Cabinet also decided upon a greatly accelerated aircraft production programme, even at the price of diminished production of other war *materiel*.[75] The final decision on the amount of expansion needed was made on 2 July when the RFC squadrons were scheduled to be raised in number from 108 to 200 (including all the additional facilities and equipment which that increase would necessarily entail) while the RNAS was also scheduled for like increases.[76]

However, long-term planning did not answer immediate needs and thus the call was made upon Haig's BEF squadrons. As Robertson stated, the defence planes then on hand in England were not up to the task. They were primarily the B.E. model fighters which had been so successful against the Zeppelins[77] but which did not have the rate of climb or speed to combat the Gothas effectively.[78] Nor was this a public secret in England — Lord Strachie had even pointed out these deficiencies in comments he made in the House of Lords.[79]

A major problem existed, however, in that Haig was already on record against sending any of his aircraft back to England. Lord French, in charge of all home defences, had already asked for such help after the first appearance of the Gothas over Folkestone. On 5 June 1917 he had written, 'I cannot too strongly impress on the Army Council my opinion that the means placed at my disposal for aeroplane defence are now inadequate and that a continuance of the present policy may have disastrous results.'[80] But, as an entry in Lord Riddell's diary clearly shows (on 10 June 1917), Haig was adamant about sending any aerial relief back.[81] In fact, the RFC requirements in machines for the

Western front as of 1917 had not been fulfilled[82] and Haig and his aerial chief, Major General Hugh Trenchard, were demanding that all available planes be sent to France at the very time that Lord French was demanding the reverse trend.[83]

This background history helps to explain the somewhat apologetic tone of Robertson's letter to Haig and the arguments which followed. Haig immediately asked Trenchard to write down his suggestions for the best defence against the Gotha raids over England. The result was another version of the 'forward defence' policy that had already been employed early in the Zeppelin era.[84] Trenchard said that the optimum answer was to capture the Belgium coast where the Gothas were based, but until that could be achieved, the next best answer would be to bomb the Gotha airfields there and that this would mean more planes and pilots in France, not less.[85]

The conflict of views between the authorities at home and in France was clearly delineated, but that conflict must be recognised as having a broader context; this was not *just* a military argument over allocation of material although it was, of course, this in part. It is quite normal practice for commanders to press for more supplies and to resist transferring their supplies to other areas. But *this* conflict was during the period when arguments in Parliament, meetings in London, articles in the press, etc., were all showing the fervent reactions to the first major airplane attack upon London in history. The Government could not let this contest over supply reach impasse. The decisive confrontation of these opposing pressures occurred when Haig and Trenchard were called back to London on 17 June to consult with the War Cabinet.[86]

In London, Trenchard repeated his previous arguments and added that the more trouble that could be given to the German air forces over France, the less would be their ability to divert their effort to London. After the War Cabinet still insisted that two squadrons of fighters must be returned to England, Trenchard warned that this would only give the Germans an advantage over the front lines which they would immediately exploit.[87] Nevertheless, the Cabinet persisted and the final result was the diversion of two squadrons to patrol both sides of the Dover Straits, one squadron to be sent to England and the other to Calais, while Haig and Trenchard were mollified by the promise that the squadrons would be returned by 5 July.[88] (It should also be remembered in this complex confrontation that Haig was already involved with his 1917 offensive in Flanders. The campaign for Messines Ridge had started, for example, on 7 June, and the plans were developed for a following campaign northeastward in the area of Ypres — that disastrous campaign remembered by the name 'Passchendaele'.[89])

Cecil Lewis, one of the pilots who was transferred back to England because of these arrangements, later wrote of the elation of the men in his unit when they heard the news. Dances, not Gothas, controlled their thoughts.[90] As it worked out, this was thoroughly realistic because the German raiders did not return during the unit's first recall to England. Haig kept pressuring for the return (as promised) of his squadrons, and they were transferred back to the front on 5 and 6 July. Due to a clerical error made in the War Office, this transfer was not reported to the Cabinet.[91] It was, therefore, a particular 'ironic twist of timing' that the Gothas made their second raid on London immediately afterwards, on 7 July 1917.[92]

This raid produced, as part of its aftermath, another letter from Robertson to Haig, written on 9 July. Robertson reported that the War Cabinet met within hours of the raid and that Lord French informed them, with great indignation, of the absence of the fighter support. 'Much excitement was shown. One would have thought that the world was coming to an end. I could not get in a word edgeways.' The result, Robertson continued, was that the Cabinet wanted the planes back again in England — and Robertson now supported the Cabinet position. 'There is no doubt that French has not got a very good force. It is mainly made of oddments, and of course oddments will not do ... The fact is we have not got enough machines to meet our requirements'.[93]

Haig once again resisted this demand. He had a campaign ready to break, and he wanted all the air support possible. Nevertheless, the Cabinet prevailed, at least to the extent of forcing Haig to send back one fighter squadron while they also held back in England twenty-four new machines that had been designated for the BEF.[94] In a later account published in 1926, the former CIGS, Sir William Robertson, drew a conclusion from this series of exchanges between London and Haig; that when the demands for home defence conflicted with those of the war fronts, the home defences, except in rare occasions, would always have first call.[95] Even as late as 1938, the former Air Ministry official, L.E.O. Charlton, remembered this transfer of defence planes to London as a forced (and detrimental) gesture to home front morale.[96]

In a broader context, this same basic conclusion became air dogma — a general principle affirmed by air authorities but also repeatedly stated by non-expert commentators — that the policy of strategic bombing compels the enemy to waste necessary resources on the relatively non-productive home defences. In 1920, for example, Air Commodore H.R. Brooke-Popham told a group of inter-service officers that this was one major plus value of long-range bombing, 'the amount of man-power and material that is absorbed in the endeavours to obtain protection from aircraft . . .', the large quantities of guns and planes that were thus not available to the battle areas, even, as Brooke-Popham pointed out, the special shelters that the Germans constructed in Belgium which used resources that could otherwise have been put to use on the front lines.[97]

Both sides used this argument during and after World War I. As has been noted, it was one of the reasons Hindenburg used to justify the raids on England in his letter to Bethmann-Hollweg.[98] The same reasoning was included in the instructions given to the commander of the specific Gotha squadron concerned.[99] When the commander of the German air services wrote his 'retrospect' about the German air effort during the war, it was one of his justifications for the strategic bombing of England; that, thereby, England 'was forced to keep a great part of its air force at home, with a corresponding loss at the front'.[100] And a recent German analysis of aerial warfare again makes the argument that defence wastage was the major gain of the raids on England in World War I apart from the effect upon morale.[101]

What is indeed involved in this reasoning is one of the key arguments for strategic bombing ever since its inception. It is also an argument which accords both with the facts and with logic (which is not the inevitable case with air power argumentation). A listing of the home defence measures existing in England by the spring of 1918 is remarkable when compared to the enemy effort which prompted them – the guns, the searchlights, the height finders and sound locators, the balloon defence system, the airfields, planes, gun stations and shelters, the number of personnel needed to build and man all these items, the tremendous amount of ammunition used, and the fantastic expense for it all . . . and only a few enemy planes (only once as many as forty-three) caused all that effort.[102] The facts thus do substantiate the principle. So too does logic. As Lieut.-Col. K.M. Loch noted, it is not just a story of a vast effort to protect against a small threat; that, rather, 'the London defences were an essential detachment in furtherance of morale. If the morale of London had collapsed, our war effort elsewhere might well have been in vain.' The basic point is one of the most fundamental in military theory, that it is essential to keep the operating base secure.[103] Just how much can be expended on defence – defence often considered relatively ineffective – is, of course, in part

Defence In The Gotha Era

a political problem. The anti-ballistic missile system is, therefore, really a new phase of this problem which dates from the earliest years of aerial warfare.

The use of more and better fighter planes was one of the major methods of defence that Britain used against the German bombers. Among the improved models used was the now famous Sopwith Camel (of 'Snoopy' reknown) equipped with two machine guns and a 110 h.p. engine which gave it a relatively fast climbing ability. The other model of the Sopwith fighter line, the Pup, flew more easily than the Camel but was less effective against the Gothas, primarily because of its slower rate of climb. Also used were the S.E.5s (S.E. standing for Scout Experimental) with a 150 h.p. engine and, later, the two seater Bristol Fighter which could attain an altitude of 20,000 feet.[104]

Fighter defence was strengthened by these various new models starting in the summer of 1917 but results were hardly spectacular for some time. During the summer daylight raids, only three bombers were shot down by British defence planes and when the Germans turned to night raids on England in September 1917 even fewer successes were recorded for months thereafter.[105]

One reason for this, certainly, was the slowness in replacing the old machines with newer models. The older fighters were usually the losers in aerial combat with the Gothas,[106] and yet it took months before the improved models comprised a majority of the available planes. However, by the date of the last Gotha raid on London (19 May 1918), almost all of the British fighters used that day were recent models, seventy-three of the eighty-four total.[107] The point is that the production increases demanded by the Cabinet took time to be effected, although aircraft production was given high priority. The Air Board were told to accomplish these increases by working with the Ministry of Production and the War Office.[108] Many people thus became involved with this production effort but the man most responsible for its final success was Sir William Weir, a very dedicated and knowing worker. He was Director-General of the Aeronautical Supply Department within the Ministry of Munitions.[109] One measure of his accomplishments is that England produced 26,685 airplanes in the ten month span from January through October 1918.[110]

Newer planes in increased numbers were hardly the complete answer to the Gotha problem however. England always had the numerical advantage over the invaders. In the three early raids by the Gothas of 25 May, 13 June and 7 July (1917), the number of defence planes averaged close to five to one over the invaders. But this advantage was really meaningless. The raiders had a flight plan and knew where they were going but the defenders were forced to fly about much more at random. Most of the fighter pilots would never see an enemy bomber while one defender might find himself surrounded by them. As always, air defence is a very complicated operation. Fighter pilots can be relatively useless unless guided by a sufficient observation and intelligence system.[111] And that system which had worked against the Zeppelins was not capable of dealing with the Gothas. Pilots, at time of take-off, could be given vague directions about the Zeppelins' course and they still had a fair chance of encountering the giant lumbering airships but far more accurate information was needed against the smaller and much faster bombers. Further, the problem of spotting the enemy became even more difficult, of course, when the night raids started in September 1917. Very often *no* sightings were made by the defending pilots during night raids.[112]

Only a ground-to-air wireless radio system would have given the defence pilots sufficient guiding intelligence. Sir David Henderson, a leading figure in the Royal Flying Corps, had long been pressing for this development but the Admiralty had opposed the idea in the fear that such radio transmission would interfere with its naval communications. This opposition was finally overcome by a reordering of priorities but a method of wireless telephonic contact was not made fully operational until after the bombing raids had ceased. (The sets were in general use by June 1918, while the last Gotha raid on England occurred on 19 May.)[113] In lieu of wireless contact with the pilots other expedients were tried such as large white arrows laid out on the ground pointing towards the bomber's position, but the results were never very satisfactory.[114]

As with the Zeppelins, so with the bombers; the new threat forced a thorough reorganisation of the defences. The need for this was strongly stated by Jan Christian Smuts, who had been assigned the task (after the 7 July raid) of evaluating the air defences of England. By then, Smuts had gained a general recognition of his abilities. The Earl of Swinton expressed an opinion which few in England by 1917 would debate when he rated Smuts as 'the greatest Empire and Commonwealth statesman, splendid – quick, penetrating, thorough and decisive'.[115] Smuts had arrived in London early in 1917 for an Imperial Conference and when the conference ended, Lloyd George and other government officials requested him to extend his stay to help on various problems that might arise – Smuts' reception in England by both the populace and establishment had been so laudatory that his assistance was expected to be most valuable. The result was that he was appointed an additional member of the War Cabinet with no specific duties assigned.[116] One indication of the esteem accorded Smuts was the uniqueness of that appointment; Smuts became 'the only full Cabinet minister of modern times to have no connexion with either house of parliament'.[117]

Thus Smuts was available when the Gothas over London made the air question of great moment and the War Cabinet appointed him to a committee to investigate the problems of home defence, of the organisation of the air services in general, and of the broader question of air power and its proper use and direction. Only the Prime Minister and Smuts were named on the committee and since Lloyd George did not actively participate, it was Smuts' 'one-man affair'.[118] Experts and representatives from the army, navy and home forces were instructed to be at Smuts' call. And in only eight days from the time of this assignment, Smuts (on 19 July 1917) submitted his first report, the one on air defences.[119]

In this report Smuts noted that London was the vital centre of the empire and that it could therefore expect increasing intensity in the air raids 'to such an extent in the next twelve months that London might through aerial warfare become part of the battle fronts'.[120] He criticised the overlapping commands and divided authority in the current defence system and said it should all be under one command. He made specific suggestions with respect to the defences; that far more anti-aircraft gunnery was needed so that a barrage effect could be achieved, and that the guns should be placed in the approach areas to London so that the barrage fire could break up the bomber formations and then, in an interior zone over the greater London area, an expanded fighter defence could attack the separated raiders.[121]

The War Cabinet quickly acted on this report and by the end of July the London Air Defence Area was established which, despite its name, included all of south-east England within its responsibility. Brigadier-General Edward B. Ashmore was placed in command of LADA, as the command became known.[122] General Ashmore was 'a truly happy selection . . .', a well-recognised artillery expert who was also interested in aerial matters and was, in fact, a pilot — an excellent leader who earned and received respect from all of his associates.[123] Air Chief Marshal Joubert de la Ferté is a witness to the excellence of this choice: 'Possessed of an excellent brain and much imagination [Ashmore] was almost an ideal commander for such an experimental organisation. He got on well with his superiors in the service and with the civilian heads of departments with whom his work brought him into contact'.[124]

Nevertheless, all reformers of systems can expect to face opposition and Ashmore received his share, especially from the War Office officers in the AA section.[125]

General Ashmore immediately set about to establish the zone defence recommended in the Smuts report.[126] The first version was an anti-aircraft barrier which ringed the eastern approaches of London some twenty miles outside the metropolis and behind this barrer (i.e. closer to and over London) was the fighter defence area. The line of demarcation (termed the green line) was strictly enforced so that these two primary methods of defence would be separated and thus not be interfering with each other as had been true in the past.[127] The barrage system of AA fire took longer to be achieved as it waited upon a sufficiency of guns. By the first of the Gotha night raids on London (4 September 1917), some 800 shells could be fired. Less than a month later (on 1 October) that number was increased to over 10,000 shells.[128]

The results of the barrage fire system were mixed at best. It has already been noted with respect to the Zeppelin era how lethal AA fire could be to the local population.[129] Obviously the shooting of thousands of shells into the air was dangerous to people and property below. A few examples clearly illustrate this point. On 29 September 1917, two civilians were killed and twenty-four were wounded by the barrage fire and 290 houses were damaged – while that same fire only hit one enemy bomber. On 6 December 1917, the barrage killed one, injured eight, and damaged 169 houses. The official historian, H.A. Jones, rarely attempted stylistic flourish but these results of AA fire did stir him to unusual rhetoric; 'most of the thousands of shells which lighted patterns across the sky exploded well away from the swift-moving aeroplanes to which they were, it seems, less of a menace than to the people and property below'.[130] Naturally, officials with a vested interest in AA gunnery (e.g. Ashmore in the First World War and General Sir Frederick Pile in the Second) would view this method of defence in a more favourable manner; arguing for its ability to frighten some invaders into turning back,[131] or, while admitting that it brought few planes down, nevertheless claiming that it had 'a strong deterrent effect'.[132] Certainly the evidence indicates that the English people generally appreciated the efforts of the anti-aircraft crews, however dangerous their fire might be – to either enemy or friend. Before Christmas 1917, a large public subscription was made to a Christmas fund for the AA and searchlight personnel of LADA. General Ashmore seems on sound ground when he claims that this shows 'the London public appeared grateful even for the moderate bag we could lay before them ...'[133]

Defence In The Gotha Era

But not *all* of the populace were grateful. Smuts' good friend in London, Miss Emily Hobhouse (the ardent advocate for reform of conditions in South Africa after the Boer War[134]), wrote him the following diatribe:

> As to your defence of London by this infernal Barrage I do trust you will stop it, as it is a remedy worse than the disease. We have lived under showers of this odious shrapnel (purely home-made) and it is costly in life and property. A woman close to me was killed in bed thereby.[135]

General Ashmore introduced one innovation in the London air defences, his balloon defence system, which gave London a wartime skyline which would be repeated in a variant form in the next World War. Ideas for balloon-supported wire barriers against aircraft had been presented at various times before, in some cases even prior to the war.[136] By late 1916, Germany had a captive balloon air defence system set up in some key areas, especially in the Saar.[137] Venice also had a balloon defence which had been examined by English air personnel.[138] But Ashmore's system was unique in that it added a wire apron effect. Balloons were anchored not just to the ground but also to each other by heavy steel cables. From these horizontal cables hung weighted wire streamers.[139]

The primary purpose of this balloon and wire apron defence was to force the raiders to fly at high altitudes, thus lessening the air space that the defence planes would have to cover. Ashmore thereby hoped to improve the chances of the fighter pilots to make contact with the enemy.[140] Twenty such apron sets were planned but only ten were ever raised, the highest altitude attained being some 11,000 feet.[141] Naturally, General Ashmore claims that this innovation of his did seriously hamper the German raiders[142] but postwar military authorities were more sceptical, stating, for example, that the apron additions made the balloons too heavy and awkward and that they had little effect at best.[143] What is certainly true is that a bomber could run into an apron, tear away two of its wire streamers, and then continue on to bomb London and fly back safely home — even though it 'much astounded the British' when it happened during a raid on 28 January 1918.[144]

The plane which performed that feat was not a Gotha but a newer addition to the German bombing fleet, the R.39 known as the Giant ('R' signifying *Riesenflugzeug*). This plane was a monstrous wonder for its age. Its wing span was 138 feet, 5½ inches or within three feet of the span of the World War II B-29, the Superfortress model. It carried a crew of seven to nine men, it was powered by four 245 h.p. engines and had an effective bombing range of almost 300 miles. A prototype model flew to London in a raid on 6 December 1917, and normal production models were ready for such missions by the end of January 1918. Fortunately for England, few of these planes were produced and made available for the English raids.[145]

The other additions made to England's defences to meet the threat of the bombers can be briefly summarised as they were extensions or changes to defence tactics established (and already described) in the Zeppelin era.

Obviously, when the Germans started the night raids in September 1917, there was a need for a more effective searchlight defence. The Royal Engineers established a school in searchlight techniques[146] and when sufficient trained personnel were thus made available, General Ashmore used these soldiers to replace the civilian volunteers who had formerly manned the lights. Naturally the resulting drain on available manpower was regretted, but the need for expert operators on the lights outweighed that disadvantage.[147] The searchlight units were also reorganised into twelve companies, each under a defence air squadron commander, so that greater cooperation was achieved between the fighter pilots and the searchlight crews. The number of companies and searchlights kept increasing, of course, and the lights themselves were increased in power, finally to 120 cm. size in 1918. By the end of the war, almost 500 searchlights were in the LADA command area.[148] Reports by German pilots indicate that the improved searchlight defence had its effect. In a raid on 6 December 1917, for example, the German commander reported that his planes were spotted 'in the cones of searchlights for minutes at a time' and that the light crews were able to pick up the planes, one light from another, as the bombers advanced inland.[149]

Observation was also aided by the appearance of sound locators in 1917. These were fairly primitive devices operating on the principle of dual reception (like human ears) wherein dual sets of sound trumpets were fixed on to the ends of two long poles, one pole rotating horizontally and the other vertically. Their efficiency was very uncertain. Miscellaneous sounds would affect them such as a passing car or gunnery fire. Also sound waves do not always follow a direct line from their source but can be changed in their course by climatic conditions. Nevertheless, they did help locate the invaders, and more importantly, they furnished England with its first early warning system because

Defence In The Gotha Era

sound locators placed on the south-east coastal cliffs could detect incoming planes for distances up to twenty miles. However, considering the relatively few planes that were downed by England's defences,[150] the official historian obviously overstated the case when he wrote (in a sympathetic context): 'It has sometimes been said that had it not been for the sound locators the German night aeroplane campaign would never have been defeated'.[151]

Mechanical aids also were produced to aid height estimations for the ack-ack guns — the Barr and Stroud Height-Finder and the Bennett Height-Finder — although their adoption was somewhat delayed by War Office conservatism.[152]

The observation system, already well organised by the end of the Zeppelin era,[153] was extended and improved upon. Observers were stationed on ships to furnish earlier warnings[154] (earlier AA fire was also provided by guns positioned on picket ships stationed off the coast[155]). The system was also made more centralised. General Ashmore had the GPO rearrange the telephone line connections so that his operations headquarters at Horse Guards received first notice of the raiders and then directions were issued to the specific subsidiary LADA defence commands as needed. It took months before the GPO could fully accomplish this task but once set up, the resulting arrangement was even closer than before to its World War II heir, the Cabinet War Room and *its* operations network.[156]

Unfortunately, no matter how elaborate the system, it was still subject to that special failing of air defence, the false alarm. An unidentified British seaplane set off the whole operation, for example, on 3 October 1917, causing a barrage fire display all along its course of flight and in midday London, for two hours, 'public business was paralyzed...'[157] On 18 February 1918, another two hour barrage occurred in the late evening hours and fifty fighters flew around uselessly when over-cautious observers heard planes off the Kent coast; planes that *did* bomb Calais which which did not come to England.[158] Ashmore's improved LADA organisation did lessen, however, the trouble caused by false alarms by dividing the country into smaller sub-control units. 'This is an important point in organising a country for air defence; the better your observation system, the smaller can be your warning districts, and disturbance is minimised'.[159]

Some suggestions for air defence were not tried. The French had used smoke screens in various areas but Ashmore was against this method because of its expense and the belief that it interfered with more effective measures.[160] Of greater interest with respect to later phases of air defence measures, a Government study was made of the possibility of evacuating the populace from a small coastal belt, but the idea was set aside as unworkable.[161] There were, however, some private efforts in this direction. Church aid groups evacuated shell-shocked children to safer refuges.[162] Schools in London for wealthy children suffered from lack of pupils.[163] Lastly in this regard, Paris furnished a precedent for England to remember. In 1918 when the city was suffering from both Gotha raids and bombardment from 'Big Bertha', its municipal officials evacuated school-age children to the provinces.[164]

With all the additions and improvements made to England's air defences, the danger to the German raiders naturally increased. The last bombing raid on England (on the night of 19 May 1918) saw the English defences have their greatest success, bringing down six of the twenty-eight bombers that came over.[165] It is certainly understandable, therefore, that the man in charge of these defences would later claim that they were, finally, the key to victory in this first Battle of Britain.[166] Nevertheless, this is an overstatement of their efficacy. Weather and the unreliability of the German machines worked far more havoc on the German English Squadron than did the English defences. On that last mentioned raid, for example, four bombers crashed on landing due to fog conditions. Thirty-six bombers crashed overall against twenty-four brought down by the British or from otherwise unexplained causes. Nor did the defences cause the Germans to cease these raids; it was, rather, that the German high command decided that the bombers were needed on the Western front.[167]

Thus, the history of the aerial defence of England during World War I. Most noteworthy, 'except for some technological refinements, notably radar, it was essentially the same air defence system which was to serve Britain so well in 1940'.[168] More ominously, however, one other phase of this story also pointed towards the future. The English reaction to the Gothas was not just to strengthen their defences. They also decided to retaliate in kind, and they created a new air service to fulfil that task — as we shall see.

4 THE BIRTH AND EARLY INFANCY OF THE ROYAL AIR FORCE

The reader will recall the 'Reprisal Map of Germany' published by the *Daily Mail* two days after the first Gotha raid on London, which indicated the German towns within 150 miles of the Western front.[1] The desire to retaliate in kind for damage received is of course a normal, if not wholly admirable, human response. England had even given official sanction to the concept of reprisals in its then current *Manual of Military Law* in which Article 452 stated:

> Reprisals ... are retaliation for illegitimate acts of warfare, for the purpose of making the enemy comply in future with the recognised laws of war ... They are by custom admissible as an indispensable means of securing legitimate warfare. The mere fact that they may be expected if violations of the laws of war are committed acts to a great extent as a deterrent ...

A following Article (459), however, limits the right of retaliation: 'Acts done in the way of reprisals ... must not exceed the degree of violation committed by the enemy.'[2]

It is hardly surprising, therefore, to see some Englishmen resort to 'the anodyne of contemplated retaliation'[3] from the time of the Zeppelin raids on; what is, perhaps, surprising is how often they felt the need to be prudent while doing so, For example, that 'freak candidate' Noel Pemberton Billing[4] emphasised that he was not advocating reprisal raids upon civilian populations when he spoke of the need for bombing strikes against Zeppelin bases in Germany,[5] even though in his first published work (which appeared during his campaign) he wrote, 'the people who object to reprisals bore me'.[6] Of similar equivocal nature was his maiden speech in the House of Commons when he specifically advocated raids only on Zeppelin sheds but talked of the need to ignore 'so-called religious scruples' in the British bombing programme.[7]

Another active spokesman about aerial matters in the House of Commons was William Joynson-Hicks (he was, for example, chairman of the Parliamentary Air Committee[8]) and, during the Zeppelin era, he showed like caution in his speeches. He spoke of the need to bomb such key German centres as Essen (his speeches of 20 July 1915, and 11

November 1915, are two instances) but (in the 11 November speech) he added, 'I say nothing of reprisals, because I want to carry the entire House with me today . . .'[9] Jix, as he is familiarly named[10], came closer to a forthright call for reprisals in an article published in April 1916, in which he pressed for raids on Berlin and stated, 'the bourgeois population of Berlin look upon security as their civil right, and their plaints will not be few if they are called upon to assist personally at the manifestation of modern frightfulness'.[11]

In that passage Jix was repeating two common assumptions in England during World War I, that the Germans had introduced the policy of frightfulness (*Schrecklichkeit*) into war and only a dose of that policy in return could make them forgo its use. This is the basic theme of a book written in 1915 by Charles Andler, a professor of the Sorbonne, which was quickly translated into English.[12] The historical record of war in the twentieth century hardly indicates that the concept of 'unlimited war' is a monopoly of the Teutonic mind, but this belief certainly made it easier for the English to justify their indulging in such activity.

In the Zeppelin era, then, calls for reprisal raids were made but often with an accompanying disclaimer concerning the suspect morality that was considered part of the whole concept. This was noticeably true in that hall of rhetoric, the House of Commons. To cite one further example, in a debate on 16 February 1916, concerning the Zeppelin raids, Mr Cecil Harmsworth argued that the English people wanted reprisals — but then he added that he did not approve of the idea (a nicety not long preserved by the Northcliffe connections).[13] But that dualistic position was hardly restricted to Parliament. In the prestigious journal *The Nineteenth Century and After*, for example, one author rued that the word 'reprisal' was impairing the ability to put into practice the activity itself. He emphasised the need for counterattacks to answer the Zeppelin raids and stated that the preacher's moralising must, therefore, go unheeded.[14]

Of course there were those who felt no need for moral disclaimers. The early air theorist Frederick William Lanchester is a noteworthy example.[15] By November 1915, Lanchester was advocating that all enemy air raids without military utility should receive reprisals in return and that 'the power of reprisal and the knowledge that the means of reprisal exist will ever be a far greater deterrent than any pseudolegal document'.[16]

Pemberton Billing's comments about 'religious scruples', H.F. Wyatt's strictures against preacher's moralising, and Lanchester's disdain of 'pseudolegal documents' indicate very important factors in the contemporary questioning of a reprisal policy. Since the opposition to that policy was so frequently based upon moral grounds, it was natural that religious leaders took a major role in that opposition. The Archbishop of Canterbury, Randall Thomas Davidson, was, indeed, one of the most outspoken opponents.[17] It was Davidson who led off a debate in the House of Lords on 2 May 1917 against a reprisal policy against Germany and moved the motion in that regard. The wording of the motion was certainly explicit:

> That this House, while fully recognising that it does not lie within its province to express any opinion on matters purely military, desires to record its conviction that the principles of morality forbid a policy of reprisal which has, as a deliberate object, the killing and wounding of non-combatants, and believes that the adoption of such a mode of retaliation, even for barbarous outrages, would permanently lower the standard of honourable conduct between nation and nation.

And, significantly, the Archbishop also said, 'We are now competitors in "frightfulness", and we feel the stain which would rest upon us if we were.' Also significant is that the Archbishop's motion had already been approved by the Upper House of Convocation of Canterbury.[18]

'Pseudolegal documents' had been produced in attempts to outlaw aerial bombardment. As early as 1899, the first of the international Peace Conferences at the Hague passed declarations 'to prohibit the launching of projectiles and explosives from balloons or by other similar new methods', and 'to prohibit the use of projectiles, the only object of which is the diffusion of asphyxiating or deleterious gases'.[19] The prohibition of bombardment from the air was renewed at the second Hague Conference in 1907, but only Great Britain, the United States, Portugal and Belgium ratified that provision, in each case compliance being conditional upon wars fought only with other states that had ratified the provision.[20] Naturally, this excludes World War I, but all the participating states in that war were signatories to the Land Warfare Convention of 1907 in which Article 25 stated 'the attack or bombardment by any means whatever of towns, villages, habitations, or buildings which are not defended is prohibited'.[21] A major problem connected with that ban was that it did not define what constituted an undefended town or village.[22] Other articles of this Convention stipulated that town authorities should be forewarned if possible before bombardment on their area commenced, and that care should be taken not to hit churches, hospitals, museums and the like.[23]

Certainly, therefore, the 'documents' were at hand, especially ones emanating from the Hague Conferences. And certainly the evidence from World War I justifies the 'pseudolegal' sneer; it justifies the historical conclusion that 'the Hague Conferences had no real influence upon the foreign or domestic policies of the great powers'.[24] But the documents did furnish useful material for the reprisal debate. The article by Sir Thomas Barclay is an example. His purpose was to charge the Germans with violating international law by their aerial bombardment of Paris (this was before the raids on England had started) and to express a rhetorical warning about the reprisal raids that such bombing might incite, although he hoped that 'neither England nor France nor Russia will debase herself by resorting to any imitation of them'.[25]

The wish to retaliate for aerial attack was naturally an emotion that affected all sides. It certainly abetted the escalation of strategic bombing incidents in the war. A classic example on the German side involved the Crown Prince, Frederick William. In June 1915, he suffered an air raid on his field headquarters just when his wife Cecilie was visiting him there. He immediately wired his father this news and said the German answer should be an air raid on London. In this instance, the Kaiser's advisers counselled him against this suggestion, stressing that 'to drop bombs on London out of revenge is completely futile'.[26]

But three months later, his advisers were pushing for the bombing of London on the basis of its effectiveness in the newspapers as a reprisal policy.[27] The officer in charge of Germany's aerial forces in World War I has justified much of Germany's strategic bombing as being reprisal raids.[28] The German aerial historian, Georg Feuchter, states that the large Zeppelin raid on London of 9 August 1915 was a retaliatory strike for a French attack on the city of Karlsruhe.[29] One can conclude that while the motivations supporting strategic bombing in World War I were many and complex, the wish to retaliate helped proponents to rationalise other motivations and the desire for revenge itself could be the uppermost motive – witness the Crown Prince.[30] And, finally, the motive of retaliation was hardly one that went unquestioned. In fact, it is *still* a debated concept in our atomic age.[31]

Of course, all the demands for bombing raids on Germany were not based merely on the desire for revenge. As early as the first month of war, the Secretary of State for War, Lord Kitchener, spoke of the desirability of bombing German industrial areas and he thought in terms of up to 100 plane raids on Essen.[32] Often the strategic bombing advocates used an early form of the 'victory through air power' argument.[33] Joynson-Hicks, for example, told the House of Commons on 11 November 1915, 'I believe with an adequate Air Service you could very largely bring this war to a speedy and victorious conclusion', in a speech which spoke of the need for long-range bombing operations.[34] In the same year, the reputable periodical *The Fortnightly Review* published an article which claimed that even a few airships employed by Britain on bombing raids 'might easily become the determining factor of the war and secure the ultimate victory for the Allies'.[35] A frequent air power spokesman in the House of Commons was A.A. Lynch and he, too, said that victory would result from an all-out effort which included intensive strategic bombing.[36]

The justification of long-range bombing as a method to attack German heavy industry was a natural conclusion considering that the emphasis on war *matériel* was so much a part of World War I. Lord Kitchener, as has just been noted, held this view right from the start of the war. By 1915, Essen had become something of a symbolic term in the press for the air target *par excellence*. Similarly, Joynson-Hicks often referred to Essen and the Krupp works there. Other similar suggestions were made in reputable journals.[37]

The evidence is clear: long before the Gotha raids of 1917, there was a significant amount of opinion in England favourable to bombing raids over the German home front, motivated by the wish for revenge and by hopes for specific gains.

Additionally, there had been some specific English military planning and actions in that aerial policy. The Royal Naval Air Service (RNAS) 'forward policy' against the Zeppelins gave that service a natural impulse towards a strategic bombing policy.[38] The naval air authorities soon jumped from bombing Zeppelin sheds in Friedrichshafen to bombing German industrial targets in the Rhine Valley from a base near Belfort, France.[39] Meanwhile, the First Lord, Winston Churchill, pushed for the development of equipment applicable to this bombing task, heavy long-range bombers and larger, more effective bombs. With regard to this, he wrote (on 3 April 1915): 'The carrying of two to three tons of explosives to a particular point of attack in a single night or day is the least we should aim at as an operation in the future ... The capacities of machines should be considered in relation to these definite tasks.'[40] The first important bomber which resulted from this RNAS pressure was the Short Bomber which was operational by late 1916 and whose bomb capacity was over 900 pounds. More significantly, the Admiralty policy eventually led, in large measure, to the development of the Handley Page series of bombers, the most important long-range bombers produced in England during the war.[41]

The RNAS strategic bombing effort was an intermittent and troubled operation because it ran counter to the aerial policy of the British Expeditionary Force (BEF). The natural wish of the Western front commanders was to have as many planes as possible devoted to their needs. Operations not directly affecting their battlefield conditions were disparagingly termed 'side-shows'. Such RFC leaders as Sir Sefton Brancker and Hugh Trenchard allowed that strategic bombing might produce hopeful results, but that it should only be pursued when sufficient spare pilots and planes were available above all demands by the BEF. These army objections finally (in 1916) forced the Admiralty to shut down the base at Belfort and some of its pilots and equipment were assigned to the BEF aerial forces on temporary duty.[42]

By then, the French had also adopted a strategic bombing policy and this was eventually to affect the English aerial history. By spring of 1915, the French high command had started to contemplate a long war wherein industrial capacity would play a key role and one result of this thinking was the establishment of a French strategic bombing group near the town of Nancy. The unit's first raid was on an explosive and gas manufacturing complex at Ludwigshafen but that their purposes were not solely limited to such industrial targets was soon indicated by later raids on such population centres as Karlsruhe.[43] This first French strategic bombing effort was abandoned in 1916, due to the effectiveness of German fighter defence.[44]

The RNAS re-enters this history when, in the summer of 1916, it again established a long-range bombing unit, this time at Luxeuil, France,[45] where they scheduled raids on Saar industrial centres, especially where U-boat supplies were produced. The French soon participated in the Luxeuil operations — but again BEF demands hindered the RNAS effort, especially after the battle of the Somme started.[46]

The French did not want to give up the joint venture and Colonel Barès, commander of the French Air Service, argued for its continuation in London meetings starting in October 1916. Hugh Trenchard, the RFC commander with the BEF, strongly protested this diversion of aircraft. He argued that the essential need was to win the battle for air superiority over the Western front and only after that battle was won should the Allies start considering long-range bombing targets. He refused to give credence to the RNAS and French claim that such bombing would hasten victory. The debate was an intense one and its decision became a political issue of major proportions. At this time, Lord Curzon was the President of the Air Board[47] and he supported

Trenchard's position. Prime Minister Asquith, on the other hand, argued that the RFC should not be allowed to dictate to the naval air service. Curzon, already sorely troubled by inter-service rivalries and by disagreements with Asquith, found this 'the last straw', resigned and, shortly after, added his influence to the movement which overthrew Asquith.[48]

These political changes resulted in a reconstituted Air Board with Lord Cowdray (Weetman D. Pearson, first Viscount Cowdray) as its new President and with Sir William Weir in charge of equipment and supply.[49] Weir, an industrialist himself, was convinced that bombing of German industry was potentially a most crippling method of war and he converted Lord Cowdray to these views.[50] Thus the RNAS stand received stronger backing from Whitehall but nevertheless by April 1917, strident demands from the Western front commanders pursuaded the Government to have the Admiralty abandon the RNAS-French joint effort.[51]

The above account gives some indication of the complex play of motives and emotions involved in the evolution of a British strategic bombing policy *before* the advent of the German Gotha raids. Significantly, it shows how military, political and popular thinking so often reflect each other, and how all three, in interaction, give this history its impulse and direction. Each domain — the military, the governmental and the popular — worked an influence on the others, positions taken in one domain were in part cause and in part effect of positions taken in the other domains. The whole story is much too complicated to justify the simple judgements that are so often made about it. The military expert and the historian are both prone to make these simplifications. For example, one Royal Air Force authority, writing about the RNAS-French operations based at Luxeuil, stated that, although steel works were their most stressed target, nevertheless, 'there is no doubt that the underlying motive behind these attacks was the feeling that the German homeland must be attacked. It was intolerable that the German people should go free while the Allied people were increasingly subjected to bombing . . .'[52] Regarding the same Luxeuil operations, the historian Edward Mead Earle states that they had the two purposes of reprisals to satisfy popular demands at home and of attacks to weaken the German will to vigorously pursue the war.[53] Both authors, surely, see part of the whole truth; both, surely, are correct in looking for underlying reasons behind the official target stipulations; both, just as surely, are oversimplifying the story.

The Gotha raids on London, starting in the summer of 1917,

brought all the above lines of development into a more intense stage and further complicated and impassioned that development thereafter. The relative prudence in the advocation of reprisal strikes that existed before these raids almost wholly disappeared thereafter. The 'Reprisal Map of Germany' printed by the *Daily Mail* in conjunction with pictures of the child victims of the first Gotha raid on London[54] represented the new climate of opinion in England. The *Daily Mail* itself maintained a continuing campaign for reprisal raids on German towns. Its articles expressed the simple desire for revenge but also stated such other arguments as that strategic raids would force the German air force to go on the defensive and thus would prevent it from continuing its raids on England, that such German defensive activity would lessen their battlefield operations, that to forbid such reprisal raids as being 'un-British' would only mean the deaths of more English women and children, and that only such offensive raids could provide an effective defence for the English people.[55]

The Times was not as strident nor as frequent in its editorial columns about the desirability of reprisal raids, but its news columns featured almost daily reports of such demands from other quarters for a period after the Gotha raids started. And it did carry editorials advocating a reprisal policy, stating that Britain must attack Germany with an equivalent form of air war consisting of raids on civil and military targets alike, that such an air policy was not reflecting merely a taste for revenge but also a more educated view of air power, that such offensive strikes provided the best defence against air attacks, and similar arguments.[56] *The Times* news and letter columns also contained many outspoken statements reflecting the new atmosphere in England. Joynson-Hicks showed that he no longer felt the need for any prudence. Immediately after the 13 June raid, he said England should 'blot out a German town' in return for every raid on London.[57] Many peers expressed their endorsement of a reprisal policy, Lords Senex, Montagu of Beaulieu,[58] Beresford and Tenterden among others. *The Times* also reported additional support coming from America. The *New York Sun* was cited for example; it was demanding strategic bombing reprisals to stop the murder of British children and it called upon America to build the necessary planes to do that task. Another column told of an American industrialist who offered $1,000 to the first US flyer to bomb Berlin and of his belief that only such raids would bring a stop to the raiding of London.[59]

Both *The Times* and the *Daily Mail* indicate a passionate public interest in the reprisal policy issue in their letter columns. Certainly the majority of the correspondents favoured the policy and there was a noteworthy emphasis on the need to ignore the bishops on the question.[60] Some bishops, on the other hand, never did give up their opposition. Even after the Government officially announced its intention to prosecute a retaliation policy — in a statement by General Jan Christian Smuts — the bishops were not silenced. On that occasion, Bishops Oxon and Ely wrote to *The Times* a statement which well represents the moral stand taken against the air reprisal policy.

> Sir — We have good reason for believing, that we represent a large body of thoughtful people in the country, who with their whole heart and conscience uphold the prosecution of the war, when we say that we are compelled to regard with profound regret and misgiving the prospect of a new policy on the part of the Government in regard to air reprisals. General Smuts describes the new policy in the following words: — 'We are now most reluctantly forced to apply to him [i.e. the enemy] the bombing policy which he has applied to us.' Such a policy we believe to be essentially and deeply wrong; it is, to use General Smuts' own words, 'bad and immoral'. We are convinced that this country will sacrifice an advantage of an importance which we cannot now estimate if it surrenders its power of entering on peace negotiations with clean hands. At the present moment, it seems to us, this country is in a peculiar sense a trustee of international morality in war-time. A practise which we adopt now 'most reluctantly' will in the next great war be regarded as the recognised custom. Further, while we are thankful to note General Smuts' words, 'We shall use every endeavour to spare as far as is humanly possible the innocent and the defenceless', experience seems to show that such a qualification will prove to be nugatory. If we allow ourselves to be led by the Germans, the descent is easy and the end certain degradation.
> Yours faithfully,[61]

Moral questioning over reprisals was not, however, the dominant mood in England after 13 June 1917. Starting the very next day, mass meetings occurred wherein the populace demanded that the Government pursue a retaliation policy. The meeting of 14 June saw one speaker demand five raids on Germany for each raid on England and he spoke of 500 plane raids on Berlin. On 17 June, the Lord Mayor sponsored a mass meeting at which many other borough mayors attended plus notables from Parliament, etc. Thousands had to be turned away even though it was held at the London Opera House. The crowd roared approval as speaker after speaker spoke of bombing the Hun. The meeting culminated with the following resolution being passed without opposition, a resolution proposed by the Lord Mayor.

> That this meeting of London's inhabitants hereby expresses its utter abhorrence of the German method of warfare by the murder of innocent women and children in air raids on open towns and cities, and is of the opinion that the only means of bringing the inhumanity and cruelty of these dastardly and criminal attacks home to the German people is by systematic and ruthless reprisals. It, therefore, calls on the Government to initiate immediately a policy of ceaseless air attacks on German towns and cities in order that their population may experience the effects of such methods of warfare and thus be induced to force the German authorities to cease this wanton and useless destruction of life and property.

The resolution also stated that copies of it were to be furnished to the Prime Minister and the members of the War Cabinet and also to the First Lord of the Admiralty and the Secretary of State for War.[62]

On the following day (18 June) there was yet another mass meeting promoting reprisals, this one sponsored by the British Empire Union. Shortly thereafter, the Corporation of the City of London petitioned the Government to take up the policy.[63]

Naturally, all that agitation found echoes in Parliament, and there too former restraints over the reprisal issue were thrown aside. Those two importunate air power advocates, Pemberton Billing and Joynson-Hicks, were, as usual, in the forefront of these debates but they had many followers now. Even before the London raids began, Billing had prophesied the occurrence of large-scale bombing raids on undefended towns – he spoke in terms of up to 1,000 plane raids – and advised that England's self-interest demanded that she start that policy before the enemy did. Victory, he predicted, would be the prize for the side which emphasised a strategic air offensive. After the 13 June raid, he returned time and again, in speeches to the House, to the need for England to retaliate with bombing raids on Germany.[64] Joynson-Hicks, in the House session the day after the 13 June raid, was already pressing for reprisals. He stressed that German townspeople must experience what their Government had inflicted upon English civilians.[65] Among the many others of like mind in Parliament, perhaps Colonel Claude Lowther was the most strident and persistent.[66]

The Birth and Early Infancy of the Royal Air Force

The correlation between the great surge of intense, passionate advocacy of this bombing policy and the instances of the Gotha raids forces the conclusion that revenge was the strongest motivating force. A strikingly clear statement of that emotion is seen in a letter by the former First Sea Lord, Lord Fisher, written during the second Gotha raid on London (7 July 1917). In this letter (addressed to Sir William Watson) Fisher wrote, 'a big raid now in progress now [sic] over my head! And these bishops want to do away now with the 58th Psalm!!!'[67] The bishops' opposition to a reprisal policy was also bitterly attacked in a series of articles by Harold F. Wyatt in *The Nineteenth Century and After*,[68] and in the first of these, written after the 13 June raid, Mr Wyatt showed that he had no doubt about what he believed was the primary emotion then being shared by his countrymen; 'a natural desire, as fierce as it is instinctive, that counter-strokes of a like kind, but on a wider scale, and with a greater destructiveness, should be dealt at the Hun'.[69] Even the quite temperate air raid observer could have a similar response. The columnist Ralph David Blumenfeld had not been very impressed by the Zeppelins[70] and the Gothas did not affect him very much either ('far more noisy than dangerous') but he too wanted to see retaliatory raids of sufficient strength to teach the Germans to cease such practices.[71]

The government leaders could not have ignored all this agitation, even if they wished to — and there is little evidence that many in the Lloyd George coalition had any desire to resist. The problem was that the Government ran up against the usual resistance of the Western front commanders to this alternative use of air power. The Cabinet discussed a reprisal policy in its first meeting after the June raid, although its major emphasis was on defence. The majority of the Cabinet favoured more home defensive measures as the best immediate answer but there was enough pressure for reprisal raids that Trenchard felt obliged to warn against such ventures in a written response to this particular Cabinet meeting's discussions.[72]

In that Trenchard later became a very outspoken advocate of the strategic use of air power, and in that he was, during the interwar years, the man most responsible for giving direction to British air development,[73] his comments on the adoption of that policy at this early stage of this history are of interest:

> Reprisals on open towns are repugnant to British ideas but we may be forced to adopt them. It would be worse than useless to do so, however, unless we are determined that once adopted they will be

carried through to the end. The enemy would almost certainly reply in kind – and unless we are determined and prepared to go one better than the Germans, whatever they may do and whether their reply is in the air or against our prisoners or otherwise, it will be infinitely better not to attempt reprisals at all. At present we are not prepared to carry out reprisals effectively, being unprovided with suitable machines.[74]

When Haig and Trenchard were recalled to London to confer (in part) on the question of defence against the Gothas, Trenchard repeated these views as part of his briefing to the Cabinet.[75] Nevertheless, at a Cabinet meeting shortly thereafter, on 2 July, discussion again turned on the advisability of ordering a retaliatory raid on Germany (Mannheim being the favoured target), and Haig was thereupon asked for his opinion. Haig, as could be expected, answered that he had no planes available for such an adventure.[76]

In sum, both Trenchard and Haig stressed the lack of surplus planes as a primary reason for their opposition, thereby continuing the pattern established by the army and its air service against the RNAS strategic operations. Not coincidentally, the June Gotha raid also caused the Cabinet to move along lines to answer that particular objection. On the very day of the raid, the Cabinet decided to expand the air services,[77] and this resulted in the specific decision of 2 July 1917 to almost double the RFC.[78] But such expansion programmes are not easily achieved. By 1917, it took some sixty-four weeks to bring an airplane engine from its experimental stage to quantity production and about thirty-four weeks for an airplane. It took up to five months to convert factory production from one airplane type to another. The quantities needed for a greatly expanded force were also beyond easy realisation. For just one additional squadron, for example, eighteen planes were needed for its first-line force, backed up by another eighteen planes for reserves and training, plus an additional eighteen planes every two months for necessary replacements – or a total of some 150 planes a year per squadron which also meant an additional 1,500 men in the manufacturing process plus 700 more personnel for maintenance.[79] It is not surprising, therefore, that as late as 25 September 1917 the new Minister of Munitions, Winston Churchill, was still warning that the 'immense programme of Aeroplane Construction' could be achieved only if it be given priority over other claims from the Admiralty and the War Office.[80]

There was, however, one great new anticipated source for aircraft;

the United Stated had entered the war as of 6 April 1917. The mounting climate of opinion about air power among the Allies was influencing thinking in this potential 'arsenal of democracy'.[81] Within weeks of America's declaration of war, missions from France and England were in Washington beseeching for, among other requests, a large aircraft production by the United States. Among the eminent personages involved in these missions were Lord Balfour (the former Prime Minister of England), René Viviani (the former Premier of France) and Marshal Joffre.[82] Then a telegram (on 24 May 1917) from Premier Alexandre F. Ribot set down some specific requests to the new ally. Ribot asked for 5,000 planes for the 1918 campaign plus a production rate by then of 2,000 planes per month for replacements and reinforcements. He also called for 5,000 pilots and 50,000 mechanics plus other personnel and *matériel*, all this necessary, he said, 'to enable the Allies to win supremacy of the air'. In that telegram, Ribot was requesting American industry to produce more airplanes in one year than France had produced in the three previous years of war. The telegram did not specify what type planes were desired (although evidence indicates that this was an unplanned omission, that Ribot had thought in terms of approximately equal numbers of bombers and fighters), so the United States was still uncounselled in that regard.[83]

Unofficial expectations in England at times expressed far larger production figures than those in the Ribot telegram. By mid-June, the *Daily Mail* was editorialising about the US air effort in terms of 100,000 planes. A more temperate estimate given in *The Contemporary Review* in September 1917 still reported that 22,000 planes would be produced in America by July 1918. It was these sort of expectations which heartened Lord Sydenham to prophesy in the House of Lords on 27 November 1917 that the planes then being produced in America would mean an early victory for the Allies.[84]

American hopes in aerial matters started out at least equally roseate. In part, as the biographer of War Secretary Newton D. Baker reminds us, the American air emphasis had a pragmatic basis: 'It was understandable that the administration should seek a dramatic way to make American might felt quickly at the front, and airplanes, the most romantic development of the war, seemed to be an appropriate means.'[85] But the air power boom in America did not need pragmatic justification; it flourished everywhere.

The first official statement on US air policy after America's entrance in the war was issued on 20 May 1917 by the Council of National Defense. It programmed a production of 3,500 planes for the first year and double that number in the following year. It also announced the appointment of an Aircraft Production Board, chaired by Howard Coffin. But less than one month later, the Council of National Defense was cited in the press as speaking in terms of 100,000 American planes of which 25,000 were to be produced in the first year. The US press also highlighted the air effort. The *New York Times* on 10 June editorialised 'First of all, Airplanes', and advised spending one billion dollars on the air programme. On the same day, it reported that the renowned Rear Admiral Robert Peary (then Chairman of the National Aerial Coast Patrol Commission) advised, 'only by sending thousands of aviators and tens of thousands of airplanes to France could we hope to help the Allies to a decisive victory over Germany'.[86] In a similar enthusiasm, the *New York Herald* headlined: 'Greatest of Aerial Fleets to Crush the Teutons.'[87]

Officials of the new Air Production Board sustained these hopes in their public declarations. For example, Mr E.A. Deeds, the assistant to Chairman Howard Coffin, said that America would build enough airplanes in its first year of war to 'fill the skies' and that this would 'assure victory' for the Allies.[88] Deed's chief, Howard Coffin, also talked of victory through air supremacy and (on 14 June) stated that Great Britain would spend $600,000,000 on its air programme in that year's budget and that America should not stint in its effort either. The same figure of $600,000,000 was used two days later by Brig. Gen. George O. Squier, Chief Signal Officer and in charge of the Army Air Service, as the amount Congress should allocate for the US air programme. As it happened, the aviation bill presented to Congress was for $639,000,000; Congress in turn quickly passed the bill at the final figure of $640,000,000 — the largest single money bill passed by Congress for a single purpose up to that time. The bill was signed into law by President Wilson on 24 July 1917.[89]

Naturally, the European allies felt heartened by the American plans. The Germans felt a corresponding concern. Germany planned an 'American Programme' of counterbalancing aircraft production, spurred on by 'the columns of the hostile press . . . filled with fantastic statements. In a short time thousands of American planes were expected to swarm over Germany and force us to sue for peace.'[90]

As it worked out, however, neither the Allied hopes nor the German fears were at all realistic. The year's span from the American declaration of war until the following April witnessed the US production of only fifteen planes total. The indignation in America was immense and, as a result, charges and recriminations were tossed about furiously.[91]

The reasons for this failure in American productive capacity are still being debated. One careful study of this problem concludes that, 'inadequate tactical and technical information was the root cause of the delays which held back the program of aircraft production in the United States'.[92] Other authors have stressed the naiveté of the popular, military and congressional expectations that the aircraft industry could expeditiously produce great numbers of planes.[93] One scholar has made the sensible observation that American aircraft production in World War I, overall, was better than its production of tanks and heavy artillery and that, although inefficiencies were certainly involved, it was primarily the inflated boasts and hopes in the early planning period which made the final record seem such a sorry performance.[94]

For the purposes of this study, however, the primary point is that English hopes in American aircraft production were not to be fulfilled. The British Government tried hard to expedite the American programme — this was one of the chief tasks of the British War Mission to the US (headed at that time by Lord Northcliffe[95]) — but it was not to be the answer to English air defence/offence problems. In fact, the immediate results for England in this regard by America's entry into the war were detrimental. Supplies formerly sent to England were now being kept in the United States and, moreover, British flyers were now being sent to America to help build its air service and train its new recruits.[96]

Meanwhile, the problems for England remained and, indeed, were intensified by the second Gotha raid on London on 7 July 1917. It is from that time forward that the various strands of this aerial history start coming together. The popular pressure for reprisals was stronger yet because of this raid. Again the Cabinet met on the problem and approval was given for reprisal raids, Mannheim again being the favoured target. And again, Haig resisted this measure, saying his current campaign operations might 'have to be abandoned' if the Cabinet insisted upon this diversionary use of his available air strength. So yet again the Cabinet put off the decision to implement a reprisal policy.[97] But the Cabinet, this time, went much further in preparing for that policy. It took cognisance of the morality issue involved and asked for a legal ruling on the bombing of cities. The memorandum written to answer that request was (as usual) somewhat vague on the question but it was, at least, permissive: 'No legal duty has been imposed on attacking forces to restrict bombardment to actual fortifications, and the destruction of its public and private buildings has always been regarded as a legitimate means of inducing a town to surrender . . .'[98]

The most significant action undertaken by the Cabinet after the July Gotha raid was the appointment of the committee headed (and monopolised) by Jan Christian Smuts to explore the questions of home air defence, the organisation of the air services, and the best utilisation of air power in general.[99] Smuts issued two reports as a result of his investigations. The first one, on home defences, has already been discussed.[100] The second report, issued on 17 August 1917, was on the broader questions in his charge. This report has been called 'the most important single consequence of the Gotha raids',[101] and 'a classic document in the history of the evolution of air power'.[102]

The tenor of the second report[103] is an emphasis of the importance of air power and the need to separate it from the restraints imposed by the older services. Smuts began by stressing the urgency of early action, that this was 'of vital importance to the successful prosecution of the war'. He then briefly sketched the history of how the air service had been subordinated to the authority of the older services. He noted the abortive Joint Air Committee (commonly known as the Derby Committee) and the existing relatively ineffective Air Board. He contended that War Office and Admiralty officials dominated all thinking regarding the use of air power. But, he emphasised, since air could be used independently of the other services (witness the recent raid on London), the time 'was rapidly approaching when that subordination of the Air Board and the Air Service can no longer be justified'.

It was the *independent* use of air power, then, that Smuts primarily used as his justification for an independent controlling air authority, indeed he gave no other justification. He emphasised that independent role of an air force in one of the most quoted statements in all air power literature: 'And the day may not be far off when aerial operations with their devastation of enemy lands and destruction of industrial and populous centres on a vast scale may become the principal operations of war, to which the older forms of military and naval operations may become secondary and subordinate.'

With respect to his recommended new air authority, he advised that it must not be just an additional body to the older air services, that this would 'make the confusion hopeless and render the solution of the air problem impossible'. There should be but one air service, under its own Air Ministry with its own Air Staff. It must incorporate within it all the older air forces and the empowering legislation for this should be enacted by the coming autumn and winter. Urgent action was necessary because airplane production was now programmed for many more planes than the army and navy would need for their operations, and by the following spring and summer, 'there will be a great surplus available for independent operations'. If the traditional military authorities supervised the use of these planes, they would be misused.

Smuts temporised by stating that his recommendations should not be taken to mean that the supporting, *tactical* use of aircraft should not be continued. At times air units should still be placed under operational control of the other services for the best attainment of such tactical purposes. But the thrust of the report was always focused on the independent nature of air power. He foresaw that the trench war might well continue in its deadlocked character, but that the air war could be carried deep into Germany and thus 'may form an important factor in bringing about peace'. Moreover, since Germany was surely planning further and more intensive strategic operations against London and England generally, it was necessary for England to beat the enemy to the punch. He also noted that the war was increasingly becoming a battle of *matériel* more than of men and that 'the side that commands industrial superiority and exploits its advantages in that regard to the utmost ought in the long run to win'.

Finally, Smuts advised a low key approach in the advised change so far as publicity was concerned. If Britain stresses a new role and emphasis in the air war, the enemy will make countering efforts. 'The necessary measures should be defended on the grounds of their inherent and obvious reasonableness and utility, and the desirability of preventing conflict and securing harmony between naval and military requirements.' But, nevertheless, the purpose was really to achieve victory in the war:

> It is important for the winning of the war that we should not only secure air predominance, but secure it on a very large scale; and having secured it in this war we should make every effort and sacrifice to maintain it for the future. Air supremacy may in the long run become as important a factor in the defence of the Empire as sea supremacy.[104]

As already noted, Smuts' prophecy in this report that strategic bombing might become the primary war technique making army and naval operations of secondary importance has become one of the most often reprinted expressions in air power doctrine. Naturally, modern day advocates of strategic bombing are prone to quote the passage as being an early justification of their position. Some RAF leaders who exemplify this are 'Bomber' Harris (Marshal of the RAF Sir Arthur Harris), the controversial commander of Bomber Command in World War II, who claimed that command proved 'the extraordinary accuracy of his [Smuts'] prophecy';[105] Marshall of the RAF, Lord Tedder, who attributed the quotation to 'a certain wise man [i.e. Smuts]';[106] and Air Marshal Sir Robert Saundby, who added that they were 'truly prophetic words, and today they have been completely fulfilled...'[107] Hugh Trenchard offers yet another example. Even though he had originally argued against Smuts' position, he later became one of its strongest proponents, and thus it is not surprising that he too reprinted the passage as being true prophecy, this in a paper first issued during the Second World War.[108] Another World War II witness who quoted the passage was Hans Rumpf who had been involved, indeed horribly involved, in the air war as the German Inspector General of Fire Prevention during the war. He too considered Smuts' concept as being the signpost for the future — although hardly praising it as were Harris, Tedder, Saundby and Trenchard. Nevertheless, Rumpf did not question its accuracy.[109] In like manner, the Canadian military expert, E.L.M. Burns, is against the aerial policy contained in Smuts' view, but he did not question its accuracy when he, also, reprinted that same passage.[110]

One obvious significance, then, of the reactions to this early statement in air doctrine, the second Smuts report, is the general tendency to accept it as an accurate description of present and *near* future conditions of aerial warfare. This is especially noteworthy because, in fact, the day has not yet arrived when strategic bombing has transformed the older military activities into 'secondary and subordinate' functions. Victory in warfare has not yet become solely the gift of the long-range bomber or the missile. It just has not *happened* that way as yet — although few would dispute that the potential in that direction has been greatly increased. The generally uncritical reaction to the Smuts report is thus a clear indication of the manner in which air doctrine is as much emotionally as it is logically received. Very few civilian writers have been more critical of the report than the military observers already noted. The official historians of England's strategic air war in World War II offer an important exception.[111] In a later work, one of these same historians wrote a very explicit criticism of the report, stating that 'much of what [Smuts] wrote in his report seemed to refer more to the age of guided missiles and nuclear war-heads than to that of Gotha bombers and 100 lb. bombs of the war in hand or Lancaster bombers and ten-tonners of the next one'.[112] One other critic of the Smuts report is worth noting. David Divine, a most outspoken voice against the course of British aerial history, has claimed that much of Smuts' presumptions about the potential of air power was based upon false (although honestly held) estimates of aircraft production given to him by his expert advisers. Divine's conclusion is, unfortunately, all too true: 'the affair is in the traditional line of optimistic air exaggeration that shows throughout the pattern of the development of air power.'[113]

There is a direct chain of events linking the second Smuts report to the creation of the Royal Air Force. There is a tendency, therefore, to view the cause/effect relationships between these two factors as being direct, simple and total. Thus, the excellent biographer of Smuts, W.K. Hancock, asserts that it was his subject who fathered the RAF[114] and, in like terms, in a lecture given at Queen's University in Belfast, he said that it was Smuts who 'had established the Royal Air Force'.[115] Moreover, due to the single stress in the report on independent bombing operations as necessitating a separate air force and also due to the circumstances which called forth the Smuts Committee — the Gotha raids and the popular and governmental pressures thereafter — there is also a tendency among many writers to view the creation of the RAF as being solely to fulfil the objective of strategic bombing.[116] But the history recounted in these pages indicates that this interpretation is too simplified; that there had been a long-building trend towards a single controlling body over the air services in England and that various reasons gave impulse to this trend.

The original formation of an English military air service had not presumed that it would be divided into separate entities controlled by the army and navy respectively. Rather, in 1912, the Royal Flying Corps was created with the intention of it becoming the single military air corps for England. But conflicting purposes, ambitions, service loyalties and the like precluded the fulfilment of that intention. Admiralty officials started talking of their 'Naval Wing' and then, very shortly after, spoke and wrote of the 'Royal Naval Air Service' and that concept eventually gained official recognition. In brief, service rivalry was not foreseen in the original conception of an air service but its inevitable presence almost immediately divided that air service.[117]

When war conditions brought an increasing sense of the importance of air power both to the Government and general populace, there was increasing concern about the waste, inefficiency and ineffectiveness associated with that divided authority over the air services. The series of air committees were established in an attempt to find remedies to those problems. The Derby Committee, as it was commonly termed, was the first such committee, and it wholly failed to lessen the problems created by the competing air services. When Lord Derby resigned in the spring of 1916, he stated that the two air wings would not be brought into proper coordination until they were joined into a single service — but he doubted the feasibility of creating that merger during the war. Thus, even in the premature Derby Committee can be seen *some* roots to the creation of the Royal Air Force. The problems which caused the formation of the former body continued and their answer was often viewed in terms of a single air service. Thus, David Divine has some logic in his conclusion that the Derby Committee was 'in a sense, the seminal instrument of the unified Air Force'.[118]

After the short-lived Derby Committee, next came the Air Board under Lord Curzon, established in May 1916. And when Curzon announced this Board's formation in a speech in the House of Lords, he clearly showed that the Government had progressed in its consideration of a single military air force. He declared that the Air Board was admittedly a midpoint on the road to such a service although the Government did not feel it could go further in that direction at that time. However, he added, the Air Board would consider as one of its duties the exploration of how the Government could proceed further toward that goal. As it happened, interservice rivalry proved to be too much to endure for the Curzon Air Board and it too had a relatively short life.[119]

Under Lloyd George, the Air Board was reconstituted (on 22 December 1916) with Lord Cowdray as President. This Board was still operating when the Gotha raids started and when the Smuts reports were issued. Viscount Cowdray was a luminary from the world of business who had been chosen specifically because he was free from any bias towards either of the military services and had the experience to run a large and difficult organisation.[120] The Board over which he presided had been given far more powers than its predecessors. Its authority over questions of supply was very great and its President was given a stature almost equal to that of a cabinet minister's. Even so, the conflicting concepts between the army and navy regarding the proper use of the available aircraft strength could not be mediated by Cowdray and his Board (although they did end much of the waste resulting from competition for supplies).[121] With respect to that debate, as has already been noted, Cowdray was a believer in strategic bombing, as was his overseer over questions of supply, Sir William Weir.[122] ('His [Weir's] ambition was to build bombers by the hundred to carry the war into Germany, for Weir regarded long-range aircraft as weapons potentially more devastating in their direct impact on the enemy's economy than the U-boat would prove on that of the British.'[123])

To sum up this point, then, long before the Gotha raids and the second Smuts report, there had been a considerable history of events which indicated a trend towards the unification of the English air forces and, moreover, various reasons prompted that trend. The Lord Privy Seal (the Earl of Crawford) was speaking the simple truth when he advised the House of Lords, upon introducing the Air Force Bill (on 20 November 1917), that this was not some new idea but 'that this question has been the central topic of political, as opposed to military, discussion about the Air Service for at least eighteen months'.[124] Nevertheless, it is almost certain that the Royal Air Force would not have been created in World War I if it had not been for the Gotha raids and the following results. These raids were the necessary catalyst;[125] without them, the general feeling that such a merger would be extremely difficult to accomplish during wartime surely would have prevented the politicians from taking that step. Politicians do not normally seek out trouble when they can avoid it — and, according to one historian, the creation of the Royal Air Force involved the most difficult interdepartmental wrangle that Lloyd George faced in 1917.[126] Also indicative that the Gotha raids produced the necessary driving emotions for the merger to be accomplished was that the impulse weakened during the hiatus between the summer daylight raids and the start of the fall nighttime series, and it was remotivated by these night raids which started in September 1917.[127]

In fact, the progress from the Smuts report to the creation of the Royal Air Force was laboured, halting and hardly assured. All *seemed* assured when the Cabinet approved in principle the report's recommendation of a separate air service, doing this only one week after its submission, on 24 August 1917. Smuts was appointed to chair another committee whose charge was to see to the details of getting this project under way, drafting the necessary legislation, etc. (the Air Organisation Committee).[128] This prompt Cabinet response can be accredited to all the motives previously mentioned — the old problems of interservice rivalry and waste, popular pressure for revenge after the Gotha raids,[129] the growing mystique of victory via air power, the hope of diverting critical German resources to home defence, etc. It is also worth considering the war conditions that then prevailed. The war of attrition on the Western front had been going on for three gruelling years. The manpower losses were horrendous and the problem of replacements was growing in importance. The distrust between politicians and generals was becoming an important emotional factor. Victory by means of strategic bombardment might well have appeared as a happy alternative and final solution to all these problems connected with the Western front.[130]

By the time of this Cabinet action, however, the daylight raids had ended and the night raids had not yet begun. As the days passed, the motives connected with the raiders thus became less urgently felt. In addition, by the end of August 1917, Lloyd George (who was the *political* champion for Smuts, this being part of their understanding[131]) was forced to retire for a time from active leadership due to overwork and depression of spirits. The nominal leadership in the Cabinet thereby went to Andrew Bonar Law who 'had little taste for strategic controversy, and no mind ever to commit himself when in doubt'.[132] The result was that the Cabinet turned to its customary palliative for indecision, the committee. Smuts found himself placed in charge of a series of these in response to his constant pressuring for action; the Air Organisation Committee followed by the Aerial Operations Committee which was reconstituted as the War Priorities Committee only to be replaced by the Air Policy Committee.[133]

Naturally, too, there was still opposition to the idea of an air service merger from entrenched military leaders. Haig and Trenchard remained in the forefront of this opposition. Haig's diary note of 28 August 1917 typifies their stand.

> The War Cabinet has evidently decided on creating a new Department to deal with Air operations, on the lines of the War Office and the Admiralty. Trenchard is much perturbed as to the result of this new departure just at a time when the Flying Corps was beginning to feel that it had become an important part of the Army.[134]

Nevertheless, the Gotha raiders had converted many in the older services to a belief that some changes were necessary. A key convert was the Chief of the Imperial General Staff, Field-Marshal Sir William Robertson. In a communication to Haig after the 7 July raid, Robertson complained about the inadequacy of the London defences and concluded, 'I am inclined to think that we need a separate air service, but that would be a big business'.[135] Another important army proselyte for an entirely separate air service was Lord French, then Commander-in-Chief, Home Forces.[136] Even the Admiralty intransigence was tempered, at least in its political representation, when Sir Eric Geddes became the new First Lord in July 1917. Geddes stated that he would accept the formation of an Air Ministry, provided that it was charged with the control of former RFC units only.[137]

The Admiralty, as it worked out, gave much more assistance than this to the creation of the RAF by its assignment of Admiral Mark Kerr to serve under Smuts as its representative. Kerr was an air power enthusiast, and was moved to an important action in this history by words of Lord Cowdray on 10 October 1917. The President of the Air Board lamented to Kerr that the Cabinet was backtracking, that it would not create the new air service and that this would end all chances for a strategic bombing force to be set up against Germany. Kerr thereupon wrote a long memorandum for Cowdray to use to help push the affair back on course. This memorandum is worth quoting at length, not only because it did have an effect upon ensuing events, but also because it further illustrates aspects of the contemporary belief in air power.

> This memo. is to point out, shortly, the extraordinary danger of delay in forming the Air Ministry and commencing on a proper Air Policy.
> In strategy it is necessary to put oneself in the enemy's place, and to decide what he will do. Information which confirms this is valuable.
> Germany will reason thus:
> 'Armies and Navies depend for existence on supplies and communications; destroy these and the enemy is beaten. Formerly we worked to outflank an enemy on land, in order to cut his communications; he then had to retire, or be surrounded and surrender. This principle remains, but modern warfare renders it impossible to carry it out round the flanks of the Allied Armies; the battle-line is too long, and there is no room for a flanking army of sufficient size on the map. Therefore we must go over the Allied line and stop them from coming over us. It is thus necessary to reduce output in the Army and Navy and make a huge bombing squadron to destroy the British and French aerodromes and machines, before they come and destroy ours, and also to bomb their factories out of existence. When we have done this, the Allies' supplies will be reduced and cut off, and their Armies must either retire or surrender.
> 'The submarines will continue the same work at sea. Our German policy is, therefore, to cut down output in everything except submarines and aircraft, and to increase our squadrons of underwater and air craft. We must strain every nerve to make our attack before the Allies make theirs. If we succeed in this, their attacking machines will be destroyed before they leave their own aerodromes, and we shall win the war. . . .'

Kerr added that evidence buttressed this prediction of German policy; that the enemy was decreasing its production of various (stipulated) armaments but that they were building thousands of bombers, larger even than the Gothas. Kerr wrote that the night raids on London were primarily training exercises for the pilots who would later be flying those massive bombers over England. He warned:

> It is a race between them and us: every day lost is a vital danger. If the Germans get at us first, with several hundred machines every night, each one carrying several tons of explosives, Woolwich, Chatham, and all the factories in the London district will be laid flat, part of London wiped out, and workshops in the southeast of England will be destroyed, and consequently our offensive on land, sea, and air will come to an end.
>
> There is no exaggeration in this, and if we are going to stop it, we must start at once with our preparations to lay their factories flat and to destroy their aerodromes . . .
>
> In short, it means the country who first strikes with its big bombing squadrons of hundreds of machines at the enemy's vital spots, will win the war . . .

Kerr then stated that only a separate Air Ministry with executive power could achieve that result for England.[138]

Lord Cowdray immediately took Kerr's statement to Lloyd George, and the Prime Minister was strongly impressed by it. Lloyd George, in turn, had reprints made of it for the Cabinet members and the memorandum became a topic of Cabinet discussion at its next sitting. Three days after that sitting, on 16 October 1917, Bonar Law informed the House of Commons that the Government would soon submit to it a bill to create a unified air service.[139]

Admiral Kerr's account is, of course, too self-serving in that he only mentions his memorandum with respect to these culminating days before the Cabinet's final decision. By then, the night Gotha raids had started and had thereby returned the air question to an urgent priority. Typically, after these raids began, the *Daily Mail* became very vehement in its attacks upon the Government about its air policy and, as usual, it insisted upon the adoption of a reprisal policy.[140] If *The Times* was more moderate in tone in its editorials, its position was, nevertheless, the same — England cannot really defend itself against such attacks except by retaliatory strikes of even greater power. And, *The Times* added, a separate air service was the way to accomplish this. In another editorial, *The Times* advised that 'there is no lack of converts now to the doctrine that the best of all defences against air attacks is a strong aggressive policy'. It also warned that the ethical argument should not apply; that the Germans attacked both war-related targets and civilian morale in their raids and the English must do likewise in retaliation.[141]

By October, the issue was of daily news value. *The Times* opened up the month with one of its strongest editorials on the matter. Insisting, as usual, that England should be bombing Germany, it argued that army and navy officers were blind to the potential of air power and all too prone to use it merely in the function of a support weapon. Therefore, it reasoned, the establishment of a separate air service which would operate a giant fleet of bombers in raids against Germany was the pressing necessity. 'When we attain that ideal, London may again know peace at night.'[142]

It was only two days later that Lloyd George informed an impromptu street gathering that the Government had approved a reprisal policy. And on the following day, 4 October, Smuts made an official announcement for the Cabinet which confirmed Lloyd George's observation. Smuts said that the decision to retaliate was made 'reluctantly' but still definitely.[143]

Obviously, Admiral Kerr's remarks had not been solely responsible for forcing the final decision; indeed, they were written *after* Smuts' announcement on 4 October. At most, they just helped the decision to be speedily implemented. But the Gothas had already (and once again) performed their work as catalysts in this history, and the well-established breadth of motivations worked out their fate. The most important of the resulting official acts were Bonar Law's announcement to the House of the Government's intentions, on 16 October; then the Cabinet's approval of the draft bill for the air service, on 16 November; the passage of this bill through Parliament, done with 'remarkable speed and smoothness' with final royal assent signed on 29 November; the establishment of a separate Air Council, on 3 January 1918; the announcement of the Royal Air Force title, on 7 March 1918; and the actual official establishment of the Royal Air Force as an operating entity, on 1 April 1918.[144]

In the postwar era, official British statements which described the creation of the Royal Air Force glossed over the part played by the desire for a strategic bombing force. For example, in the Command Paper 'Synopsis of British Air Effort during the War' presented to Parliament in April 1919 it was claimed that the separate air service was established to end the duplication of effort and *matériel* and wasted energy which resulted from having two air services.[145] That explanation not only neglects the strategic bombing factor but also the role played by interacting influences from the general public, the press and the Government. That latter error of omission was not made in a later public statement (in 1928) about the creation of the RAF, but nevertheless, the basic explanation of the 1919 version was still repeated; this in a speech to the House of Commons by the Secretary of State for Air at that time, Sir Samuel Hoare. He stated that the Royal Air Force:

> came about, not as a result of the claims of theorists, but as the result of a most definite demand from public opinion. It was demanded, first of all, for the purpose of avoiding the duplication of organisations, and to entire unity of command and unity of effort in meeting the menace of air raids in London and in this country.[146]

Another list of factors which had brought the RAF into being was presented in an excellent Belgian military journal. This 1921 view also stressed the wasteful rivalry between the RNAS and the RFC, especially for industrial output; it added the problems caused by poor liaison between the air services where their authorities overlapped, for example with respect to air defence of coastal areas; and it claimed that the discipline demanded by the older services proved too rigid for air personnel relationships. But, again, no mention was made about the pressures and purposes connected with the Gotha raiders.[147]

The contemporary evidence is very clear, nevertheless, that strategic bombing hopes were connected with the birth of the RAF. Moreover, this was well realised at the time. *The Times* editorial after the announcement of the Air Force Bill, for example, celebrated the fact that air power would no longer be just a support weapon for the older services.[148] Indeed, an essential continuing characteristic of the RAF was established in its very creation, it was an *offensive* service arm which was created to deal with *defensive* needs.[149] As shall be analysed in Chapter 6, RAF leaders in the interwar period were often to justify the services's independent and offensive role in terms of its fulfilling defensive needs. The entire aerial history covered in this study is substantively of a piece.

Although the intent was present right from the beginning to make the new Royal Air Force fulfil a strategic bombing role, that purpose was not, for the most part, achieved. Troubles plagued the new air arm from the very start. Naturally, former opponents to the merger were not instant converts after its accomplishment. Much cynicism prevailed.

The Birth and Early Infancy of the Royal Air Force

Trenchard found the official formation date of the RAF to be very apt – All Fools Day. And Trenchard was the man selected to be the service's first commander (Chief of the Air Staff). Also indicative of his continuing scepticism was his comment to Haig on the day he was returning to London to assume his new duties; he then complained that the Air Ministry politicians were wholly irrational in their hopes that air power could end the war.[150]

The political leadership of the service also started out in a very restive state. The general expectation was that Lord Cowdray, the chairman of the body which was being replaced by the new Air Ministry, would be appointed as Air Minister.[151] The well-informed civil servant, Sir Almeric Fitzroy, saw the Air Bill as 'in substance the tardy attempt to fulfil the pledge given to Lord Cowdray on accepting his office'.[152] Nevertheless, Lloyd George offered the new post to Lord Northcliffe, the press lord who controlled both the *Daily Mail* (the paper with England's largest circulation) and *The Times*.[153] Lloyd George later claimed that he did not believe Cowdray's health was strong enough for the added responsibility.[154] The 'official' historical account would have this decision based primarily on the fact that Cowdray had expressed doubts about the wisdom of the air merger during wartime and that the Government wanted a man fully in accord with its policies – although Cowdray's doubts reflected an earlier viewpoint.[155] A more cynical interpretation states that Lloyd George was punishing Cowdray for an article in Cowdray's *Westminster Gazette* which told of the Prime Minister fleeing London because of his fears of air raids;[156] another account states that Lloyd George was seeking to profit from Northcliffe's public popularity.[157] There was also Whitehall gossip of the time which speculated about potential trouble over Cowdray's Mexican oil dealings.[158] For whatever reason the decision was made, Lloyd George was to rue it; Lord Northcliffe, in a dramatic *coup de théâtre*, answered the Prime Minister's approach with an open letter declining the offer, which he published in *The Times* on 15 November 1917. Northcliffe gave no prior notice to Lloyd George, nor had Lloyd George yet said anything about his plans to Lord Cowdray. The letter thus was something of a bombshell. The launching of the Royal Air Force therefore started in a prophetic manner, that of passionate charges and countercharges and generally heightened emotions.[159]

The Prime Minister next offered the position of Air Minister to Lord Rothermere, Northcliffe's brother, who accepted.[160] But this selection also proved to be an unhappy one, primarily due to the friction that developed between Rothermere and his choice[161] for Chief of the Air Staff (CAS), General Trenchard. The appointment of Trenchard began with high hopes, as evidenced by a letter written by Evelyn Wrench, a key aide of Rothermere's: '... everyone thinks General Trenchard's promotion as Chief of the Air Staff is a good move'.[162] But Trenchard was far less sanguine, and in less than a month after his appointment during a visit in France, he was complaining bitterly to Haig about his new political chief and 'the rascally ways of the politicians and newspaper men'.[163]

One key disagreement between Rothermere and Trenchard was their opposing views over strategic as against tactical use of air power, Rothermere pushing for the independent air role against the German heartland.[164] Soon after his appointment, Rothermere had made his position clear on this matter in a speech given at Gray's Inn wherein he stated that he and his co-workers were 'wholeheartedly in favour of air reprisals. (Loud cheers.) It is our duty to avenge the murder of innocent women and children ... we shall slave for complete and satisfying retaliation.'[165] Trenchard's emphasis on front line needs first has already been described. After the final break between the two men, Rothermere justified his opposition to his CAS in part because Trenchard 'was insisting on the ordering of large numbers of machines for out-of-date purposes'.[166]

There were, of course, other reasons which exacerbated their relations. Trenchard was especially sensitive to Rothermere's disdain of the formalities of chain of command. The minister sought advice wherever he wished while Trenchard believed that he should have been the sole source of expertise to his political chief. The outcome of the matter was inevitable; Trenchard resigned and his resignation was accepted on 10 April 1918.[167]

Sympathy for Trenchard was evidenced from the royal family on down.[168] The RAF pilots started wearing black ties as symbols of mourning.[169] The reaction of the military correspondent Repington was representative of the general feeling: 'A real bad business as Trenchard is a tip-top man.'[170] Naturally, the matter was brought up in the House of Commons where the debate was an emotional one and favourable to Trenchard almost in entirety.[171] The upshot of all this was that Rothermere was also forced to go[172] and (the point of major importance here) that Trenchard was given a new and important assignment – almost, one might say, by popular demand.[173] Thus it happened that Trenchard was given the command over the newly created Independent Air Force, the strategic bombing force that the

RAF was, in such strong measure, created to bring about.

The Independent Air Force evolved from a strategic bombing unit that was forced upon the RFC the previous October (1917). The Government, concerned at the time by the Gotha night raids, requested Trenchard (still the RFC commander in France) to spare planes for such a unit. The result was that three squadrons of bombers were thereupon sent to the same general area near Nancy where the RNAS had already furnished precedents for such procedures. Lieutenant Colonel Cyril L.N. Newall was given this command, and he made his headquarters at Ochey, France. On 17 October 1917, this force had made its first raid — on an iron foundry near Saarbrücken — and the RFC had entered the age of strategic bombing. The first RFC night raid of a strategic character followed only a week later.[174]

Trenchard was characteristically sceptical about this new venture, as usual pointing to the lack of available surplus machines. However, Lloyd George stood firm and told Trenchard that he expected successful results to assuage public opinion at home. Gradually, then, Trenchard shifted in his attitude as Newall's accomplishments both surprised and impressed him. By December 1917 Trenchard was pressing for more support to the Ochey effort — although the Government was convinced that the RFC could be even more helpful in that regard.[175]

There were reasons for Trenchard to be impressed. Even though Newall's force was too small to engage in more than very limited activity, the reaction in the German home front was similar to that in England from *its* limited suffering. A delegation of Rhineland mayors, for example, pleaded to the higher authorities that Germany should stop its policy of strategic bombing so that the reprisal raids might also cease; this in February 1918. In the same month, the King of Spain sent word that he was willing to intercede for the Germans to arrange limits upon aerial bombardment. Then the Bavarian and Baden Chambers of Deputies passed resolutions expressing their desire to have bans placed on the bombing of home front areas.[176]

The Cabinet leaders were heartened in their aerial policy by these reactions in Germany. They felt that the advantage in the air war must now be on their side and so they should persevere and profit from that advantage. Thus the indications that Germany might be willing to stop strategic bombing only gave additional impulse to British bombing plans — so much for the formerly stated objective of having reprisal raids in order to stop the German raids! The Cabinet also reasoned that 'in the interest of future peace, it was undesirable that the civil population of Germany should be the one population among the belligerents to enjoy immunity from the worst sufferings of war'.[177]

The bombing operation offered to Trenchard for his command was, therefore, already of proven value and of increasing emphasis. The original three squadrons had been increased by two additional squadrons; the 41st Bombing Wing had become the VIII Brigade. Its further expansion into the unit to be designated the Independent Air Force was already planned — that unit officially constituted as of 5 June 1918. The Air Minister who replaced Rothermere was Sir William Weir, a fervent strategic bombing enthusiast, and he added yet more pressure for the strengthening of that force — as did Trenchard after he took over command. By the end of May 1918, Trenchard wrote Weir, for example, that 'as you know, we are pushing the bombing'.[178]

Nevertheless, problems and hindrances still dominate the story. The very name 'Independent Air Force' caused friction. As one RAF officer later acknowledged, that term *independent* 'gave rise to a fear that the newly created Air Staff wished to engage in a private war of its own, separate from, and quite unconnected with, current land and sea operations'.[179] And well might that fear have arisen with respect to the aerial authorities; indeed one essential distinction between the Air Minister (Weir) and the IAF Commander (Trenchard) was that only the latter kept front line needs in mind as a continuing priority.[180] Moreover, those fears were made manifest by actual opposition. As the new CAS, Major-General Frederick H. Sykes, later reported, the BEF commanders continued to oppose this independent bombing activity, as did the French military officials by the time of the IAF.[181]

The French opposition forced the matter to come before the Supreme War Council, where it remained an issue for weeks. Marshal Ferdinand Foch was especially insistent that the IAF should either be placed under his supervision or else it should get out of France — and since he had been (as of 26 March 1918) officially 'charged by the British and French governments with the coordination of the action of the Allied Armies on the Western Front',[182] his position was a strong one. This problem was finally resolved by the adoption of an idea already promoted by Weir some time before, the expansion of the IAF into an Inter-Allied Independent Air Force to be composed of British, French, American and Italian bomber forces. This new unit was placed under Foch's overall supervision and Trenchard was appointed its Commander-in-Chief on 26 October 1918.[183]

The official brief for this Inter-Allied force included this stated 'object of the Force. To carry war into Germany by attacking her industry, commerce, and population.' The brief also stressed that it be a large air force whose raids were to be carried out 'with tenacity'. However, it also stipulated — the French and BEF opposition clearly involved here — that Western front line needs must have first call on its services.[184]

The war ended before this Allied venture ever progressed further than being just a paper organisation, but, nevertheless, it also was involved in early difficulties. American participation in the joint force would have been a very doubtful factor. The American Secretary of War, Newton D. Baker, strongly protested against the force's purpose when he was informed of it. He stated that the United States would never 'war upon a defenseless civilian population', and he wrote to the Army Chief of Staff, General Peyton March, that the United States should not participate in any bombing that 'has as its objective, promiscuous bombing upon industry, commerce or populations in enemy countries disassociated from obvious military needs to be served by such action . . .'[185]

Weather was another trouble that plagued the British effort in strategic bombing; 'by the time the force really got going the weather was unfavourable, poor in September, and really bad in October'.[186] And there were still more problems. Enemy fighter defence forced Trenchard to direct one-half of his effort against German airfields — a programme he planned to continue through the summer of 1919. Trenchard was also forced to provide his day raiders with fighter protection because of the rate of loss. From all this, England should have learned the lesson of the need to attain air supremacy in order to fulfil a strategic bombing programme, but this factor was later neglected in the interwar RAF planning. The informed critic, Noble Frankland, believes that this neglect was the 'cardinal mistake' of the Allied air effort which culminated in the bombing of Germany in World War II.[187] Trenchard's strategic bombing effort was also diminished because Marshal Foch called upon IAF help on the Western front in September and October.[188]

Finally, time did not favour the IAF record. Although plans for its complement fluctuated — in July 1918 plans called for 104 squadrons in the force by the fall of 1919 — the IAF never *attained* anywhere near the strength that its mission presumed. By the armistice it only contained nine bombing squadrons plus one fighter squadron, the bombing squadrons divided between four long-range and one short-range night bombing squadrons and four daylight raiding squadrons.[189] But had the war continued into 1919, then the strategic bombing of Germany would have reached a very intense effort indeed. The planes for that task had really just started to appear by the armistice. By then, three of the Handley Page H.P. 15 v/1500 were in service, a 'gargantuan' bomber powered by four 375 h.p. engines with a 126ft. wing span and a bomb capacity of 7,500 pounds.[190] The bombs were also appearing in ever improved models. A 1,650 lb. model was at hand by war's end and in 1919, Berlin among other cities was scheduled to feel its power.[191] Berlin and other German centres were also to receive poison gas (especially Lewisite, a new blister type gas developed by the United States) and incendiary attacks from the air.[192]

In sum, Lloyd George wrote accurately of the anticipated results when he described the 'devastating' strikes on German cities that were planned for the year 1919.[193] It is important to remember that those anticipations were part of a history with various formative roots but nevertheless they were substantively the result of a *defensive* response to the Gotha raids on England. It is also important to remember that the Royal Air Force was primarily created to achieve just those types of hoped-for results. As shall later be described, the RAF's continued status as a separate military service in the interwar years greatly depended upon its insistence that the independent action of strategic bombing was necessary to English *defence* needs and that this military purpose could only be achieved by a separate air force. But just as the creation of the RAF needed the support of non-military people who held similar conceptions of air power, so too did the continued existence of that force need continuing civilian concepts of air power which would allow acceptance of that RAF argument. Thus, to understand the later development of RAF air doctrine and its ever present connection to defensive needs, it is necessary to analyse the development of aerial concepts and theory in non-military circles. This will be the subject of Chapter 5.

5 THE FORMATIVE YEARS OF NON-MILITARY BRITISH CONCEPTS OF AERIAL WARFARE (TO 1931)

Since humankind has long been concerned, and often fascinated, by the potential threat of attack coming from the skies, modern authors have had no difficulty finding applicable statements from the past which reflect that fascination. One commonly used source is the Italian Jesuit academician, Count Francesco Lana de Terzi (1631-87) who, in his book *Prodromo overo saggio di alcune invenzioni nuovo* (1670), foresaw the invention of aircraft that would thereupon be used in war using bullets and bombs and 'aritifical fire' among other means.[1] Erasmus Darwin affords another prescient vision in his poem *The Botanic Garden* (1791) which predicted the advent of aircraft and their military effectiveness, aircraft in which

> ... warrior-bands alarm the gaping crowd,
> and armies shrink beneath the shadowy cloud.[2]

Lord Tennyson's poem *Locksley Hall* (1842) furnishes another often cited passage, one which provided the title for the notable Royal Air Force leader Sir John Slessor's autobiography.[3]

> For I dipt into the future, far as human eye could see,
> Saw the Vision of the world, and all the wonder that would be;
>
> Saw the heavens fill with commerce, argosies of magic sails,
> Pilots of the purple twilight, dropping down with costly bails;
>
> Heard the heavens fill with shouting, and there rain'd a ghastly dew,
> From the nations' airy navies grappling in the central blue.[4]

From such past evidence one can conclude that if it is true that 'nations have tended to regard flight as a prerogative of war',[5] this would only be in accordance with the workings of human imagination long before the age of flight. Certainly, the joining of the idea of flight with warfare typifies the writings of H.G. Wells, surely one of the most influential literary prophets of air warfare before World War I.[6] As soon as Wells started 'anticipating' aircraft, he foresaw their use in war, and his thinking in that regard grew more cataclysmic. These prophesies began in 1901 when he presented his 'Anticipations' in the *Fortnightly Review*, which were then published as a book in 1902.[7] This book represents Wells' first major essay into the field of non-fiction prophecy and it received a very welcome public response. '*Anticipations* established the renown as a seer from which Wells could never extricate himself, nor did he wish to. The Prophet gloried in it and practised his clairvoyance with brash confidence all the rest of his life.'[8]

In this work, Wells was remarkably restrained in his aerial forecasts compared to his later writings. He forecast the appearance by 1950 of heavier-than-air craft which would be war-related. There would be air combat to achieve 'command of the air', the victor of which would have immense advantages in aerial observation, in troop morale and in the ability to subject the enemy troops and also targets deep behind the enemy lines to a bombardment of explosives and incendiaries. Thus Wells vaguely implied the practice of strategic bombing, but he did not pursue this concept beyond that single intimation.[9] Wells' description of the opening phase of an air war as one consisting of a clash to achieve 'command of the air' is worth special note as this would later become semi-official military doctrine during the interwar years — as will be noted in the following chapter.

In 1908, two more works appeared by Wells in which he envisioned future air war in a much more awesome manner. In another work of non-fiction prophecy, *First and Last Things*, he described:

> great towns red with destruction while giant airships darken the sky ... For the first time in the history of warfare ... the rear of the fighting line becomes insecure, assailable by flying machines and subject to unprecedented and unimaginable panics. No man can tell what savagery of desperation these new conditions may not release in the soul of man.[10]

Wells thus states what will become a continuing fear about air war, that it would involve mass panic and an aftermath of violent behaviour. We have already seen how the question of panic became a major concern during the First World War in connection with strategic bombing,[11] and we will later see how, in the interwar years, that concept of panic and resulting violent behaviour became associated with the spectre of mass revolution.[12]

Also in 1908 appeared Wells' famous fictional account of *The War in the Air* and in this work the future horrors were even more unlimited. Great cities were destroyed (including New York in a transatlantic raid by German aircraft), battleships proved defenseless against the bomber,[13] and the belligerents exchanged staggering air attacks with the result that all the powers were too weakened to gain a positive victory.[14] This was Wells' fictional view as of 1908. After the Great War, his journalistic descriptions of present reality were in close accord with that earlier fictional presentation. In 1921-22, for example, he reported for various newspapers on the Washington Conference on the Limitation of Armament, and his reports were published in a book in which he describes the next war as one involving air raids upon capital cities within hours of its start, with gas, incendiary and high explosive bombs all as likely possibilities.[15]

H.G. Wells furnishes an excellent example, in sum, of the manner by which human imagination foresaw the use of flight as a means of bombing urban populations and how, by the interwar years, that earlier vision was replaced by a fixed presumption of the present-day plans for air power. It is worth noting that the prewar writings by Wells which were cited in the above analysis were all published before the occurrence of the first successful flight across the English Channel. Naturally, that flight heightened apprehensions in England about the military implications of the new invention.

Louis Blériot was the man who accomplished that dramatic flight. Flying a plane of his own design and given impulse by a *Daily Mail* offering of £1,000 prize money, Blériot traversed the Channel on 25 July 1909, making a thirty-seven minute trip from Calais to Dover.[16] The English reactions to this feat were mixed. Crowds gave Blériot a hero's reception and the Lord Mayor of London officially received him. Nevertheless, there were definite signs of disquietude. The press was prone to see the event as signifying that England had lost its moat-like security. In both England and France appeared notices that England was 'no longer an island'.[17]

The connection between the age of flight and the loss of insular security was one that was commonly made. The early enthusiast, Louis P. Mouillard, whose researches had been closely studied by the Wright Brothers, had forewarned that the invention of aircraft would mean 'no more frontiers! No more insular seculsion [*sic*]! No more fortresses! . . . Will society perish?'[18] Naturally, Englishmen would be especially prone to react with this sense of lost security because they had long been relatively secure behind their naval-protected surrounding waters. H.G. Wells, by then enjoying his new role as the prophet of the modern age, published an article in response to the Blériot flight in which he stated that a continental power could build within a year hence hundreds of planes capable of performing bombing raids on London of wave-like continuity.[19]

Gradually the phrase that England was no longer an island became a cliché. Major T.L. Baird, Parliamentary Secretary to the Air Board, used it in his speech to introduce the Air Force Bill in its second reading.[20] Viscount Templewood (the former Sir Samuel Hoare) used it in the 'Foreword' to his memoirs of his years as Air Minister.[21] Winston Churchill used it in a speech on 15 March 1922, wherein he was emphasising the importance of air power to Britain's future.[22] Other examples abound but the point is that, as usual, this cliché represented a generalised viewpoint; in this case, that England's traditional defensive security was lost with the development of the airplane and that England existed thereafter in grave jeopardy. This mental shift in England from confidence to insecurity about its defensive position was of major consequence during the interwar years. It naturally affected budget priorities among the armed services – the air advocates being quite outspoken in their use of this argument. For just one example, Lieut. Col. Alfred Rawlinson concluded his study (published in 1923) of London's wartime aerial defences with a warning about the island's present and future needs. He told his countrymen not to 'emulate the asinine antics of the ostrich and ... bury our heads in our traditions of naval defence ...', and in a later passage of clarification he added, 'it is necessary for all to recognise that, as the responsibility for the first line of the defence of these islands gradually changes from the SEA to the AIR, adequate provision shall be made to enable us to maintain our legitimate position in the comity of nations'. Thus, he argued, England must spend less on naval defence and more on air defence.[23]

It was not just the press and public who evidenced this early fear of lost defensive insularity. According to Sir Walter Raleigh, the official historian of England's fledgling years in the air, the highest officers in the English military services resisted the concept of aerial warfare just because that threatened loss was involved.[24] On the other hand, the British Government showed a like concern but with more positive results. Even before Blériot's flight, a British governmental commission 'reported that the peril of aerial bombardment to England's insularity was a justified concern'.[25] The Blériot event (and, presumably, the general progress in aerial development by 1909) moved the Government from enquiry to action and an Advisory Committee for Aeronautics was appointed with members included from the fields of science, engineering and military service. Science was especially stressed, seven members being fellows of the Royal Society.[26]

There was a like response in Parliament in 1909 which resulted in the establishment of the Parliamentary Air Committee with members from both Houses. Its original chairmen were, in order, Arthur Lee (later Lord Lee of Fareham), Colonel Wilfred Ashley, and then, of most interest, William Joynson-Hicks (later Viscount Brentford of Newick) whose role as a parliamentary air spokesman we have already observed. This committee proved to be an active ginger group in prodding the service ministers for more air-mindedness by their respective arms.[27]

The publishing world naturally also reflected the heightened interest in aerial matters. By 1911, there had already appeared over 300 books and 25 periodicals devoted to this area.[28] The popular press naturally played its part. Lord Northcliffe of *The Times* and the *Daily Mail* prided himself upon his activities in sponsoring English aerial interests. His large prizes for aerial 'firsts' were in that context, such as the £1,000 award for the first cross-Channel flight. And consistent with the trends already discussed, Northcliffe was both advocate and Cassandra about aerial affairs. Before the World War began, for example, he was already pressing for England to build up a four-fold air superiority to Germany as he warned that Germany would be attacking England with her opening blow being an attack by air.[29]

The Northcliffe press was hardly unique. Two interesting and parallel reports appeared in early summer 1914 in the *Morning Post* and the *Daily Telegraph* after their respective reporting teams had visited Royal Flying Corps installations. Both articles gave the usual courtesies — comments about efficiency and careful preparations and the like — and then both speculated upon the future of aircraft in warfare ending with a similar verbal shudder about that future; the *Morning Post* foreseeing the possibility of 'the use of the plane as the most awful engine of destruction the world has seen . . .' and the *Daily Telegraph* describing the airplane as 'the deadliest weapon of man's invention'.[30]

By that year specific wartime events had furnished some concrete evidence to bolster human imagination. In the Italo-Turkish War of 1911-12, a certain Lieutenant Cavotti, on 1 November 1911, dropped four 4½ lb. bombs upon some very surprised Turkish troops and thus initiated the age of aerial bombardment. Italian flyers kept up the practice and the outraged Turkish protests show that this war practice occasioned a far more violent reaction than did much more damaging, but more traditional and expected, forms of bombardment.[31]

So trends were already being set in the worlds of both thought and action that would be extremely intensified during World War I, that 'forcing house' for aviation as Lloyd George so aptly described it.[32] One enlightening way to observe the effects of that 'forcing' action is to follow the trend of thought during this period of that frequent British commentator upon aerial concerns, Claude Grahame-White (often writing in collaboration with Harry Harper). In a book published in 1911, Grahame-White's major stress was on the observation and messenger uses of aircraft in wartime. He reported that military authorities were discounting the airplane as a bombing instrument although he did foresee future improvements in aircraft design affecting that presumption. He mentioned that there was some current thought that bombing would produce panic far beyond what would be justified by the actual physical danger involved but his discussion assumed such bombing would only be directed at troops in combat, and he personally believed that the troops would react sensibly to such attacks once they became accustomed to them. In essence, then, this 1911 work presents a very temperate vision of the air war to come.[33]

A year later, Grahame-White, writing in collaboration with Harry Harper, was still not making very dire prophecies but he was by then certain that bombing would definitely be part of the warplane's function and that 'the next great war will begin, not on earth, but several thousand feet in the air' where opposing aerial fleets would struggle 'above all else, to obtain supremacy of the air'.[34]

In an article written immediately after the start of the war, Grahame-White predicted that airplanes would be used primarily for 'scouting' rather than for offensive, destructive attacks. He commented that the press preferred to emphasise the bombing danger but that this was not realistic, that the number of planes on hand were too few to risk them in missions other than reconnaissance. (Grahame-White was obviously one who presumed that the war would be over before that supply situation greatly changed.) He suggested that airships might well be used on bombing raids, especially in that so many believed that the effect on morale would be so much greater than the actual physical results justified. And he did agree that if the war had come five years later, then certainly there would be aircraft at hand especially designed as bombers and that the war would have opened with an air battle for aerial supremacy. Lastly, with respect to that future probability, he warned that an international ban against the bombing of civilians must be arranged or the alternatives would be horrible.[35]

In 1916, Grahame-White and Harper wrote an article summarising the preceding two years of war in its air phase. They claimed that it was now commonly accepted military doctrine that the first task of an air force is to achieve 'a clear and definite command of the air'. The goal, however, is still to gain a monopoly in air reconnaissance — although they saw much greater advantages accruing from that supremacy in future wars when the offensive power of aircraft would be so much more effective. 'Any nation which falls behind in the struggle for air power may, in years to come, be defeated in a campaign lasting not a year or a month or even a week, but as the result of a blow delivered and completed within a few hours.' Future war would mean hundreds of planes attacking 'the very nerve centres and vital arteries' of the enemy, hitting governmental, military and industrial targets alike, and even 'blotting out whole cities'. Both gas and bacteriological bombardment would occur.[36]

A year later, in an article primarily concerned with the great peacetime contributions that flight would bring, the two authors again warned of the horrors that it would add to warfare. 'After one or two staggering blows, in which its chief cities are destroyed, and its means of communications paralysed, a country may find itself so helpless that there will be nothing for it to do but sue for peace.'[37] Also in 1917, they published a book which illustrates that they had become firm believers in a great many assumptions about air power which would characterise the interwar years. Its preface states that 'the greatest lesson of the war' was that air supremacy would ultimately guarantee to its fortunate possessor 'the dominion of the world'. Within the body of the book they argued that victory in war would henceforth depend upon victory in the opening clash of opposing air armadas; that a state's preparations for war should thereafter include placing the important governmental, administrative and industrial potential targets underground while shelters should also be provided (probably) for the civilian populations; that strategic air strikes against enemy 'vitals' using all means available (gas, high explosives, incendiaries and bacteriological inundations were all mentioned) were the only logical future war tactics because only thus could a state achieve quick victory and no state could endure another extended war; and that such a strategic air strike could well precede a formal declaration of war. They had no doubt of the quickness of victory by such methods, indeed, victory might be achieved on the opening day. Therefore, all depended upon prewar preparations. Air forces must be kept supplied with the latest and best equipment, government funds must be used to support the aircraft industry, and the people must be educated in all the portents of the air age now coming about. Lastly, after examining past efforts to restrict air war methods, they concluded that all such efforts would always be unavailing.[38]

This analysis of the evolving thought of one of England's most active early aeronautical commentators shows how he quickly adopted many concepts which came to dominate much of interwar aerial dogma. These concepts include the beliefs that future wars would be undeclared affairs that would either open with air strikes against enemy 'vitals' or with aerial contests to achieve dominance over the skies, or both; that the strategic raids would leave reeking snake pits of the urban centres, stricken from incendiaries, high explosives and a vaguely defined complex of chemical and biological warfare methods;[39] and that such raids were the only realistic means to victory, and that it would be a very quickly gained victory. Also of note in this mental odyssey, is Grahame-White's change of mind with respect to legal restrictions — at first he stresses the need for such restrictions and later he denies that they would have any real restraining effect. The interwar years also evidenced much ambivalence over this question. Lastly, it is obvious from the chronology of Grahame-White's writings that it was the Great War which did so much to change his early patterns of thought into much more fervid forms.

One other example should well illustrate this point. H.G. Wells was certainly among the most unrestrained prophets about the potential of air power, but he hardly envisioned world dominance and total victory by such means: 'The War in the Air kept on through the sheer inability of any authorities to meet and agree and end it, until every organised government in the world was as shattered and broken as a heap of china beaten with a stick.'[40] But world dominion via air power was just what Grahame-White and Harper derived from the war as its 'greatest lesson'.[41] Nor were they alone in that idea. In 1916, the prestigious Alexander Graham Bell, in an address to the National Convention of the Navy League of the United States, had already made a similar prophecy. 'We may therefore look forward with certainty to the time that is coming, and indeed is almost now at hand, when sea power and land power will be secondary to air power, and that nation which gains control of the air will practically control the world.'[42] In sum, World War I represents a great watershed in the development of air war concepts.

The peacemakers at Versailles were moved by yet another fear about air power that would characterise much thinking during the interwar years, the fear that commercial aircraft could easily be transformed into strategic bombers. Nor was this fear very illogical at the time, as it was only in later years that aircraft design would become ever more specialised.[43] An early indication of this mental linkage between commercial and military aircraft was made by the first British Civil Aerial Transport Committee, established in May 1917. This committee recommended (in July 1918) that commercial aircraft design in Britain be purposely oriented to easy military conversion. They also advised that wartime utility should be one of the qualifications for future planned civil air routes and landing fields.[44] These recommendations closely followed the position taken by Chief of the Air Staff, Sir Frederick Sykes, in his evaluation of the 'Air Situation and Strategy . . .' presented to the War Cabinet on 27 June 1918. Sykes admonished the government that 'in developing our commercial air force, war use must be considered. Our air routes and aircraft bases must be designed to help to meet the requirements of war as well as those of peace', and that 'commercial aircraft could be quickly converted into battle craft'.[45]

Just after the armistice was arranged, the commander of England's air defence system made a similar warning to the Commander-in-Chief, Home Forces. Therein, General E.B. Ashmore advised that that defence system should be maintained because commercial flight would soon be a reality and each plane involved would be a potential bomber.[46]

Thus, concern about future civil aviation was well established by the opening of the Versailles Conference, and as General Sykes led the British Air Section, it was certain that this concern would be voiced.[47] However, in the event, the French at Versailles evidenced more anxiety about German commercial air development than did the English.[48] Therefore, discussions and papers at the conference frequently expressed that theme. General Duval, the French representative on the Aeronautical Commission, warned that there was no way to distinguish unequivocally between a German civil airplane and one designed for military purposes. A.J. Balfour agreed that the growth of a German civil aircraft industry would constitute a danger and that it might just be a front for the rebuilding of a military air force. General Groves concurred: 'the military brain of Germany', he charged, 'intends to develop German air power under camouflage of civil aviation.'[49]

Like positions were taken when the peace arrangements with Turkey were discussed. When international officials concerned with these negotiations met at the British Foreign Office (on 22 March 1920), the Aerial Navigational Committee reported to them that 'the technical committees had unanimously recognised that there was no practical means of preventing civil aviation from being rapidly transformed into military aviation'.[50]

Naturally, the belief that all commercial planes were potential bombers increased British anxieties with respect to its aerial defences during the interwar years. In a work published in 1922, General Sykes made the point very explicitly:

> In the event of a war of short duration that power will win which has the greatest preponderance of machines, service or civil, fit to take the air. The asphyxiation of a large enemy city, if within range, can be done by night-flying commercial machines, and it would require a defending force of great numerical superiority for its successful defence.[51]

This belief also complicated and inhibited another approach to the air defence problem, the attempt to achieve an international air limitation.[52] Naturally, statesmen would feel loath to deprive their countries of military air strength if they envisioned their neighbouring countries' commercial aircraft as being potential raiders. It was primarily such thinking which prevented an air convention from being achieved at the Washington Conference of 1921-22. Too often this conference is only discussed with respect to its efforts concerning naval limitations.[53] The original invitation issued by Secretary of State Charles Evans Hughes to the world powers stated that its proposed goal was for a limitation of arms in general.[54] It was officially called the Conference on the Limitation of Armament[55] and one of its various subcommittees was dedicated to aircraft. General William 'Billy' Mitchell was one member of that committee and his stance was 'that the only practicable limitation as to the numbers of aircraft that could be used for military purposes would be to abolish the use of aircraft for any purpose'.[56] The subcommittee's final report echoed that stance and thus concluded that no practical way to limit air power was possible at that time. This conclusion was then endorsed by the conference delegates in general, in a statement issued on 9 January 1922, which announced that it was 'not at present practicable to impose any effective limitations upon the numbers or characteristics of aircraft, either commercial or military'.[57]

This pattern of thinking and its resulting impotence was to continue throughout the interwar years. Akin to General Mitchell's expert testimony to the Washington Conference, for example, was that of Chief of the Air Staff Trenchard's to the British Cabinet in 1925 when that body was discussing the prospect of aerial disarmament. Trenchard advised that all aircraft would have to be banned before any restriction would be effective, 'that commercial aircraft, as at present constructed, can be very rapidly adapted for use in war.'[58] Meanwhile, various publications were presenting the same argument to the general reading public. The outspoken Labour MP (and former wartime consultant to Lloyd George[59]), J.M. Kenworthy, wrote that any attempt to abolish aerial warfare would be in vain due to the fact that civil aircraft was so easily adaptable to war usage. Likewise, the air defence expert, General Ashmore, warned that England would always be in danger of an air attack, no matter what Europe did about disarmament and for exactly the same reason.[60]

How justified were these assumptions about commercially styled airplanes? Expert advice during these same years was unsure. In an address in 1924 to the Royal United Service Institution, one RAF officer said that such convertibility was no longer practical (although future designs might find a way to close the gap).[61] But in an article seven years later in that institution's journal, another military expert was certain that civilian airplanes were still able to be used as bombers. ('Herein lies the danger of masked preparation for unexpected attack and for acts of international brigandage.'[62]) Chief of the Air Staff Trenchard was not a trustworthy guide on this subject either. In 1925, in his evidence to the Cabinet, he was fearful of commercial aircraft but just two years before, he had told his Air Minister (Sir Samuel Hoare) that commercial and military design would not develop along similar lines and that the idea of their interchangeability was a misleading delusion.[63]

Edward P. Warner, then professor of aeronautics at Massachusetts Institute of Technology, gave a sensible assessment of the problem (in 1926) in an address given before a seminar studying 'Disarmament and Security'. Warner noted that commercial planes were not efficient war machines – that, for example, they were not able to defend themselves effectively – but that nevertheless they *could* be used as bombers, any plane could drop bombs, because, accuracy aside, no special equipment was needed to drop a bomb because gravity provided the single essential requirement.[64]

Also pertinent to this subject is that design was not greatly differentiated between commercial and bombing aircraft during most of the interwar years. In 1928, for example, the Junkers aircraft concern of Germany started producing a passenger plane of large size for its time and concurrently brought out a military version of that plane for the Japanese army, equipped to be a heavy bomber. Even World War II evidence is not wholly negative in this regard as the early Dornier and Heinkel bombers used by Germany had been originally used in commercial transport. *Somme toute*, air power mythology is never wholly disassociated from air power realities.[65]

Given the generally held idea about the war potential of commercial aircraft, then naturally the interwar development of the German commercial air services made its contribution to English feelings of insecurity about air defence.[66] This German development was surprisingly robust considering the Versailles settlement which not only banned all military aircraft to Germany but also limited her commercial strength at the time to 140 planes plus 169 additional engines while forbidding further construction of commercial planes there until 1922.[67] Of course, the Germans found these provisions objectionable. They said that they needed seven hundred planes for air mail service and that in various other ways the rules were far too strict with respect to their commercial prerogatives. But for a time the Allies remained adamant against such complaints.[68]

Nevertheless, the German air industry managed to keep operating, primarily by turning to neutral countries for location of their construction facilities. Many aircraft firms of World War II fame used such expedients in the immediate postwar years including Dornier, Junkers and Heinkel.[69] Soon German controlled corporations were making a large impact upon the world air scene. To cite just one example, 'Scadta' in Latin America was such a corporation and it became so powerful as to cause official concern in the United States, the American Secretary of War pointing out that its aircraft were potential bombers which threatened the Panama Canal.[70]

The English were quick to show anxiety about this German activity. Lord Northcliffe by 1920 was already warning of German air prospects and intentions and Lloyd George, according to a report by Lord Derby, was in agreement that there was a 'dangerous development of civil aviation in Germany'.[71] The following year saw General Masterman of the Inter-Allied Military Commission in Berlin expressing his concern over the burgeoning German civil air industry and its potential war danger.[72] One year later, in 1922, the para-official military source, the *Journal of the Royal United Service Institution* reported on the great progress being made in German commercial aviation.[73]

In 1922, the Versailles ban on German construction was lifted, and naturally, this caused even greater progress in German development. More names of future fame appear; the Focke-Wulf Company in 1924, Messerschmitt in 1925.[74] The next major change in the situation occurred in 1926 when, as part of the Locarno concessions, what restrictions had remained on civil air construction were removed; restrictions such as maximum allowance on motor power, wing span, load capability and the like. In turn, the German government promised to limit its financial aid to the industry.[75] The ban on military aircraft still applied, of course, but Gustav Stresemann, the German negotiator at Locarno, had no doubt about the military usefulness of the air concessions because 'civilian aviation might be converted into military aviation'.[76] 1926 was also important in this story as being the year when the German civil air lines merged to form the Deutsche Lufthansa which operated thereafter as a state-controlled monopoly.[77]

All this activity kept British anxieties very much alive. In 1928, for example, the former RNAS leader, Murray F. Sueter, wrote about his foreboding of the German civil air service which had, in just the previous year, doubled its passenger totals and quintupled its freight load, whereas 'all their civil machines can be turned into war machines at short notice'.[78]

Certainly to regard the German commercial activity in just military terms and not to recognise the profit motivations and civilian communication needs and the like which would prompt it was unrealistic. But nevertheless the British fears were perceptive, as hindsight informs us. Many of the Luftwaffe leaders during the Nazi era had participated in these developments. The first managing director of Lufthansa, for example, was Erhard Milch who was later to become Deputy Air Minister under Hermann Göring – and Milch and Göring were already in mutual contact over commercial aerial matters by 1927.[79] If Dr Emme is perhaps overly harsh in concluding that 'it was this aeronautical development, blessed with legality by the peace mechanism, which for the most part enabled the Nazis to so rapidly rearm Germany in the air after 1933',[80] nevertheless the anxiety about that development could be justified then and can still be justified now.

Also contributing to the worry over German air power was the fact of the clandestine rebuilding of her military air service. This illegal activity (in terms of the Versailles restrictions) has been analysed often and at length[81] and need only be briefly noted here. The greatest activity in this regard took place in Soviet Russia. There is some evidence that the German military attempted to arrange warplane production in Russia as early as 1919[82] but serious negotiations only began in 1921. By the following spring, the Junkers firm was in production there.[83] In April 1922, the Treaty of Rapallo between the two states included a secret clause which broadened their scope of aerial cooperation. Thereafter, among the German officers who were actively engaged in training programmes in Russia were Albert Kesselring, Hans-Juergen Stumpff and Hugo Sperrle, the three leading commanders of air operations against England in 1940.[84] However, the overall results of these measures were not very dangerous in any immediate sense[85] — although they did have a long-range influence ('German rearmament after 1933 would have taken considerably longer without the experience gained in Russia . . .'[86]).

Of more importance for this study is that knowledge of this illicit activity was, to some degree, known and publicised in England. The *Manchester Guardian* printed articles on these developments in December 1926, and the works by Kenworthy and Sueter give evidence that this Russian programme was fairly common knowledge in the following two years.[87] In sum, then, developments in both German civil and military aerial spheres in the 1920s added to the climate of opinion under discussion in this chapter.

Yet more attitudes that typify English aerial conceptualising between the wars may be seen in an analysis of the Government's efforts in air raid precaution planning undertaken during the 1920s.[88] The Government naturally had to concern itself with the prospect of a future air war and its attendant civil requirements. In 1923, the Air Ministry suggested that the Home Office was the department which should develop ARP plans and the Home Office concurred. Consequently, in January 1924, the Committee of Imperial Defence (CID) established a subcommittee to investigate ARP needs (the Air Raid Precautions Committee) under Home Office direction, the chairmanship being assigned to the Permanent Under Secretary of State for the Home Department.

That office was then held by the remarkable Sir John Anderson. Another admirable civil servant, Lord Salter, says of Anderson that he was 'the greatest administrator of his age', and notes the 'qualities which . . . brought his greatness; the sure judgement, the personal authority, the intellectual grasp and the capacity to bring unity of policy out of conflicting personalities and interests'.[89] Another civil servant has recalled John Anderson's era at the Home Office during which all his associates considered him as 'Jehovah, the all wise, the all knowing, the all powerful'; a man who was wholly unflappable, ever sound, uncommonly filled with common sense.[90] Anderson's biographer has also commented on his austere image and the cognomen 'Jehovah' and the genius of the man who so often made the right choice between alternatives; this man, who during World War II, would be named by

Winston Churchill (under impulse from George VI) to be Prime Minister if he, Churchill, and Eden both died during their trips overseas.[91] The point then: the chairman chosen to investigate air raid precaution necessities was hardly a man prone to wild exaggerations and panicky thinking. It is thus very convincing evidence of the all pervasive air fears of the day and that these fears fitted with the 'logic' of contemporary conditions that Anderson was one of those who was thoroughly convinced of the justification for these fears. And it was Anderson who was most influential in the formation of the ARP committee's guidelines, its first and trendsetting report being largely the product of his thinking.[92]

The committee's assigned task was 'to enquire into the question of Air Raid Precautions other than Naval, Military, and Air Defences',[93] and to report yearly to the CID on this question. As the official history observes, this purview was extremely broad and difficult and what essentially was involved was the examination of 'all means by which the civil authorities could co-operate to make the policy of the Fighting Services effective'.[94] This is a doctrine which has been accepted throughout the air age; as one RAF authority recently expressed it, 'it should be remembered that although civil defence can never win a war, the lack of adequate civil defence could well lose it'.[95] Air warfare, in other words, has added more complications to a problem as old as war itself, the necessity of keeping one's home base secure. This, then, was the basic assignment given to Anderson and his committee.

Under Anderson's guidance, the enquiry was parcelled into eight major divisions: early warning, prevention of damage, repair of damage, maintenance of necessary services, the possible need of the removal of the Government to a safer location and the details connected with that possible action, education of the public with respect to the threat involved and the best measures for their self-protection, legislative powers that would be needed, and, finally, the specific individual departmental responsibility for all the actions recommended. Each of these divisions covered considerable areas of investigation. The official history notes, as an example, that the topic 'prevention of damage' embraced such large problems as lighting restrictions, camouflage, shelters, gas masks and the evacuation of the civil population.[96]

This chapter is concerned with attitudes more than actions. Therefore, its attention will now be centred on the assumptions and conclusions taken by the committee. Analysis of the specific measures taken or recommended will be made in the following chapter.

One primary assumption that was clearly held by the committee was the *measure* of the air threat that they foresaw. The Air Ministry had provided the base for measurement in estimations that it gave the committee in its second meeting. These 1924 presentations estimated that the next war could bring 200 tons of bombs down upon England during the first 24 hours, 150 tons during the second 24 hour period and 100 tons per day thereafter. Defence could perhaps (only in the most hopeful prospect) prevent up to half of this bombardment from occurring. London was presumed to be the primary recipient of that tonnage.[97] By complex calculations based upon London's experiences during the past war, the Air Ministry further stated that each ton of bombs would produce 50 casualties, one third of them fatal. This pounding of London would continue *at these rates* for at least one month. Thus, in the opening thirty days of war, Londoners could expect to suffer, at a very conservative estimate, almost 80,000 casualties, one third of them fatal. These estimated figures were accepted with alarm but without question by the ARP committee.

That fifty casualties per ton estimated rate had a tremendous and lasting influence. While the passing years saw the estimated tonnage raised, the fifty multiplier remained for the casualties – easy to use and remember – and it became the accepted truism.[98] And yet this figure was based upon very dubious calculations. The sociologist Professor Richard M. Titmuss has analysed this point:

> The use of this multiplier of fifty casualties per ton can be criticized on several counts. First, it might reasonably have been argued that such a casualty rate could not continue to operate for long. With several thousand of tons [*sic*] of bombs being dropped every twenty-four hours, the population of London was, in such circumstances of damage and destruction, bound to diminish as a result of (a) evacuation (b) the number killed and (c) the number injured and removed to hospitals outside London. Thus, within a few days of the first raid the population would be smaller and the density of population per acre thinner. The ratio of casualties to tons of bombs would continue to decline as the ratio of population to the land area of London declined. For this reason, it was not valid to use a fixed and constant ratio of casualties per ton. A second criticism of this ratio derives from the sketchy and unreliable character of the statistics on which it was founded.

Professor Titmuss then proceeds to give specific examples of questionable statistics, both with respect to tonnage and to resulting casualties during the war, plus other statistical faults as, for instance, errors in transcribing figures.[99]

Nevertheless, it has only been since the Second World War that such arguments have been pressed against these extrapolations from the Great War's experience. *Now* it is easy to recognise that the sample base was far too small for good statistical evidence.[100] We now know that the true casualty figures lay between 15-20 per ton with the one third fatality estimate still holding fairly true.[101] But the interwar years in Britain were not ones of very logical and dispassionate thinking about the future air threat. Counter-evidence to current assumptions and expectations was just not given much credence. And the results of that pertinacity were great. As another of the 'official' histories states:

> This exaggeration of the number of casualties which strategic bombing was likely to produce was a major factor in all strategic thinking before the Second World War and exercised a profound influence on the minds of the services, the political chiefs, and being translated into even more sensational language by journalists and publicists, on public opinion at large.[102]

The ARP committee's first report was issued in July 1925, ('it remained the basic document on the subject until some twelve years later...'[103]) and it shows that many other arguments by the Air Ministry were fully accepted by the committee. The report agreed that London would be the key enemy aerial target and by that was meant the London area in general rather than specific military targets within it, that the possibility of effective defence was small and that public morale would therefore be a major concern, one which might well be the decisive factor in the war, and that *control* of the civil population must therefore be an important consideration.

The committee looked at the possibility of abandoning the city and moving its vital functions elsewhere but they concluded that this was beyond practical achievement and that England could not carry on if London could not maintain those functions. Nevertheless, partial evacuation was deemed necessary, primarily to lessen the *probable* incidence of civil panic. ('The serious moral consequences of air attack ... was to preoccupy every subsequent committee concerned with air raid protection up to 1939...'[104]) The committee's recommendation thus traced the fine line between what they considered essential in London's maintenance and its parallel terrifying vulnerability; those who worked in the required activities should be aided, perhaps even forced, to stay while those not needed (especially, of course, women and children) should be encouraged to leave. The report further recommended that the respective ministries and boards over health, transport, trade and education should start planning the necessary operations to achieve these goals. Of major importance is that these conclusions within the first official interwar ARP report came to represent continuing and basic governmental assumptions: 'the Government thought that a large exodus from London and other cities was inevitable; panic would send the people out and unless the Government took firm control chaos and confusion were bound to ensue'.[105] A major concrete step which reflected this viewpoint was the creation in 1931 of the evacuation subcommittee of the Committee of Imperial Defence and this committee so believed in the probability of a panicked outburst of the populace from London that they talked in terms of a police cordon surrounding the city to prevent this from reaching extreme proportions. They debated on 'how control of the population was to be exercised'. Control by police units became the recommended method after talks with the Commissioner of Police and the Home Office. To this extent, then, had this aspect of air defence proceeded by the start of the 1930s.[106]

Control of the civilian population was an important prospect in air raid planning, not only to help (or enforce) their continuance at work and their orderly evacuation, but also to maintain order where otherwise panic and chaos was expected. As early as 1908, H.G. Wells had envisioned the need for military aid to preserve law and order among urban masses after air attack.[107] In 1922, the famous Geddes Committee[108] advised the use of territorial army troops for such home air defence activity.[109] The ARP committee was of like persuasion throughout its sessions during the 1920s and into the 1930s. In 1930, for example, this committee advised that three battalions of troops should be prepared for action for home air defence duty. The chairman, Anderson, wrote that if such troops were not on hand, it was his belief that 'such a state of panic would be produced as might bring about a collapse, certainly of the community in London, if not of the whole country'.[110] (Military concepts are generally long-lived. Of interest in this regard is this comment by an RAF authority in a recent work: 'The army will play their greatest part in home defence in the event of severe air or rocket attack, whether nuclear or otherwise, by maintaining morale as well as supporting local authorities and essential public services and utilities.'[111])

A point to remember in connection with these British fears about panic and the need for troops to maintain public order is that these years were not only ones of the Air Scare but also of the Red Scare. Historians have no trouble finding evidence of extreme, often neurotic, fears over the 'Bolshevik' threat among Europeans in general and politicians in particular during the whole interwar period. Professor Sontag found this a major element worth his treatment in his review of Europe 'between the wars'.[112] James Joll has claimed that 'the fear of revolution was a real one for all European politicians in 1919. They believed in the possibility of world revolution as firmly as Lenin himself . . .' and given the events of the time, Joll finds that belief 'a plausible one'.[113] Maurice Baumont, in his outstanding study of the interwar years, has devoted a chapter to those events of 1918-19 which prompted so many Western fears, events he summarises as 'la contagion révolutionnaire en Europe (1918-19)'.[114] Max Beloff has also described 'the fear and horror' that Communism provoked in the West.[115]

Certainly, English history gives evidence of that 'fear and horror'. In 1919, for example, Winston Churchill at the War Office circulated to his army commanders a 'Secret and Urgent' questionnaire which basically was an enquiry as to the trustworthiness of the troops for such duties as preserving popular order, strike breaking service, and possible military service against the Soviet Union.[116] (Churchill's fear of public disorder was greater than his fear of possible troop disloyalty. It was also in 1919 that he pressed for a rebuilding programme for the Territorial Army, primarily for their potential use in maintaining order in England.[117]) When the hunger marches started in England in 1922, they were often seen as sinister manifestations of some revolutionary programme and, as Arthur Marwick perspicaciously observes, these marches were then indexed in *The Times* under the category 'Russia: government propaganda'.[118] Also in 1922 one finds one of the better pieces of evidence of how paranoiac the Red Scare fear could become − a speech to the assembled military personages at the Royal United Service Institution by one famous Russophobe of the day, Mrs Nesta Webster. Of added significance is the fact that all the floor comments after her address were in full agreement with her statements, which included such warnings as: 'British Christian civilisation is being undermined whilst we sleep. We must awake. We must awake and act before it is too late.'[119]

All this bears upon the purposes of this chapter because the Red Scare could so easily be tied to the Air Scare. Given this fear of popular revolt plus the belief that air war would be directed primarily at urban centres wherein the masses were basically without protection, then popular panic might well be foreseen as a prelude to mass revolution — and this link provided an incentive for governmental planning in terms of population control by police and/or military troops. In a contemporary comment by one British military authority, England had to concentrate upon its air defences because it was 'the great cities and towns of England where Bolshevik upheaval might be hoped for by the enemy'.[120] In sum, interwar fears in England were not only extremely strong but were also often interacting and reciprocally reinforcing.

Also worth notice in the committee's report was their presumption that aerial defence measures were woefully inefficient and that this forced them to a conclusion that we have seen as a continuing theme in early British aerial history, that only an offensive strategy could answer defensive needs. They reported that only an aerial policy which stressed offensive operations might provide security at home. 'Adequate protection for the civilian population from aerial attack could only be secured by the vigorous prosecution of an active offensive campaign which would carry the war into the enemy's country and subject his people to conditions even more trying than those which the people of Britain would be called upon to endure.'[121] No international agreements would prevent these exchanges of air blows. Therefore the basic quality of a people's willpower might well become the final arbiter. 'In the next war it may well be that that nation whose people can endure aerial bombardment the longer and with the greater stoicism, will ultimately prove victorious.'[122] The calm, Jehovah-like John Anderson and his committee had seen the future in as cataclysmic a vision as was described by the most outspoken gloom and doom air prophets of the time.

A great many of these prophets were members of the Royal Air Force, but there were also many who were not then in military service (although usually of past military service) who had much to do with the creation of a common mode of thought about aerial matters. Chapter 6 is concerned with contemporary military air doctrine. The following account will look at these other prophets who may, perhaps, be usefully considered as the major paramilitary theorists.

Among these, the late Basil Henry Liddell Hart had certainly achieved the most respected reputation. The acclaim given to Liddell Hart upon the appearance of the first volume of his memoirs gives impressive evidence of his honoured status.[123] He has been mentioned in high esteem by President Kennedy, he has been called 'the greatest military thinker of the twentieth century' by General Chassin of France,[124] and another authority has described him as 'this century's most brilliant and creative thinker on military subjects'.[125] Obviously,

Liddell Hart no more than John Anderson represented an extremist, illogical persuasion. That his views on air power were of the category generally described throughout this chapter is yet another solid indication, therefore, of how pervasive those views were.

B.H. Liddell Hart was a history student at Cambridge who joined the army in the First World War to serve as an infantry officer. He showed early talent in military theory and in 1920 (at the age of 24) he was assigned the duty of writing the first postwar manual on infantry training and then, shortly after, the first task of editing the official *Small Arms Training Manual*. But then the army removed him from its active list in 1924, ostensibly because of his wartime disabilities (a concussion and exposure to phosgene gas) although Liddell Hart claims the true reason was that the hidebound army leadership found his adventurous mind not sufficiently suitable. The following year he became military correspondent for the *Daily Telegraph* and in 1935 he filled the same role for *The Times*.[126]

Also in the year after he was 'invalided' out of active duty, Liddell Hart published 'his first important popular work . . .',[127] *Paris: or the Future of War*, which is, among other things, a remarkable testament of faith in the new military weapon of air power. World War I furnished his starting point of reasoning; Europe, he claimed, could not repeat that kind of military fiasco without thereby destroying the essential fabric of Western civilisation. The problem was, therefore, to find a war method which would bring a quick victory, to produce a weapon which would seek out your enemy's most vulnerable point just like Paris' arrow found Achilles' heel — thus the title and basic theme of the book.[128]

Liddell Hart contended that the Great War had proved that confrontation between opposing armed forces was no longer feasible and that a *new* target must be emphasised, the enemy's *will* to continue the war. 'The aim of a nation in war is, therefore, to subdue the enemy's will to resist, with the least possible human and economic loss to itself' (p. 19). There were various means at hand to attack that target — propaganda, blockade, etc. — but air attack was superior to all others. To support this view, the author quoted Marshal Foch (the italics were added by Liddell Hart): 'The potentialities of aircraft attack on a large scale are almost incalculable, but it is clear that such attack, owing to its crushing *moral effect in a nation, may impress public opinion to the point of disarming the Government and thus become decisive.* (p.14).

The Achilles' heel of modern war is, then, the populace's ability to endure. To attack that vulnerability from the air also meant gaining an additional advantage of hitting the enemy economic resources at the same time. But the priority target must always be the enemy morale: 'The enemy nation's will to resist is subdued by the fact *or threat* [Hart's italics] of making life so unpleasant and difficult for the people that they will comply with your terms rather than endure this misery' (p. 31). This dislocation of the populace is best achieved by air raids on the centres of its most vital activities – thus morale and economic targets were not distinctly separate categories. The author argued against the previous war's lessons with respect to strategic bombing, that such raids were far too weak in character to be valid guides. But weak as they were, they still caused panic and dislocation. That they were not decisive was not a valid objection. Nor do moral arguments against such tactics have any validity. Strategic bombing, he claimed, would result in far fewer fatalities than such lengthy bloodbaths as the previous war. Moreover, if gas were used as the primary aerial weapon, then war would become even less lethal. Gas offers much greater chances for survival. (It is worth remembering the author had been a gas casualty of the First World War.) Gas was, in fact, an ideal weapon in many ways. Its effective range was extensive so that accurate bombing techniques were not necessary, it was invisible in many of its forms and it was an excellent surprise weapon because it could be made under secret, peacetime cover activities.[129]

In sum, Liddell Hart argued that future victory in war might well be decided by which side has the better air arm and the stronger will to endure. The endurance factor would not be long in contention; victory might come within hours, very probably within days. As for England's chances in such a war, he was very explicit and pessimistic:

> Imagine for a moment London, Manchester, Birmingham and half a dozen other great cities simultaneously attacked, the business localities and Fleet Street wrecked, Whitehall a heap of ruins, the slum districts maddened into the impulse to break loose and maraud, the railways cut, factories destroyed. Would not the general will to resist vanish, and what use would be the still determined factions of the nation, without organization and central direction? (pp. 41-2).

One saving hope for England was that such a war might not occur if the contending sides were approximately equal as to air strength and vulnerability; then each might refrain from such strategy in fear of retaliation. In that light, the author then turned his attention to more traditional forms of warfare, on land and sea.

Paris, or the Future of War had an immediate and influential impact. Liddell Hart correctly states that the work 'fitted the developing trend of the Air Staff's views on the subject [of air power] '.[130] The Chief of the Air Staff, Sir Hugh Trenchard, took care to see that the book was widely read by his RAF officers and by leaders in the other services. The civilian air minister, Sir Samuel Hoare, was also a reader and believer.[131]

The commentary by Marshal Foch that was quoted by Liddell Hart has its own separate interest. As Robin Higham has observed, the 'conversion' of Foch to this air power stress was a very influential event; his statement of the potential of air power (that quoted by Liddell Hart) was used to give an aura of absolute authority to strategic air war doctrine ('the equivalent of a papal seal of approval').[132] One finds it requoted again and again as though his view was the final blessing needed. Some indication of the frequency of that usage is furnished by the *Journal of the Royal United Service Institution* wherein one finds three different military authorities giving that same passage by Foch in less than a two year span.[133] And Rear-Admiral Sueter saw fit to present those comments in full twice in the same book.[134]

Although *Paris* was Liddell Hart's first book on these themes, he had already indicated in earlier articles that his thinking was developing along these lines. In 1922 he wrote that 'the conquest of the enemy nation's will to resist is the fundamental principle ...'[135] and he reaffirmed that target concept in March 1925, adding that air power (especially using gas bombs) was the ideal means of attacking that target.[136] Furthermore, this pundit continued to believe that air power was the supreme military means throughout the period under study here. By 1930, he does express some doubt that future war will start by an exchange of strategic terror air raids, believing that the belligerent states might refrain from this in order not to alienate neutral opinion — but there will be air strikes, he claims, against economic targets right from the start, these will include targets within home front urban centres, and that 'in this future warfare, economic in aim, the air will be the dominant partner'.[137]

Another paramilitary authority on aerial matters of major influence was 'the fanatic of air power',[138] P.R.C. Groves. Groves had joined the Royal Flying Corps in 1914 after having served some years with the land forces. He became a friend of General Sir Frederick Sykes who, when made Chief of the Air Staff in 1918, appointed Groves his Director of Flying Operations. This brought Groves into a supervisory role over the Independent Air Force in France. After the war he was

British Air Representative at the peace conference and later a member of the military team overseeing German disarmament. He then served as Air Advisor to the Council of Ambassadors in Paris and to the Council of the League of Nations. After he retired from military service in 1922, he still held influential positions as air correspondent for *The Times* and as an active promoter of the Air League of the British Empire.[139]

As an air power publicist, Groves is most famous for a series of newspaper articles he wrote in 1922, articles which brought forth much debate and eventually legislative action in Parliament and which also helped to confirm popular attitudes on the subject.[140] In these articles[141] Groves stated that the Great War had furnished a lesson which must be heeded by England or all its expenditure on defence would be for naught. That lesson was that war had become an exercise in 'areas' rather than in 'fronts', and that henceforth the primary areas to be attacked will be those of the home front, via air raids.[142] 'Each side will at once strike at the heart and nerve centres of its opponent: at his dockyards, arsenals, munition factories, mobilization centres, and at those nerve ganglia of national *morale* — the great cities.'

These air raids would involve that unholy trinity of aerial weapons, high explosives, gas and incendiaries, delivered by thousands of planes. All this is 'beyond doubt'. The *sole* practical defence lay in long-range bombing strikes and these should be directed at both the enemy air force bases and depots and his vital urban centres; that is, attempt to inhibit his ability to raid your centres while you attack his public morale. Purely defensive measures will be of almost no value. Thus, 'the Independent Air Force is not only the striking arm, but also the shield. It is the first line of national defence.' And since one's reprisal threat was the only possible deterrent, it must be built up during peacetime. Government money must be used to support civil air services and aircraft industries as part of that build-up.

It was in that regard that Groves created an immediate sensation because he warned that the French civil and military air force was far stronger than the British counterpart.[143] Groves reminded the English that they could no longer depend upon the Channel as a moat, that it now only added to the threat by limiting the ability to sight the oncoming raiders while, at the same time, easing the enemy return flights.

Typical of the air spokesmen, Groves was prone to reiterate his warnings, even to the point of verbatim restatement. A case in point is a 1924 article of his which warned about the French air menace and in which he repeated the passage quoted above concerning the air targets of future wars. In this article he also used the oft-quoted statement by Foch (already noted) to bolster his argument. He stressed that air war was becoming ever more terrifying because improvements kept appearing in planes and bombs while defensive abilities remained futile. He also emphasised the peculiar vulnerability of England to this type of war, being so urbanised a society that London was 'the largest aircraft target in the world'. He repeated his concept of a war, now, of 'areas' and prophesied that such a war was a near possibility, perhaps within the next decade. The result of that war could be utterly dreadful: 'In such a contest each of the belligerents would be in a position to inflict upon its opponent destruction on a super-cyclonic scale. If they were fairly equally matched, devastation might continue until the collapse of the entire social and industrial system of both combatants.'[144]

Yet another paramilitary writer on air power – a prolific one – was James Moloney Spaight. Spaight was a civil servant who evidenced early interest in aerial matters and who became a key official in the Air Ministry during the interwar years, rising to the rank of Principal Assistant Secretary before he retired in 1937. He was a close friend of the air force's chief, Trenchard, and has been called 'one of the most authoritative writers on air power . . .'[145]

Spaight was more moderate than the usual air prophet, perhaps because of his official position within the establishment. Indeed, he believed that 'the hot-gospellers of air power' hurt their own cause.[146] But still, before the war's outbreak, writing in the spring of 1914, he described terror raids on cities as a possible air war strategy although holding out the hope that current standards of morality plus fears of reprisal would prohibit such activity. However, he advised that steps be taken to reduce the attractiveness of urban centres as targets; remove all military activities from these cities, place all railway stations underground, make the defensive capability as effective as possible – suggestions that governmental leaders may well have found somewhat impractical.[147] Near the end of the period discussed in this chapter, Spaight still preserved some hope that the cities would not be targets *per se*. He had no doubt that they could be completely demolished[148] but still felt that air staffs might prefer other targets, especially armament factories. He disagreed that terror raids on cities would be an infallible way to break the enemy will, they might bring just the opposite result, a people only the more determined to endure. 'Besides', he added with more of the aura of an Air Ministry publicist than of an accurate reporter, 'air power does not particularly wish to smash the cities. If it has the impatience it also has the chivalry of youth.'[149]

If Spaight can be described as a moderate air power advocate because of such views, further examination of his writing shows just how superficial moderation about air power could be. For example, in an attack on the League concept of collective security written in 1928 (the title of which clearly denotes its primary thesis, *Pseudo-Security*) he specifically states that air attacks against urban areas will be the basic character of warfare in the future and, significant to this theme, that this precludes any security that the League claims to afford. His argument was that the outbreak of war would entail devastating attacks against cities and the League machinery would move far too slowly to prevent those raids. Thus, 'our cities might have suffered irreparable damage meanwhile. It is even conceivable that the war might have been decided by then.' Spaight produced a novel description for the urban raids. Recalling the Japanese earthquake of 1923 which caused over 150,000 casualties, he wrote of 'stupendous aerial earthquakes' which could possibly bring like results.[150]

Many other aerial clichés of the day were included in this book: gas would be a probable weapon, both gas and airplanes were peacetime products easily converted into wartime horrors, the time of England's insular security was over and her period of appalling vulnerability had begun, neither disarmament nor legal prohibitions would stop the threat of strategic air raids, etc. He expresses more faith in fighter defences than many do, but admits that air defence is really an untested hope. Yes, strategic bombing also represented an untested gamble — but he had little doubt about its effectiveness; he described what might happen to Geneva if it was made the headquarters city of a collective security military force, that it 'would stand an excellent chance of being bombed sky-high in the first great war of aggression'.[151]

The prevailing concepts about air power brought Spaight to the conclusion that League collective security was futile. The same concepts helped bring David Davies to the conclusion that only a League police force guaranteeing collective security could preserve Europe from those horrors.[152] Reactions to air war fears could greatly vary, the fears themselves were generally of a piece. Davies wrote of air raids upon cities including the use of gas and incendiaries and he also believed that no prohibition would stop the use of these means — unless the League had at hand his advised collective security police force. Davies was a total believer in the air menace potential; it could mean the end of civilisation. 'The destructive potency of existing weapons is sufficient to ensure the annihilation of peoples and the total destruction of their cities. There is no need to exaggerate: the facts are clear and it only remains for us to realise their true significance.'[153]

The fervent British champion of the League, Lord Robert Cecil (later Viscount Cecil of Chelwood) was another who believed that the menace of aerial warfare furnished a pressing justification for a League police force — in fact he called for an International *Air* Force under League control to serve as a deterrent and retaliatory striking force.[154] Not only the mildly turncoat Conservative among the Cecils, Lord Robert,[155] believed this, so too did that 'high Tory' of the Cecils, Lord Hugh,[156] as seen in his passionate address to Commons on 21 March 1922.[157] The air scare could — and did — bridge all positions in the British social and political spectrum.

But before turning to political manifestations of the air scare, a few other phases of the story are worth mention. A work of history, for example, should not ignore the fact that historians of the day were no more immune from these fears than were other members of society. The Vere Harmsworth Professor of Naval History at Cambridge, J. Holland Rose, derived like conclusions from the Great War as Liddell Hart. He also believed that that war had proved that defensive means had the effective advantage in land and sea engagements and that such tactics could only reproduce, therefore, the lengthy horrors of 1914-18. He also realised that strategic bombing offered an alternative to this military stalemate, but Rose saw this alternative as being the final absurdity.

> Wearied by these long-drawn-out and ineffective struggles, inventors are turning their attention to the civil populations . . . As armies only slowly wear each other down, aircraft are to pass them by and deal with the large cities — direct action with a vengeance. These developments open up hideous vistas, unless mankind regains its sanity and comes to see that the whole ghastly business is a colossal farce. The climax of scientific warfare, in practise, to be a *reductio ad absurdum.*[158]

Present-day historians do agree that one major reason why England pushed a strategic bombing policy in the interwar years was that it did represent an alternative to war in the trenches.[159] The evidence in this paper does not contravene this interpretation but it does indicate that defensive concerns played a key role in the formulation of the policy also. But certainly the horrors of the trenches would be remembered and a war method that could avoid those horrors would be welcomed by many — if seen by others (e.g. J. Holland Rose) as an absurdity. And in this regard, since this chapter concerns thinking within civilian circles, it is worth remembering that six million British non-professional soldiers served duty on the war fronts of World War I; the memory was a deep and widespread experience.[160]

Naturally, British political opinion was not immune from the catastrophic views of air power. Of special importance because of their influential position within the 'corridors of power' was that the civilian air ministers were notable examples in this regard. The three important officials in that office in the era concerned in this chapter are Winston Churchill from 1918 to 1921, Sir Samuel Hoare from 1922 to 1929 excluding the Labour Government hiatus in 1924, and Lord Thomson who served in the Labour Cabinets of 1924 and 1929-30.

Winston Churchill, in an article written in 1925, showed that he had no doubts of the horrors to come via the air. He stated that future wars would witness air raids against civilian populations with gas and other unspecified chemical bombs predominating. He stated that this would have been the character of the last war if it had extended into 1919. Mankind, he said, had the means *at hand* to 'accomplish its own extinction'.[161] Later, in 1929, Churchill published his analysis of the postwar era (actually the final volume of his history of the Great War) in which he repeated — in some cases verbatim — his fears expressed in the 1925 article. Air war using chemical means could end urban life, he warned, and future war could well mean the end of civilisation itself. The best hope for Western man, he advised, was in League security.[162] Churchill's public speeches could take similar lines; for example, in a campaign speech in 1924 he stated 'I say that if another war is fought, civilisation will perish.'[163]

Sir Samuel Hoare was probably the government official most dedicated to the RAF strategic bombing concepts. His commitment is clearly denoted in his memoirs of his years as air minister. He was captivated by Chief of the Air Staff Trenchard ('the first really great man that I had ever met in my life . . .') and saw his role as Trenchard's public spokesman, both within Cabinet and government circles in general and also to the populace at large.[164] That he was such a spokesman in Parliament will be noted shortly and his speeches on the stump often took the same line, as for example at Lincoln in October 1925 and at Southend in November 1927.[165]

Lord Thomson, the Labourite air minister, held like views on these matters as his Tory brethren, and he too has left relevant written testaments — as stated, aerial concepts bridged all parties, all sections of society. In 1927, Thomson published a book whose title and preface both stressed that he would present the 'facts', not exaggerations. And the 'facts' of aerial warfare were explored in a chapter which opened with that often repeated quote by Marshal Foch and which then followed with Thomson's attempt to make explicit the implications in Foch's statement. He suggested that all defence expenditure might be for naught if Britain was not made secure against aerial attack because defeat could then be suffered before the older services had any time to go into action. He had no doubt that wars between industrialised countries would inevitably begin by opposing air fleets engaging in strategic raids against the enemy vital centres, these raids perhaps taking place before any formal declaration of war was made. Future war therefore would allow no time after its beginning for preparations and mobilisation. Nor would any international agreement or ban be effective in preventing these opening blows — the passions of war would not be restrained by prior rules. Lord Thomson was another of those who held out little hope for air defences *per se*, only offensive blows could provide defensive security. ('Since the best form of the defensive is the offensive, the bombing airplane is a shield as well as a sword.') But even that form of defence would not prevent enemy raids from taking place and thus both winner and loser would end up with 'ruined cities, widespread distress among the masses of the people, hospitals filled with the maimed and mutilated of all ages and both sexes, asylums crowded with unfortunate human beings whom terror had made insane'. Thomson's only real hope to avoid that catalogue of 'facts' was the achievement of the outlawing of war altogether.[166]

There were two former RAF leaders who also became political appointees to the Air Ministry, the successive chiefs over its civil air section, and these experts were also ready and forceful advocates of the prevailing air gospel. The first of these appointees, Sir Frederick Sykes, expressed his ideas in an early work published in 1922. Therein, future war is again described as starting off with 'huge' aerial armadas setting out on strategic bombing raids against the enemy heartland with gas (and other CBW horrors as developed) as part of the payload. Such use of air power would be the decisive factor of the war and no way to ban its use was possible. 'Just as gas was used notwithstanding the Hague Convention in the Great War, so air war, in spite of any and every international agreement to the contrary, will be carried into the enemy's country, his industries will be destroyed, his nerve centres shattered, his food supply disorganised, and the will power of the nation as a whole shaken.'[167] Considering these views, it is worth noting that Sykes had another base for influence besides his position in the air ministry; as of 1920 he was the son-in-law of Andrew Bonar Law and his wife served as Bonar Law's official hostess when he entertained as Prime Minister during his tenure of 1922-3.[168]

Sir Sefton Brancker followed Sykes in the civil air post and he, too, made his beliefs very clear, above all as a frequent speaker at public functions and in frequent press articles and open letters. To enumerate some examples: by March 1919, he was already stressing that 'aerial strength will mean everything in the next great war'; in 1920, he urged that the major defence expenditure should be in air power since it represented the offensive weapon of the future; by 1921, he was yet more outspoken — 'No efficient defence against determined air attack has been evolved, and the only apparent method is to be so strong in the air as to be able to defeat the enemy's aerial attacks and to carry the aerial offensive against his very vitals.' In that piece he also predicted that the older services would become purely defensive in their functions while air power would be the *sole* offensive striking force. Later that year he also predicted that the near future would see the strategic bombing payload capability raised to 1,000 tons per day. In a 1928 speech given less than two years before his death, Sir Sefton took as axiomatic the horrors of the air war to come, but hoped that this might prove to be the needed boon to achieve lasting peace, for who would venture to start such a *Götterdämmerung*? Lastly, Brancker (like Sykes) was a strong advocate of civil air development, in part because it represented war potential. Of course, this position fitted both Brancker's and Sykes' roles as officials over such development.[169]

The opinions of these political appointees within the Air Ministry reflect the general political rhetoric about air power during the 1920s. A good source to explore for this rhetoric is the House of Commons debates over the yearly presentation of air supply estimates and this evidence will be examined here very shortly. However, some personages deserve special attention and these will be looked at before undertaking that other review.

Stanley Baldwin, in 1932, furnished what was to become a cliché in aerial discussion when he said 'the bomber will always get through',[170] but he was deeply concerned about air warfare long before. For just one example, Baldwin in 1927 gave a speech to the Classical Association in which he warned that aerial gas attacks would constitute part of the next war and that these would be beyond the ability of defensive measures to prevent. His conclusion was wholly pessimistic: 'Who in Europe does not know that one more war in the West and the civilisation of the ages will fall with as great a shock as that of

Rome?'[171]

Baldwin's implication that the general view throughout Europe was in accord that another great war would mean the downfall of Western civilisation can be supported. The military historian, Ian Hay, has written that there was general agreement in the 1920s among the European peoples that a future war 'might ultimately lead to the disintegration of civilised life in the world'.[172] As prestigious a militarist as Major General Sir Frederick Maurice was in basic agreement. In 1926, he made a like prophecy of war's future results but added that he did not see another great war as a near possibility.[173] Or, to return to the present political emphasis in this chapter, Lord Grey, the former foreign minister, gives yet another statement of that doomsday prophecy in his memoirs, when he turned to some final reflections on the postwar world. He spoke of the changed conditions in warfare: 'War ... used to imply a contest between armies; it will henceforth, by common consent, mean the destruction by chemical agencies, of the crowded centres of population; it will mean physical, moral, and economic ruin'.[174]

A noteworthy political convert to the air power proponents was A.J. Balfour (Earl of Balfour, 1922), a leader among the Tories ever since his premiership of 1902-5. Balfour, as First Lord of the Admiralty, was a naval champion against the claims of air power advocates during the First World War. At that time, he was a most bitter opponent to the aspirations of the air power spokesmen.[175] However, by 1921, when he served at the request of the Committe of Imperial Defence as an arbitrator over divergent positions about air power being taken by the three armed services, his decision was firmly with the position taken by the RAF. A year later saw Balfour repeating as gospel fact that RAF estimates as to the available bombing tonnage involved in the current air threat and Balfour added his agreement that this could annihilate the London population and that no defence existed except in the retaliatory threat of one's counterattack potential.[176] Of similar interest is Viscount Haldane of Cloan who was also a former champion of an older service who moved in the interwar years to an air power emphasis.[177] Two other converts within the political milieu who were quite influential in forwarding the role and the doctrines of the RAF were Lord Salisbury and Sir Maurice Hankey (later Baron Hankey). Salisbury, the high Tory personified, 'became one of our strongest supporters'. And according to the air minister Sir Samuel Hoare, Hankey, who was an essential lynchpin linking Cabinet thinking with that of the Committee of Imperial Defence 'came down decisively ... on the side of the Air Force'.[178]

Of course, the long-persevering political air enthusiasts continued to propound their views. The one who reached the greatest heights of power among this group was Sir William Joynson-Hicks (buffoon though he was[179]) who was Home Secretary from 1924 to 1929. His interwar rhetoric about air power is typified in a piece published in the *Daily Telegraph* in 1922 where he described the manner in which the next major war would open . . . and close: 'The Navy and Army will almost certainly be impotent spectators of an air battle or series of battles which will determine the issue by the destruction of the enemy capital before even the older Services can get into operation.'[180]

The interwar evidence is conclusive; the great, the near great, and lesser lights alike, within British political circles, were in agreement over the potentiality and tendencies of aerial warfare. The single best source for this evidence is *Hansard*. Each year the House of Commons had its seizure of air power bombat when the air service budget requirements were presented. In the following review of these debates, repetition of theme after theme will be inevitable, indeed the fact of this repetition is very much to the point of this chapter. A mental climate is not made up of a few striking statements from notables of some vanguard position but of an ever mounting reiteration of specific concepts and their spreading pervasiveness.

The debate in the year immediately following the war did not, naturally, evidence a great amount of concern about a coming war and its aerial threat. The Great War was over; nobody was prone to think in terms of another such challenge in the immediate future. This was the year of demobilisation disorders and soldiers demonstrating with signs reading, 'We want civvie suits'.[181] The major questions emphasised during this debate,[182] then, were on peacetime issues; the fact of the dual political assignments of the army and air posts being under one minister (Churchill) and on the justification for a continuing separate status for an air service. Both of these matters will be analysed in the next chapter. However, there was an aspect to the separate air force question which is of present moment – the question of the RAF's share of the overall defensive budget and the parallel question of the air service's effectiveness compared to the other services. In these regards various speakers made comments which indicate the mental climate which this chapter has delineated.

Major-General John E.B. Seely remarked upon and agreed with the opinion of Lord Fisher, the former naval leader, that air power had already made some types of warships obsolete. Mr G. Lambert was of like mind and even more specific when he stated that the battleship's day was over — and he too had a former naval great to cite as a bolstering authority, Admiral Sir Percy Scott. Another blow at the navy was struck by Lieut. -Col. J.T.C. Moore-Brabazon in his contention that the British navy could no longer furnish England's security because the future danger lay with strategic aerial attacks. (Moore-Brabazon had served with the RAF and would soon serve some years in the Air Ministry and would later be a wartime Minister of Aircraft Production under Churchill.) His argument included the contention that industrial capacity was now the *major* military target and that air power would thus be directed at that target from the outbreak of war on. He also placed himself in the paradoxical position of stating that only the air service (again, not the navy) could provide defence against that menace while also maintaining that 'there is no defence against aircraft to-day... The only answer is to have a bigger Air Force so as to have the potential power of hitting back'.[183]

Major R. Glyn, like Moore-Brabazon, saw future wars opening with strategic air strikes and Glyn included national morale as one of the targets for these strikes. He referred to the past war to support that argument, mentioning the effect of the Gothas on England's morale, and warned that the future would entail not just thirty-five bombers, but, possibly, 3,000 and that gas bombs would be part of their cargo. (General Seely also referred to the war in his eulogy to air power when he claimed that 'the Independent Air Force had a great effect on . . . its final issue'.) Both Glyn and the future Labour Cabinet minister W. Wedgwood Benn, argued for more support of civil air development because additional military bombing strength was inherent in such development. For the same reason, Glyn questioned the wisdom of allowing the Germans to develop a civilian aircraft industry which would only result in their becoming a military threat once again.[184]

The following year (1920) still saw Winston Churchill holding both the War Office and Air posts and this still received much commentary in the air estimates debates.[185] A new topic of interest and controversy was the use of aerial bombardment against native uprisings within the Empire or mandated territories. Such tactics had been used in their first important instance in January 1920 in Somaliland[186] although two incidents in May 1919 may be described as precedents, when airplanes were flown over Kabul, Afghanistan as a threat to force the Amir into negotiations (successfully)[187] and when airplanes were used in small operations in Mesopotamia.[188] This is another subject which will be looked at in the following chapter.

Otherwise, the trend of repetition was being established as various spokesmen in the previous year's debates repeated their positions. Major R. Glyn again worried about German civilian planes becoming attacking bombers and Wedgwood Benn repeated his desire for greater support to British civil air development (in part because of its military utility), Major-General Seely and Noel Pemberton Billing concurring. This debate saw one contrary view on this position; Lord Hugh Cecil not being in favour of using government funds to support the civilian air industry and also questioning the ability to use civilian planes efficiently as war machines. A new voice, Mr A.B. Raper, joined Glyn, however, in fearing German civil air progress on the same basis, that it represented a military threat.[189]

Attacks on the older services in reference to budget shares for the RAF was another repeated theme. Wedgwood Benn spoke of the 'obsolescent' older services, Lieut.-Col. A.H. Burgoyne foresaw the eventual abolishment of the army and navy (Lord Fisher again being cited for supporting evidence) and both Burgoyne and Pemberton Billing used the parallel argument that it would be air, not sea power, which would provide England's future security. However, this basic theme of readjusted defence priorities received its most impressive support in the traditional speech of ministerial reply by the Air Minister, Winston Churchill, who said that yes, the RAF 'is going to grow, expand, and develop at the expense of — and to the advantage of the public — of [sic] both the other great services'.[190]

Another theme reaffirmed in this year was that the air service had greatly contributed to the victory of the late war, Pemberton Billing making that assertion this time. Speaking for the government, Major G.C. Tryon tied many repeated ideas together when he said 'if war comes to this country, no longer an island, it will have to meet attacks on its towns, its factories, its communications, and its shipping, which no armies can stop and no battleship avert'. Lastly, with respect to this debate, Lieut.-Com. J.M. Kenworthy added a new line of thought which would be repeated often in later debates; that the League should possess a strong international air force and that this would allow a parallel reduction in the British air force. Major H. Barnes joined Kenworthy in that viewpoint.[191]

The 1921 debates[192] again involved questions which will be of concern in the following chapter — a growing feeling of jeopardy for the independent status of the RAF and the double ministerial control still held by Winston Churchill. (At this time he was both Colonial Secretary and Air Minister.) Otherwise, those who argued for more government support of civil air development were more numerous and emphatic this year and as usual spoke in terms of military as well as civil benefits from such support. New and noteworthy speakers among that group were Sir William Joynson-Hicks and Sir Oswald Mosley. The latter was that famous political chameleon who had started with the Tories, then moved to the Labour Party and then ended up founding the British Union of Fascists. He had been a fighter pilot during the war and found that the maintenance of the image of an aerial expert was a useful one during the interwar years.[193] In two speeches during this debate, Mosley obviously attempted to project this image of authority in aerial matters. Besides the stress on more civil air expenditure, he spoke about air power eventually sweeping warships off the seas and went further to prophesy that the future of warfare would be in the air.[194]

A new doubting Thomas about the convertibility of civilian aircraft to military purposes was Mr A.B. Raper[195] but otherwise he declared himself an 'air enthusiast' who foresaw the RAF providing the first line of security in England's next war. Winston Churchill also emphasised the role of the RAF in future air defence and used this argument against those who spoke against the independence of the air service. He also restated his contention that the air force would keep taking on more functions formerly performed by the older services.[196]

The 1922 air estimates debates[197] evidenced far more concern over the potentials of aerial warfare than had the previous debates — even though the earlier ones were impressive enough in this regard. Two occurrences which happened before the debates seem to have had the most influence in heightening this concern; the appearance of the first of the newspaper articles by P.R.C. Groves which had described future air war in such awesome terms and had keyed on the weakness of the English forces compared to the French air service,[198] and the American tests conducted by General 'Billy' Mitchell the previous summer wherein bombing planes sunk various warships including the battleship *Ostfriesland.*[199]

The debates were led off by Capt. Frederick E. Guest who had replaced Churchill for a short-lived tenure as Air Minister, and he set the tone for the debates by making the most fearsome speech on air power up to then given by an official Government spokesman. Guest's speech shows the influence of those events mentioned above; he spoke about Grove's article and he was very outspoken with respect to the air versus sea power question: 'In 10 years' time I believe that a combat between the forces of the air and the forces of the sea will have become a grotesque and pathetically one-sided affair.' He was just as positive with respect to the course of future war, it would 'certainly' begin by opposing aerial strikes. Guest was only somewhat unusual with respect to the defensive capability against such air strikes in that he saw some usefulness and hope in a fighter defence force but even here he agreed that air attacks upon enemy territory would be the 'first line of defence'.[200]

One new and prominent participant in this debate who also held out hope for fighter defences was Colonel Josiah C. Wedgwood, a Cabinet member in the 1924 Labour Government. But other personages – Seely, Joynson-Hicks and Churchill – took the more common stand that offence was by far the most reliable form of aerial defence.[201]

The speaker who followed Guest was Major-General Seely, who had participated in all the preceding years' air estimates debates, but this year he felt cause to be far more outspoken than ever before. He, like Guest, commented favourably on the accuracy of the statements in Groves' article. Then he turned to his version of the next war. Hundreds of bombers would come with bombs ten times heavier than those used in the last war and any *effective* defence plans against this threat would be 'impossible'. Retaliatory strikes to 'destroy' the enemy towns would be the only useful activity and Seely had no doubt about the ability to *efface* urban areas. But England was woefully lacking in the necessary air power for that task of retaliation; even the Germans now led them in aerial expertise. (Here he repeated his usual call for more support to civil air development.) He concluded with a passionate warning: either humanity finds a way to achieve permanent peace or the 'peoples of the world' could easily 'destroy each other altogether'.[202]

The tendency to downgrade the older services compared to the new RAF was still evident and in fact strengthened by the Mitchell tests. Guest, J.T.C. Moore-Brabazon (who had spoken in these debates since 1919), Colonel Wedgwood and Rear-Admiral Murray F. Sueter (the naval air service leader) all referred to these tests. On the other hand, parliamentary spokesmen for the navy tried to minimise the tests. Captain Viscount Curzon stated that the tests were not made under realistic conditions and Commander C.W. Bellairs claimed that the battleships currently under construction were superior to the *Ostfriesland* and could not be sunk by aerial bombardment. The traditional argument that air war methods would progressively replace the activities of the older services was again given voice by such 'regulars' in these debates as Joynson-Hicks, Wedgwood Benn and Churchill with a new debater, the RAF veteran C. L'Estrange Malone, adding his support as he claimed the older services would play very secondary roles in future warfare.[203]

Like Seely, Joynson-Hicks and Lord Hugh Cecil also gave more impassioned speeches than in their previous performances in the air debates. 'Jix' contended that air war against civilians was absolutely inevitable, popular morale was a target that would not be overlooked. 'Vast fleets of aeroplanes' carrying awesome weights of high explosives, gas and other horrors (e.g. cholera bacteria) would exchange blows in an unannounced opening to the next war, 'a war of the most cruel and destructive character, in which the women and children will be bound to suffer just as much as the men'. Lord Hugh Cecil was just as pessimistic. He too had no doubt about aerial inundations of high explosives and gas, a threat from which there was no defence and the concept that one had the ability to retaliate in kind was no consolation at all. War was really now a question of who struck first, and since the British were not a people who would do so, the future was terribly grim. War under these conditions could only mean a horrible regression for man ('we should be back to a stage of barbarian development') so enduring peace was man's only hope. For Lord Hugh, the alternatives were peace through the League of Nations or 'absolute ruin and destruction'.[204]

Clearly, by 1922, the hysteria over air war which was one characteristic of Britain's interwar years was very evident in the august debating chamber of the House of Commons. The clichés were still being repeated – Churchill for example reiterated that 'we are no longer an island' and warned that England was more vulnerable to air attack than the continental states – but a new passion was also evident. There were some doubters still, but they were equivocal in their opposition. Field-Marshal Sir Henry Wilson spoke against air's infringement into army prerogatives but he also affirmed a need to further civil air progress for future war needs. Major J.W. Hills believed that the air threat was being exaggerated – he did not accept air's ability to obliterate capital cities for example – but he agreed that, yes, the menace was 'formidable' even though he thought that effective defence measures would come to hand.[205] But all in all, the debates indicate that England was well into its interwar military fixation.

The following year's debates [206] were relatively calm in comparison. The disparity between the French and English air strengths was the key theme; some arguing that a readjustment of that imbalance was necessary while others took a pacifist-tinged stance against starting up another arms race on this issue. Moore-Brabazon was one who saw a readjustment as necessary to allow Britain to take independent and equally strong (compared to France) positions in foreign diplomacy.

The Air Minister, Sir Samuel Hoare, used a less clearcut argument. He contended that the French air emphasis put into question the weaker British policy; he denied that France should be envisioned as a possible enemy but that, nevertheless, national prudence demanded a more equitable military status.[207]

The issue which best illustrated that the preceding year's fears were still very much present was over the efficacy of air defences. Hoare admitted that this was the greatest problem which faced the air ministry. The former Chief of the Air Staff, Sir Frederick Sykes, presented his authoritative view that 'attack is not only the best but almost the only form of defence'. Thus his conclusion that strategic bombing functions must be the major purpose of the RAF and not any type of tactical support effort. Lord Hugh Cecil also repeated his doubts about air defence, as did Moore-Brabazon.[208]

Also of note in this debate is Wedgwood Benn's appeal for an international air convention similar to the Washington naval agreement as this was to become another often repeated theme in these debates. Lastly, the disclosure by Hoare of the fact that there were five squadrons in England devoted to home defence, only one of which was a fighter squadron (the rest were all bombers) is of much interest as it indicates the general pessimism then prevailing about fighter defences and that this pessimism was reflected in RAF planning and actual operations.[209]

The 1924 air estimates debate[210] is of special interest because the Labour Party was in office and thus its spokesmen for the first time represented the Government position. Significantly, that position was basically the same in the presentation speech. The Labour Party still accepted the need for an increased air force, especially because of that continuing imbalance of strength compared to France. This debate showed that all parties were in general agreement as to the nature of the next war and that it represented a Pandora's Box of airborne horrors that England would have to face. Few speakers took a really divergent stance from that general consensus. The most contrary view was expressed by Lieut.-Col. H.M. Meyler, in his maiden address to the House. He argued that the Londoner's response to bombing during the last war hardly supported the present day air power claims; that people were just made more firm in their anger by that war method and, therefore, no power would use that strategy in the next war. Others, however, were quick to argue that air technology had advanced greatly since the war and much greater efforts in that war strategy could be expected (Oswald Mosley and General Seely the most notable speakers in that context).[211]

Perhaps the passionate statement of F.D. Acland may be considered as an oppositional stand in that he strongly objected to the fact that all parties were ensnared by the same presumption that the future strategic bombing of cities was inevitable and he pleaded for some new approach that would bring the world out of that impasse Nevertheless, he obviously accepted the basic concepts of the day as he admitted that the end of Western civilisation would be the result if such air war came — a result brought just as easily by converted passenger and cargo planes as by warplanes *per se*.[212]

An important new participant in these debates was Sir John Simon, notable because he would later be an influential Cabinet Minister during the appeasement era. Simon had been on Trenchard's staff during the war and was one of Trenchard's fervent supporters[213] so it is no surprise to find him among the air power proponents. He had no doubt that the next war would see high explosive and gas bombs raining down upon the contestants' cities; that to expect just military targets would be bombed was 'the height of nonsense', and that attempts to ban indiscriminate bombing would not be successful: 'the truth is that the reprisals of one war become the normal practise of the next.' Nevertheless (this is an excellent example of the sense of futility which permeated British interwar thinking on aerial matters), Simon suggested that Britain should try to attain just such a ban. 'The only hope in this matter, really, is that there should be an attempt made, however difficult it may be, to get some form of international limitation.' Otherwise, Simon repeated an idea expressed earlier by Mosley that the German war concept of 'frightfulness'[214] had finally arrived at the full possibility of realisation through air power. He also agreed with the general belief that civil planes were convertible into bombers. Only with respect to air defence was he more optimistic than the prevailing opinion in that he believed that daylight defence was definitely feasible and that anti-aircraft fire would be able to keep the bombers at high altitudes. But these ideas hardly led to very consoling conclusions. Given these defensive abilities, the bombers would be most apt to raid at night from great heights — which meant just an indiscriminate dropping of bombs upon the cities as any attempt at accurate bombing under these conditions would be unavailing.[215]

Simon's future fellow 'Appeaser', Sir Samuel Hoare, was also pessimistic about air war restrictions and suggested that the only possible way to achieve such limits was to attain an international ban on *all* military air power. Thus this debate included the peculiar situation of the Conservative 'shadow' Air Minister requesting the Labour Party to push for a total ban on aerial warfare.[216]

This debate included many perennials repeating their customary views, such as Mosley, Seely, Kenworthy and Moore-Brabazon. Even among these, the subject of air defences created some disagreements. Mosley saw no hope in air defence while Kenworthy believed effective daylight defence was possible but agreeing that the law of *lex talionis* was still the only defence against night raiders. Moore-Brabazon and Seely both had some favourable comments to make for fighter airplanes but Seely really believed that fighters were just a better alternative to *no* defence at all.[217]

New speakers iterated stereotyped arguments: Sir Geoffrey Butler, soon to be the parliamentary private secretary to the Air Minister, was pessimistic about fighter defences; Mr W. Leach, then Under-Secretary for the Air Ministry, affirmed that civilian planes could easily become night raiders (though possibly not day raiders due to a limitation in their ceiling capacity); Mr G.D. Hardie pressed for a ban on strategic bombing and its attendant horrors; Lieut.-Col. C.K. Howard-Bury joined Moore-Brabazon in emphasising the need for research in the development of new defensive techniques; Lieut.-Col. Sir Philip Richardson believed that the deterrent striking threat was the only real defence and that, therefore, the single major question was whether the country had sufficient air power to feel confidence in its deterrent capability.[218]

There was one other major interwar political leader who made, in 1924, a forthright statement on air power — Anthony Eden in a maiden speech in a debate held before the air estimates debate, on 19 February 1924. In this speech, Eden showed that he, too, was another 'believer' in the air power assumptions of the day. He viewed aerial bombing as 'the greatest peril of modern war'; he stated that defence mostly depended upon one's retaliatory ability — while also stating that London 'is especially vulnerable to attack from the air' (why was not the basic conflict between these two ideas more readily recognised by these thinkers? Is logic avoided when one catches at straws?); he pressed that England should start right then preparing for the next war, that there would be no time to build up the nation's strength when the need arrived.[219]

The 1925 air estimates debate[220] took place after the Conservative Party had regained the government and this fact seemed to release a more fervid attack upon aerial armaments by the pacifist left than had ever occurred before. It has been generally recognised that the British interwar pacifist movement was very strong and influential. In great part based upon the experience of the Great War, the movement was also strengthened by the vision of the next war producing an airborne holocaust for one and all.[221] Certainly, it was these future horrors from the air which were stressed by pacifists during this debate.

Mr E. Thurtle was the first among these speakers in 1925 and his address set a notable precedent in impassioned oratory. He forecast figures similar to those of the air ministry[222] for the future bomb tonnage that threatened England – in this case, 170 tons per day at first and 75 tons for the daily and extended rate thereafter – and he claimed that *no* defence against that tonnage was possible. The result would be a 'smoking and reeking' London with its populace lying 'mangled' on the streets. These foresights, he claimed, 'are not vain imaginings, but are really the things which we can seriously look forward to in the next air war . . .' He went so far as to jar the members with a description of a direct hit on the House of Commons itself. His conclusion was that England must lead the way out of that suicidal course by abolishing its air force. These terrors described by Thurtle received confirmation from various following speakers, firstly from J.H. Hudson, who cited a number of authorities in support, including Foch (as usual) and Thomas A. Edison who had predicted that the whole of London's population could be gassed in the space of three hours. The prominent Labourite, J.R. Clynes (he was Lord Privy Seal in 1924 and was to be Home Secretary from 1929-31), also concurred in the statements of Mr Thurtle and he, as did Hudson, also agreed that disarmament was the only sane alternative. That pacifist personified, Mr George Lansbury,[223] followed and with his impressive forensic ability he argued for the cause of unilateral disarmament. The Communist Party member, Mr S. Saklatvala, agreed with that stand for unilateral disarmament as he reminded the House that unarmed states did not suffer in the previous war. All together, this debate contained an impressive number of fulminations by the 'disarmers' within the House.[224]

This debate also witnessed some new men of prominence taking part. Philip Snowden, for example, was then the shadow air spokesman for the Labour Party (he was Chancellor of the Exchequer in 1924 and would be so again in the Labour Cabinet of 1929) and he now disassociated the Labour Party from the RAF expansion programme, stating that England was thereby getting itself involved in another armaments race. He pushed, rather, for universal disarmament, reasoning that since no limitation on the future conduct of air war was possible, then disarmament was the *only* rational approach. The Liberal Party also had a new spokesman of future importance, Sir Archibald Sinclair (he would be Air Minister in the Churchill coalition from 1940-45), and he too showed the same pattern of thought that typified these debates. Air power, he said, was by far the major threat of the day and thus the old services could no longer offer security. Cities would suffer a rain of death from high explosive, incendiary and gas aerial bombs.

> Therefore, I think there is no hon. Member in any quarter of the House who will dispute this statement, that it is quite possible for our industries to be entirely deranged, our great cities to be shattered by incendiary and high explosive bombs, our people demoralised by incessant bombardments and the terrors of gas, and reduced to the verge of starvation by the destruction of communications, our Government paralysed, and all this while the Navy still holds command of the seas, and yet incapable of firing a shot in our defence. [Remember, these are the thoughts of the future wartime political air chief.] [225]

Sinclair concluded that Britain must maintain a sufficient counterstrike force while pressing for international disarmament. Indeed, disarmament was a major continuing theme throughout the 1925 debate. Besides the speakers already noted, Captain G.M. Garro-Jones, Commander Kenworthy and Mr Morgan Jones all spoke in that cause. The horrors of the air war to come was also, of course, a continuing theme, and besides the comments in that regard already noted, the most extreme statement came from Captain A.S.C. Reid, a former wartime pilot, who predicted that 'a colossal hostile fleet of bombing aeroplanes' could end the next war within the opening hour by a raid on London.[226]

The Government spokesman, Sir Samuel Hoare, was hardly more optimistic. He too agreed that cities were now dreadfully vulnerable, that air defence was not worth much, that international restrictions were very difficult to achieve (especially because of the civil air complication), and that, yes, something must be done or 'air warfare in the future may well mean the destruction of civilisation as we know it today'. Hoare recognised – as later in the debate did Sinclair and Commander C.W. Bellairs – the need for more research, especially in developing new methods of air defence, but Hoare explained that finances were extremely short for such additional expenditure. Lastly, and of interest with respect to future developments, Hoare agreed with a Mr T. Henderson that a shelter programme was a sensible idea and that the Government was, in fact, looking into that possibility.[227]

With one specific exception, Hoare was just as much in agreement with the comments made in the 1926 debate.[228] He again agreed that air defence was of outstanding importance and that it was an extremely difficult problem, that the horrors of air war were beyond imagination, and that Britain's only hope which could then be foreseen was that its counterstrike strength might keep these horrors from happening.[229]

The issue over which Hoare was assailed by both rearmament and disarmament advocates was the relevance to the air programme of the recently concluded Locarno Pact. (The Locarno arrangements had been announced on 16 October 1925.[230]) Locarno, as is well known, had brought a new sense within Britain of an easement of tension in European affairs.[231] Thus, Hoare announced that the Locarno accomplishment allowed the Government to retard its air expansion programme. But others were not at all satisfied by that announcement. The 'disarmers' were unhappy that Locarno had not brought an actual reduction in Britain's air strength rather than a delayed expansion. As Clement R. Attlee, the future Prime Minister, expressed it, 'there is no echo of the Locarno spirit in these Estimates'. Capt. Frederick E. Guest was another who would use Locarno as the wedge to seek more aerial disarmament. Mr Rennie Smith repeated that admonition. Mr E. Thurtle also rued that Locarno had not resulted in a reduction of Britain's air service. Other spokesmen for disarmament were Josiah Wedgwood and J.H. Hudson. On the other hand — but starting from the same premise of the threat involved in air warfare — Admiral Sueter, Commander Kenworthy, Capt. G.M. Garro-Jones and Mr F.G. Renny were unhappy that the expansion programme was not being hurried to completion.[232]

Clement Attlee was the most important new participant in this year's debate and he proved to be of the general mood. Air power meant horrors for which neither defence nor international limitation was possible. He also stated that civil aircraft were potentially an equal menace as war planes and he advocated placing all aircraft, both civil and military, under League control.[233]

Defensive abilities received a bit more hopeful expectation this year than usual. Capt. W. Brass was one, for example, who believed that fighter planes would seriously beset the incoming raiders. However, he asked the disturbing question whether England's fighter squadrons were prepared; what would the RAF be doing, for example, if enemy bombers were on their way right then. (Some House members must have squirmed most uncomfortably at that idea!) In another speech, Capt. A.S.C. Reid repeated the awesome image he presented the previous year of an air war achieving victory on the first day but nevertheless, he gave a somewhat favourable report on fighter defence, stating that although the fighters would not have time to get airborne before enemy bombers arrived over London, they would still be effective against the enemy on his return flight home. He then added the hope that such potential losses might possibly deter the enemy from attempting such raids in the first place — surely an illogical expectation given his viewpoint that strategic bombing could produce instant victory. Lastly, with respect to fighter defence, Mr F.G. Renny was more hopeful yet in that he spoke in terms of the defence's ability to actually 'repel' the raiders. But he qualified this by stressing the absolute necessity of repelling 'the first attack'. Nobody was ever *very* hopeful in the interwar air debates.[234]

Captain Reid added one other grave thought to this debate. The threat presented by civil aviation had often been mentioned but Reid made this threat very concrete by reminding the House that Germany would soon have more civilian aircraft (all potential bombers) than England had in military aircraft. Thus France was no longer the only potential immediate air foe within the context of these debates; a new and, certainly to many, more logical enemy had been introduced. Then the former Air Minister, Capt. Frederick Guest, added to this fear by warning his fellow members that planes were relatively inexpensive and speedily assembled. England, he said, would not have a prewar warning next time; it could, rather, find itself awaking to a morning of urban ruins. All in all, terribly grim forebodings were again the story in the 1926 debate.[235]

The 1927 debate[236] evidenced a heightened emphasis in the longstanding theme of disarmament due to the forthcoming spring meeting in Geneva of the Preparatory Commission for the Reduction and Limitation of Armaments.[237] Speaker after speaker pleaded with the Government to take the lead in this Geneva meeting with a strong programme for disarmament. The most emphatic in this regard were H.B. Lees-Smith, Miss A.S. Lawrence (an under-secretary in the Labour Government of 1929), Mr Rennie Smith, Hugh Dalton and Alfred Duff Cooper. Other 'disarmer's' in the debate included Commander Kenworthy, Capt. G.M. Garro-Jones, Arthur A. Ponsonby (a leading Labourite by the late 1920s) and A.L. Shepherd, the latter two pressing for British unilateral disarmament with Shepherd wanting to ban the RAF entirely.[238]

A.S. (Susan) Lawrence added a feminine and extremely pacifist voice to the proceedings. As so many others had done through the years, she used General P.R.C. Groves as her major authority for her cataclysmic view of the coming air war from which no effective defence was possible. The major targets would definitely be the urban centres with the purpose being 'to produce such terror, disorganisation and disorder among the civil community and the country generally as will make it impossible for a government to carry on a war . . . ' Millions would die, especially from gas attack. Just two to three hours would be sufficient to 'break up' such centres as Paris or London. To avoid all this, Miss Lawrence would have Britain pursue a policy of banning all aircraft in the coming Preparatory Conference. Let man not fly at all. (Lees-Smith and Rennie Smith were only slightly more tolerant of civil flying and proposed that all aircraft be put under League control.)[239]

The long-term Under-Secretary of State for Air (he was to hold this office for twelve years), Sir Philip Sassoon,[240] answered for the Government with respect to their disarmament policy at the coming Geneva meeting. That answer was on balance less than adequate in logic. Sassoon asserted that the Government shared the same concerns about future aerial warfare as did Miss Lawrence and her fellow pleaders. The Government leaders, he affirmed, 'are seriously anxious to see a limitation of air arms'. However, the problems in that pursuit were extremely difficult. The question of civil aviation had to be faced and for Great Britain, civil aviation represented a great boon as well as a great threat; an imperial power should not attempt to impede faster communications. And with respect to military aircraft (here is where Sassoon's logic seems less than satisfactory), since Britain was not the largest air power, it could not 'take the initiative on the question of air disarmament', though it would support other nations who did take that initiative. The non-sympathetic members might well have wondered what difference it made which state advanced the first proposal at Geneva.[241]

Two important new participants in this year's debate were Hugh Dalton and Alfred Duff Cooper, future Labour and Tory Cabinet members respectively. Typical of how the air phobia united otherwise diverse viewpoints, their speeches were of a piece. They both specifically prophesied that future war would be an air-wrought disaster that could bring about the virtual end of Western civilisation. Dalton did dwell more on the coming horrors, among which he included chemical and biological air war means. He noted that some were proposing gas masks for babies and the young as one solution and that this only indicated just how horrid present day conditions were. (Earlier, Commander Kenworthy had recommended the issuance of gas masks to everyone south of the river Trent.) But basically, little difference can be found between the speeches of Dalton and Duff Cooper – and this holds true for the entire debate; the differences in viewpoints were generally very few.[242]

In 1928, the speakers and scripts represent little change from the previous year.[243] The desires for greater future effort at Geneva in the cause of disarmament were now bolstered by the charge of insufficient effort there during the previous year. There was an increased number of speakers who argued the need to expand the air force but their position still echoed that taken by the disarmers – total disaster was the alternative to the programme they proposed. The list of speakers on each side of that two-faced coin are as follows: those arguing for more aerial disarmament or else *Götterdämmerung* included A. Barnes, W. Wellock, Rennie Smith, H.B. Lees-Smith and J. Beckett. The reverse stand was taken by the novelist-politician John Buchan, Admiral Sueter and Dr A. Vernon Davies. Only that doughty defender of the Admiralty's interests, Commander Carlyon Bellairs, argued against the feasibility of the concept of future strategic air warfare, claiming that the wastage of aircraft on such strategic missions would be too great for any belligerent to risk losing.[244]

Otherwise, perhaps the 1928 debate need only be noted for two expressions of an assumption that the strategic air warfare concept was one of general acceptance. Mr A. Barnes in a major address for more disarmament action at Geneva – he moved an amendment in that regard – had this to say regarding that consensus of thought: 'No matter in which direction we look, whether we take the views of statesmen or of experts, of military men or of naval men, or the view of the man in the street, the immense disastrous consequences of aerial warfare are known and it is impossible to exaggerate.' And shortly after, Mr W. Wellock quoted a statement by the British air representative at Geneva, Group Captain H.F.M. Foster, which attested to that consensus among aerial experts: 'I do not think any airman nowadays in high position would guarantee that under favourable weather conditions to the enemy, immunity could be insured against a great city being flooded with gas, set alight with incendiary shells, and bombed with high explosives.'[245]

In the 1929 air estimates debate,[246] there were again attacks on the Government air policy from those who would cut back and from those who would expand the RAF, but the temper of the House had definitely shifted in the direction of the 'expanders'. Sir Philip Sassoon, speaking for the Government, noted that shift as he commented that it appeared that the Air Ministry had not asked for enough money on this

Non-Military British Concepts of Aerial Warfare 153

occasion. The estimates amount, in fact, had been questioned in many speeches. As usual, no member was pushing for an increased defence budget overall, but many desired a readjustment in the defence budget so that the air service received a larger share. As part of their argument, they stressed the concept of 'substitution'; i.e. using the air service in more places and in more ways where the older (and more expensive) services then held sway. Among this group were Wedgwood Benn, Moore-Brabazon, F.E. Guest, Cunningham Reid, Sueter, Kenworthy and the Communist C.J. L'Estrange Malone — all by then familiar participants in the air debates.[247]

Of course, the request for a readjustment in the defence budget did not logically exclude the desire for general disarmament and some of those mentioned did take both stances, above all L'Estrange Malone who also supported an amendment which took the Government to task for not making greater effort to achieve aerial disarmament at Geneva. Mr A. Bellamy, in a maiden speech, moved that amendment and his talk cited authority after authority (including P.R.C. Groves in two separate references) to bolster his Damoclean view of the horrors to come. Others who focused on these horrors to justify disarmament pleas were James H. Hudson and Hugh Dalton. (Dalton took pains to scare his fellow members with the vision of the destruction of the buildings of Parliament at the outset of the next war so that England would even be without a forum to discuss peace terms.)[248]

The terrifying vulnerability of London in general again received emphasis. Sassoon (for the Government) was as ready as, say, Cunningham Reid to admit that its geographical position made it the major city most vulnerable to air attack. Dalton used this assumption to make the sensible point that it made England's retaliatory air strength quite irrelevant. In that context, he mockingly quoted a speech by Lord Halsbury to the League of Nations Union (in December 1928): 'It seems to me', Halsbury had stated, 'the only possible solution is to have enough force in this country so that, if Germany were to attack London and obliterate it, we should have enough force to make reprisals on Germany'. Cold comfort indeed! Here, clearly brought out by Dalton, was the awesome dilemma of the English, given their prevailing axioms about air power. As for the ability to 'obliterate' cities, L'Estrange Malone declared that 'every town in this country is liable to annihilation within five minutes of the declaration of war . . .' and both he (Malone) and Bellamy envisioned gas attack as the main instrument of that annihilation, Malone mentioning diphenylchlordarsine while Bellamy spoke of lewisite gas. In this debate as in the previous year's, only the Admiralty spokesman, Commander Bellairs, remained obstinate about 'all the rubbish that is talked nowadays of wiping out London...'[249]

The 1930 air debate[250] occurred when the Labour Party was again in power and, just as in 1924, their air programme was essentially a continuation of the one presented the year before by the Conservative Party. This basically reflects the quandary presented by the prevailing air war concepts. Given the almost all-pervasive fear of the nature of future warfare, the belief that Britain was the most vulnerable to that type of war, the assumption that no defence was really effective except possibly the deterrent threat of a bombing force, and the added conviction that the achievement of a guaranteed restriction on air war would be extremely difficult, then there were few alternatives open to the government in power, of whatever political philosophy. The Government spokesman, Frederick Montague, Under-Secretary of State for Air, attempted, of course, to put the best gloss possible on his proposals. He emphasised that the Government was pursuing a moderate policy and that England was below France, Italy and the United States in air strength. He also stressed that the RAF was a relatively inexpensive means to police various imperial outposts. In sum, Montague delivered a typical 'Conservative' speech and this fact was not ignored by the 'shadow' spokesmen, Sir Samuel Hoare and Sir Philip Sassoon. Both complimented Montague on his speech and policy while expressing disappointment that the home defence squadrons were not even further strengthened (the Labour estimates had added one regular and four auxiliary squadrons to these home defences). On the other hand, some Labour backbenchers (still free from the responsibility of power) were appalled that their party leaders were supporting a policy that entailed the purposeful killing of civilians.[251]

In addition to these enlightening exchanges during the 1930 debate, a lengthy contribution by Capt. Harold Balfour is of consequence in that he was Under-Secretary of State for Air in 1938. In that Balfour was an RAF veteran, it is hardly surprising that he, too, was an outspoken air power advocate. ('The air is the power which can command in the future.') He urged that the air service be given a much larger share of the defence budget and that a drastic substitution policy should be undertaken, stating, for example, that air power had already, in fact, ended the era of the capital ship.[252]

The last debate to be surveyed here, the one for 1931,[253] was even more than usual a restatement of the previous year's speeches. The Labour Party was still in power and again presented a policy of continuity in its air service proposals. The Tory shadow ministers repeated their approval about that continuity although (still and again) wanting more increases. And once again, some backbenchers voiced their dismay that England was not voluntarily indulging in unilateral aerial disarmament.[254]

Sir Samuel Hoare's speech is also noteworthy in that he had by then become one of those who pressed the Government to push harder in the Geneva talks for international cutbacks in air power. The Labour ministers must surely have found that ironic! Hoare's argument here was that aircraft had become much more efficient over the last few years — he furnished specifics — and that, therefore, the air threat was becoming ever more fearful.[255]

This chapter has attempted to survey the non-military thinking in Great Britain concerning air warfare through the year 1931. This thinking largely coincided with (and surely had much influence on) the military doctrines being developed at the same time, a development that will be described in the next chapter. This chapter is, then, a practical exercise in the practice advised by John Lukacs; that 'in the twentieth century ... the historian must face the difficult problem of delineating the mental climate of the majority'. It also furnishes evidence to justify Lukacs' belief that public and official opinion on a given matter often have a symbiotic relationship; that 'the history of public opinion is often inseparable from the history of opinion-making'.[256] And, just as much to the point, *vice versa*.

An excellent indication that the interaction of beliefs about air power had already achieved a profound impact by 1931 is presented by the book *What Would be the Character of a New War?* published that year on behalf of the Inter-Parliamentary Union.[257] The Union, originally formed in 1859 and by 1931 consisting of members from about forty parliaments, i.e., from most of the world's parliamentary bodies, had decided to promote a written symposium by an international body of experts on such questions as security, armaments and the nature of warfare in the immediate future, and to publish the results before the disarmament conference met in Geneva in 1932. An indication of the seriousness of the enterprise is that simultaneous editions were published in English, French and German. Nineteen authorities from nine countries contributed articles, presumably fulfilling the Union's desire to establish their enquiry on 'a solid scientific basis'.[258] Of moment to the purpose of this chapter is that most of these articles contain the same presumptions about air power that have been witnessed throughout this study.

The experts represented either academia or the military services. On balance, the academicians were the more extreme with respect to their prophecies respecting air power. For example, both General von Haeften of Germany and Professor André Mayer, then Vice-President of the Collège de France, wrote articles on 'Protection and Defence Against the New Methods of Warfare'. Professor Mayer foresaw no such measures as being effective and thus concluded that if mankind again decided to go to war, it would mean mass 'suicide' (pp. 225-49). General von Haeften, on the other hand, granted some usefulness to fighter and ground AA activity . . . though not much. For the English, he was even less consolatory with respect to a country's dependence upon a deterrent bombing force as he sensibly argued that it not only had to be the greater of the contending air forces but also that the country's own vulnerability to strategic bombing should be less — or certainly not greater. Such vulnerability depended upon geographic, demographic and economic factors. (On all points, England was extremely vulnerable. Her capital was too close to the Channel, her population density was very great, and her economy was primarily industrial and urban-oriented.) Finally, the general had no hopeful comments to make on civil defence measures so in sum his reflections on 'protection and defence' offered very little of either to his English readers (pp. 208-25).

Even more bleak were the predictions of Professor Joerg Joergersen of Copenhagen University who dealt with the psychological effects of terror bombing (pp. 250-73). He presumed such bombing would start with the outbreak of war in an effort to break the civil 'will to war' and that:

> This phase of the war will produce the greatest panic and be characterised by the utmost barbarity. All moral principles, all education and discipline, will be forgotten . . . The instinct of self-preservation will involuntarily oust all other emotions, and human existence will degenerate into wild chaos . . . A complete moral collapse is to be expected.

This writer also predicted revolution or civil war and a potential mass mental breakdown as other probable consequences. And, in a following article, Professor Liebmann Hersch stated that if Europe engaged in another war, it would be one of inevitable aerial horrors (especially CBW) which would result in the end of Western civilisation and the future control over Europe by another civilisation and/or racial stock (pp. 274-314).

The last selection that shall be noted here from this astonishing book is by Dr Gertrud Johanna Woker, then head of the Institute of Physico-Chemical Biology at the University of Bern, who wrote on 'Chemical and Bacteriological Warfare'. Her article quotes one authority after another who affirm that the next war would involve air raids on civilians consisting of high explosive, incendiary, gas and bacteriological inundations which would be horrible beyond imagination. Millions would die in the opening hours of warfare and a city as large as Paris could be 'annihilated' in the very first hour. She presents page after page of such prophecies with never a glimmer of hope, a possible chance of defence, or a possible means of restriction (pp. 354-91).

What Would Be the Character of a New War? clearly shows that the thinking in England as described in this chapter was part of an even more general climate of opinion about aerial warfare. The particularly vulnerable situation of the English would naturally make them hypersensitive to these concepts which were so generally accepted. Now it is time to turn to an analysis of military developments in thought and deed in these aerial concerns and to see how they correlated with the trends described in this chapter.

6 THE FLEDGLING YEARS OF THE ROYAL AIR FORCE IN DOCTRINE AND DEVELOPMENT

The dominant problem for the Royal Air Force during the early postwar era was whether it would in fact survive as an independent service. This question of survival profoundly influenced the formulation of air doctrine among the RAF leaders and, as will be seen, along lines parallel to popular thought.

Before the armistice, the question of survival did not appear to be a future problem; indeed in 1918 the new service seemed to be thriving and ambitious plans were at hand for its future use. The most ambitious projects were for the Independent Air Force under Trenchard's command.[1] By the spring of 1919 it was programmed to make 'devastating' raids (Lloyd George's term) on the German homeland which would leave 'shattered' (Winston Churchill's term) gas-ridden cities, including Berlin, in their wake.[2] Special long-range bombers were being produced for this campaign, most prominently the 'massive' Handley Page bombers, the H.P. 15 v/1500, which could carry bomb loads up to 7,500 pounds, possessed a range up to 1,100 miles and carried a seven man crew. Another plane designed for this purpose was 'the gargantuan Tabor F1765', larger even than the Handley Pages and the final British designed triplane. (It crashed on the takeoff of its maiden flight, both pilots losing their lives, and no further attempts on any triplane were thereafter made.)[3] Trenchard had already prepared, prior to the armistice, accommodations for sixty squadrons of long-range bombers and more bases were also planned to be established in England and the Balkans.[4]

These spectacular campaign plans may obscure, by their obvious prominence, the much wider scope of the air service's growth. In 1918 there was a strong shift in the overall emphasis in Britain's military effort towards its air war phase. The shift was so noticeable that British Expeditionary Force leaders complained about it to Churchill, then Minister of Munitions, when he visited France in September 1918. And Churchill was a justifiable target for their complaints. His provisional munitions budget for 1918 (submitted on 1 November 1917) clearly foreshadows the shift. He wrote in terms of a three-fold increase in aeronautical supplies; he presumed an expected output of 2,700 airplanes per month by mid-1918 for the planned new squadrons; and he stipulated that the great majority of the added labourers in munitions work for the coming year be allocated to the aircraft industry.[5]

Churchill's projected figures were more than fulfilled. In the ten war months of 1918, 26,685 airplanes were produced (and 29,561 engines) and by the armistice, the monthly rate of aircraft production was up to 3,500 planes. According to Lloyd George, there was an additional 40,000 planes on order. Total outstanding orders for the RAF by the armistice amounted to £165,000,000 or more than one half of the expenditure on aerial equipment during the war. More than 100 large factories were making airplanes and engines while over 3,000 more plants were producing related parts. At least 250,000 workers were connected with the air industry — one estimate places this figure at almost 350,000. And for one final indication of how much all this represented a shift away from the other military services, over one half of all outstanding orders issued by the Ministry of Munitions at war's end were for the air service.[6]

Indubitably, the new air service had grown to surprising proportions, but, unfortunately for its postwar peace of mind, it was not allowed the time to prove (if indeed it could have done so) that this new strength was justified in its war effectiveness. The spring strategic bombing campaign of 1919 was, of course, the most important missed opportunity in proving the capabilities of air power. The Independent Air Force of 1918 was not equipped for such campaigns and its efforts were of little effect on the war. Later, one strategic bombing expert of the RAF summarised these operations: 'It cannot be said that it [the IAF] produced decisive, or even very positive, results.'[7] A historian of British aerial concerns recently reached a similar conclusion, that the air service's contribution to the war effort was disproportionate to its consumption of the war budget.[8] RAF leaders at the time rued their lost opportunity for conclusive performance. A striking example is a comment made by Sir Frederick Sykes about the fact that the armistice precluded the use of a small number of H.P. 15 v/1500 bombers which would have been available shortly after that date for a bombing run on Berlin. Sykes, then Chief of the Air Staff, later recalled that 'it was a great blow to all of us who had worked hard to bring this small force into being that on instructions necessarily given by the War Cabinet the order [for that Berlin raid] had to be cancelled'.[9] One listens here to the *cri de coeur* of the air power professional — oh, give us just one chance to prove what we could do with one raid on Berlin!

The regret for what remained unaccomplished naturally did not deter RAF spokesmen from claiming as much as possible for what they had accomplished. One such claim was by the Air Minister, Sir William Weir, who stated that the strategic bombing activity produced the best results of all war tactics for effort expended. He based this conclusion above all on the amount of resources the Germans were forced to divert to home front defence due to such activity.[10] The former IAF Commander, Trenchard, stated (in January 1919) that only the lack of sufficient means prevented him from thoroughly devastating many German industrial centres — although he conscientiously added that this might have taken up to five years.[11] Mark Kerr, the former Royal Naval Air Service authority who had transferred to the RAF, was much more outspoken. He avowed that 'if it had not been for the formation of the Air Ministry and the Independent Bombing Force that brought the war to an end much sooner than was expected, the war would not have concluded with victory for the Allies'.[12] Nor was the official RAF 'Synopsis of British Air Effort during the War' very unassuming. This was a document presented to Parliament in April 1919. It stated, 'the effect, both morally and materially, of the raids on German territory carried out during the summer of 1918 can hardly be over-estimated'.[13] Nevertheless, the impartial observer — and some partisan observers — could see that the reality lay elsewhere; that the RAF had contributed very little to the Allied victory.[14] It was this lack of conclusive evidence of military capability that made a considerable contribution to the threat to the air arm's separate status in the immediate postwar years.

That this separate status would be questioned was inevitable. With the coming of peace and the consequent evaporation of the emotions that had engendered the Royal Air Force, it was only natural that the army and navy would attempt to regain control over their former air arms. Furthermore, with the advent of postwar budget paring, the utility of having a third military service would certainly be debated. From a note in Field Marshal Haig's diary (of 26 January 1918), it is clear that Trenchard, then Chief of the Air Staff, had foreseen these problems. 'Trenchard', remarked Haig, 'thinks that the Air Service cannot last as an independent Ministry, and that the Air units must again return to Army and Navy.'[15] Shortly after the armistice, this threat to the RAF became public knowledge when the *Daily Express*, on 18 December 1918 editorialised that the Air Ministry would shortly be abrogated and that the air force would again be split into navy and army subunits.[16]

Nevertheless, the Air Ministry and RAF leaders proceeded in their planning and actions as though their separate autonomy would be permanent. While most were highly motivated in this regard by their ardent belief in air power, it would have been a normal reaction anyway by any new bureaucratic assemblage. The first major attempt to delineate a postwar programme for the RAF was by Chief of the Air Staff (CAS) Sykes in his 'Review of Air Situation and Strategy for the Information of the Imperial War Cabinet' dated 27 June 1918.[17] The opening section of this 'Review' concerned 'The Future of Air Power' in which the first sentence reads: 'In the next war the existence of the British Empire will depend primarily upon its Air Force.' Consequently, Sykes called for the postwar creation of an *imperial* air force, under one overall command, whose task would be to protect the entire Empire. For that role, it must 'always' be maintained at war-ready efficiency because surprise was the 'essence of air tactics' and only an air force kept on a 'war footing' could protect the Empire against surprise attacks. He envisioned the development of enemy 'aerial dreadnoughts' whose great range and speed would allow them to circumnavigate and threaten all the British dominions. But the danger did not wait upon such developments; even present day aircraft could guarantee victory at the very outset of war if a belligerent state possessed 'undoubted aeroplane superiority'.

Concerning the makeup of this proposed imperial air force, he typically advised that it be a strategic 'striking force', one that would be 'undoubtedly our surest means of defence'. In addition to this force there should be a 'ready reserve' force, presumably to be converted from the available commercial aerial activity. He thus stressed a need for postwar governmental support to the civil aircraft industry as well as the governmental establishment of civil air routes and bases on the basis of their future wartime utility. In sum, Sykes proposed a very ambitious postwar air programme and he warned that if Great Britain followed a weak policy of 'drift', her fate would be 'swift and final'.[18]

Shortly after the armistice Sykes expanded his suggestions into a very detailed programme presented (on 9 December 1918) as the 'Memorandum by the Chief of the Air Staff on Air-Power Requirements of the Empire'.[19] In this paper, as part of a warning against the Government making too drastic a cutback in imperial defences, he stated:

> Before a formal declaration of war it may be possible to deal a paralysing blow at some vital nerve centre; the Air Force must be the first line of defence of the British Empire, and in the next war, however near or distant, the existence of the nation will depend largely upon air power.

Among his specific recommendations was that the air defence planning should be formed on an all-inclusive imperial scale and he listed suggested air strengths to be maintained for all areas then under imperial control. The total air units thus suggested added up to sixty-two fully operating squadrons plus ninety-two more operating on a cadre skeleton level. The immediate cost estimate for this programme was twenty-one million pounds — or about the cost of two dreadnoughts.[20]

This was hardly the approach to please a nation tired of war and of war costs and the Cabinet rejected the programme. (Sykes long remained bitter about that rejection; in 1942 he wrote, 'we threw away, with one stroke of the pen, our hardly won supremacy in the air'.) Sykes thereupon proposed an alternative 'small' (in his terms) programme of a sixty-two squadron 'Home Air Force' which would be 'kept constantly at war pitch'. He continued his advice for the Government also to sustain a strong civil aircraft industry and air traffic activity.[21] Sykes, in his civil air emphasis, was very much in accord with his Air Minister, Sir William Weir, who had expressed his views on this matter at the Mayor's Banquet at the Guildhall just two days before the armistice. According to Weir:

> The future of Aviation — the future, perhaps of the Air Force — will come, I hope, from the need of peaceful commerce rather than from the tragic necessities of war and one of the duties of the Air Force will be to maintain and still further develop that degree of technical superiority which we have now achieved.[22]

Weir and Sykes were both to be wholly frustrated in their plans for the postwar air service and its fight for survival was about to begin. Weir soon resigned as Air Minister to return to his business concerns and, the 'coupon election' behind him, Lloyd George had this office among his problems in selecting the new Cabinet. The cavalier attitude with which he filled the post reflected the Prime Minister's new indifference to the face of the RAF. The mercurial Welshman[23] who had been so influential in creating the service[24] was now ready to disband it. According to the report by Winston Churchill who was offered the ministry, Lloyd George told Churchill: 'Make up your mind whether you would like to go to the War Office or the Admiralty, and let me know by to-morrow. You can take the Air with you in either case; I am not going to keep it as a separate department.'[25] In the event, Churchill was persuaded to take the War Office as the Prime Minister wanted him to handle the developing crisis over demobilisation. As promised, the Air Minister's office was tied on to the arrangement as something like a short-term gift.[26]

Thus was created the dual ministry role for Churchill that received so much debate and criticism in Parliament and elsewhere. For example, during the House of Commons debate on the 1919 air estimates (on 15 December 1919), J.E.B. Seely, J.T.C. Moore-Brabazon, Wedgwood Benn, C. Lambert, Oswald Mosley and R. Glyn all spoke against Churchill's dual portfolios and one common aspect to their attacks was the assumption that this double posting threatened the separate existence of the Royal Air Force. Andrew Bonar Law answered for the Government and, given the terms of Lloyd George's offer to Churchill, he was less than truthful in stating that 'there is no idea in anyone's mind that the Air Force is not to be an independent Force'. Nor was Churchill's contribution to the debate very exact when he claimed that the Government was doing everything possible to ensure the permanent autonomy of the RAF.[27] Churchill was, in fact, very helpful in the RAF's future struggle to stay separate and intact, but a helpful attitude was hardly typical of the governmental purpose of 1919 in that regard. And just one year earlier, even Churchill had been ready to accept the idea of closing down the Air Ministry as is evidenced in his letter to Lloyd George which concurred with the PM's approach on the controversial Cabinet offer.[28]

As usual, parliamentary indignation reflected contemporary attitudes. RAF leaders were among those opposed to the dual ministry. For one, Major-General Sir Sefton Brancker publicly campaigned against it, stating that it represented the road backwards to the former dividend control over tactical air arms. The press reaction to Churchill's joint tenure was also distinctly negative. The strong public response to this threat, as it was considered, to the air force, prompted the Government to respond with an official announcement that Churchill's double appointment did not alter the status of the Air Ministry at all; that its independence was complete, just as before.[29] Thus the governmental leaders found that prudence was at least necessary in their dealings with the air force; that there were political risks in their actions against the new service that had captured the popular imagination.[30]

The Churchill controversy produced a yet more explicit manifestation of political disquiet with the well-publicised resignation of the Under-Secretary of State for Air, J.E.B. Seely. In so doing, Seely stated that Churchill gave the great share of his attention to War Office business, that he (Seely) should have been chairman of the Air Council rather than Churchill (the Air Council was a body analogous to the

Army Council, a guiding body composed of both civilian and service members), that the air service had had to suffer too many economy cutbacks, and finally that the service should be *wholly* separate under a *full-time* civilian service chief.[31]

The charge that Churchill gave most of his time to War Office duties undoubtedly was true. According to the Marchioness of Londonderry, whose husband succeeded Seely as Under-Secretary of State for Air, the Marquess 'practically controlled the Air Ministry' because Churchill's interests lay elsewhere.[32] Nevertheless, Winston Churchill's period over the Air Ministry was a fortunate one for the air service. This is true despite the fact that one prominent authority, Sir Frederick Sykes, has stated that Churchill's tenure was something of a disaster for the RAF.[33] Sykes' malice is easily explained as it was Churchill who replaced him with Trenchard as Chief of the Air Staff and that shift alone justifies Churchill's term of office. The Trenchard appointment was one of the most salutary events in the early RAF history. The Trenchard era is also very important for the focus of this chapter as he was a key factor in the establishment of the doctrine emphasising the strategic bombing policy as a defensive deterrent measure.

It was apparently the former Air Minister, Sir William Weir, who first suggested to Churchill that Trenchard would be the best choice for CAS. He held that a strong personality was needed to preserve the young service in its postwar years of peril; that Sykes was not the necessary leader and that Trenchard was. Churchill was easily persuaded. He had already been cool towards Sykes; he had already been impressed by Trenchard (who had, among other things, been very helpful in Churchill's efforts to quiet the demobilisation unrest); and (a most necessary item) Trenchard had proposed a far more modest and economically feasible programme for the RAF than Sykes' 'small' alternative. For a variety of reasons, then, Hugh Trenchard regained the CAS title on 15 February 1919. He was to hold that position for the next ten years, years which he greatly helped make into ones of major importance.[34]

Trenchard's prudence in setting down postwar goals may well have been conscious hypocrisy; he was not only desirous to regain the air leadership but also eager to force Sykes out. The two men had been bitter rivals for some time, and would continue to be so until their deaths. Trenchard much later remembered Sykes' tenure as CAS in extremely acid phrases.[35] The history with which this study has been concerned is, typically, not just one of logical occurrences in understandable sequence but also, and inevitably, one of irrational human emotions and motivations.

Trenchard soon formally expanded his ideas for the peacetime RAF into a 'master plan' that he submitted for Cabinet approval in November 1919.[36] Trenchard maintained his realistic prudence in these specific recommendations and also demonstrated a sound pragmatic sense in his ordering of priorities. He recognised that the state could anticipate peaceful years in the near future except for possible imperial brushfire incidents. Thus there was no immediate need for the expenditure of large sums for the maintenance of a large number of service squadrons. What was needed was the establishment of the foundations by which a large force could be quickly provided when needed. Training facilities were a related area of immediate importance. So Trenchard proposed to spend more on buildings and other permanent equipment than on machines and other war equipment, his list of priority needs including an air force staff college, a cadet college, training bases, permanent barracks, etc. He also advised establishing part-time auxiliary squadrons and university squadrons. In sum, Trenchard's specific suggestions were sensible and his eulogist and one-time air minister, Viscount Templewood, is correct in stating that Trenchard had set down the model for the basic permanent foundation of the RAF and for many other, later, independent air forces among the world powers.[37]

But there is much more of importance to be found in Trenchard's paper. In its early paragraphs he stated his underlying principles of the functions of the air service and in these passages there is much that is controversial. The pertinent paragraphs are as follows:

> The principle to be kept in mind in forming the framework of the Air Service is that in future the main portion of it will consist of an Independent Force, together with Service personnel required in carrying out Aeronautical Research.

> In addition, there will be a small part of it specially trained for work with the Navy, and a small part specially trained for work with the Army, these two small portions probably becoming, in the future, an arm of the older services.

> It may be that the main portion, the Independent Air Force, will grow larger and larger, and become more and more the predominating factor in all types of warfare.

At the centre of Trenchard's concept of the future role of the RAF, then, was the same emphasis upon air power's independent role that was at the centre of the purposes for its wartime establishment and, likewise, the same de-emphasis of air power's tactical support role.[38] Trenchard was later to rue his comments about the two 'small' tactical units 'probably becoming, in the future, an arm of the older services'. These words gave added force to the arguments of the War Office and Admiralty leaders when they demanded the return of control over their air arms. Thus Trenchard later regarded this statement as 'a fatal lapse of judgement'.[39] Nevertheless, his basic position was unequivocal — the air force was primarily to be a separate service over a separate, *independent* function of warfare — and, as all knew from the late war's experience, the words 'Independent Force' were equivalent to the concept of a strategic bombing force.

It would seem, then, that by 1919 Trenchard had converted to the concept of air power that he had strenuously resisted in 1917.[40] Nevertheless, the timing of his conversion has received some scholarly debate. Nobody would deny that he did finally become one of the most fervent exponents of the strategic bombing doctrine. By the time of the Second World War, he even complained about airplanes being diverted to anti-submarine duty. Strategic bombing, by then, was his sole answer for all military questions. He argued against the forming of a second front, against *all* campaigns that would diminish a joint allied air campaign against Germany. Churchill banteringly observed that his nickname should be 'Bomb' rather than 'Boom' Trenchard.[41] What is in debate is whether Trenchard had become a true advocate of strategic air war by the time of his 'master plan'. The foremost American historian of British air power, Robin Higham, is one who questions this and he notes in his argument that Trenchard only proposed two squadrons in his White Paper for the immediate needs of the Independent Air Force and then further observes that the Air Ministry placed no orders for true long-range bombers until 1932. However, this latter point is very ambiguous evidence for his argument in that Higham also contends that Trenchard had turned to 'wholehearted support of strategic bombing by 1921'.[42]

This conflicting evidence can be resolved by keeping in mind the situation of the time. Trenchard's emphasis on the independent role of air power was pragmatically sound when the very existence of the air force was in debate. This was not only the function which had justified its original formation but also the one most in accord with the current public image of air power (as described in Chapter 5). But to emphasise this strategic role for the purposes of survival was one thing, to advise the spending of large sums on the formation of a sizeable strategic striking force when no present need could be seen for it was something else. That surely would have been the road to ruin. It can therefore be argued that Trenchard was throughout the White Paper really being consistent with respect to his uppermost priority – the survival of the Royal Air Force.[43]

What is beyond debate is that by the time he was CAS, Trenchard was proclaiming the independent function of air power as its most important attribute and was depreciating its tactical, supporting role. In the very first briefing that he gave Churchill, he took this same tack, disdainfully alluding to pilots either being chauffeurs for the other services or working independently in a manner 'that will profoundly alter the strategy of the future'.[44] In his 1919 White Paper, he again repeated that analogy of the misuse of pilots as tactical chauffeurs.[45]

The threatened position of the postwar RAF thus led to a renewed stress upon its strategic bombing role, and this function was always preferably described in its *defensive* image as a deterrent force. The argument was already at hand – Whitehall and mass opinion had spawned it – and one of the most persuasive interwar leaders, Trenchard, made it his own when the survival of the independent RAF became his life's major dedication. Above all, the need was to promote a *separate* function for the air service to justify its separate autonomy.[46]

Historians of later phases of British air history have noted the profound importance of this early phase of the story. The official study of the strategic air war against Germany in World War II includes the observation that strategic bombing was the rationale for the birth and continued independence of the Royal Air Force.[47] The historian of the first major strategic stike against Germany in that war reached the same conclusion, that 'strategic bombing was fundamentally the RAF's reason for existence ...', and this from its birth on.[48] And for a final example, an historian of the near-contemporary British military defence scene has noted that the strategic bombing orientation of the RAF established in its early history, especially by Trenchard, continues on to the present, fostered by Trenchard's acolytes. He described these men, now the senior officers in the service, as follows: 'their deification of Lord Trenchard and belief that air power is a separate and somehow special instrument of strategic policy spring partly from the experiences of the twenties'.[49]

It was not just by doctrinal theory that Trenchard helped preserve the RAF. As has been seen, he also had the sound realisation that economic prudence was a necessity.[50] Churchill, after his installation as Air Minister, had treated the Sykes 'small' scheme with 'extremely withering' scorn and thus greeted Trenchard's relative parsimony most favourably, although advising his new CAS to go further yet, to be sure to remodel the service at a price the market would bear.[51] Even so, the Churchill-Trenchard team evidenced more expensive tastes than the Cabinet would allow, as is seen in the Cabinet meeting of 14 April 1919 where the air force budget was discussed.[52]

The resulting final reductions in the RAF were impressive by any standard. Within two years of the armistice, its personnel complement had been slashed by over 90 per cent, from almost 300,000 to less than 25,000 men. (The women's branch had been disbanded.) The aircraft totals had been reduced even more drastically, from over 22,000 planes to about 200.[53]

Another event in 1919 had compounded the budget problem; this was the establishment of the infamous[54] Ten Year Rule. The history of the imposition of this rule goes back to a query posed by Admiral the Earl Beatty, then First Sea Lord, to his civilian service chief, Mr Walter Long. Beatty in later summer, 1919, wanted guidelines as to how long it was to be presumed that the Empire need not fear the outbreak of another major war. Long, in turn, put this question to the Cabinet, asking for an official ruling as to how long the service could assume that Britain had 'immunity from war with a great power or a combination of small powers giving an equivalent enemy force'. The Cabinet answer came on 15 August 1919 in the form of a memorandum to all three service departments instructing them to revise their budget estimates 'on the assumption that the British Empire would not be engaged in any great war during the next ten years'.[55]

Andrew Boyle has suggested that Lloyd George promoted this rule among his colleagues as part of his postwar self-image of a leader in the move towards world peace.[56] Partly through that image, Lloyd George had initiated a major change in English (and to some extent in European) diplomatic practice by assuming the major responsibility in international negotiation rather than leaving this in the hands of the foreign office, an example that Neville Chamberlain would later make infamous.[57] Whether or not the Ten Year Rule belongs in this wider context, it is certainly true that its imposition was 'wholly empirical and was not based on any scientific analysis . . .'.[58] Of course, in 1919, the assumption of ten years of grace from war was correct but the problem was that the rule was continuously reaffirmed until, in 1928 (under the impetus of Churchill, then Chancellor of the Exchequer) it was given an even more permanent status:

> The basis of the Estimates for the Service Departments should rest upon the statement that there would be no major war for a period of ten years and that this basis should advance from day to day, but that the assumption should be reviewed every year by the Committee of Imperial Defence.[59]

As one historian has aptly commented, England's peaceful future was extended by each new rising of the sun — and this remained official policy until 1932.[60]

Trenchard immediately realised that the Ten Year Rule would worsen the budget rivalry between the three services in that by further limiting the available defence money it would intensify the struggle over the relative shares.[61] It was this financial competition which was probably most responsible for the interservice rivalry directed against the new service. The army and navy had long been accustomed to their own competition over the budget but now a new — and to them unnecessary — service further complicated the financial division. The problem of the defence budget has long been a central governmental question and in the modern age it is an ever increasing problem. Michael Howard's statement of its nature in the late 1950s applies equally well to its nature in England during the interwar years.

> How much of the national wealth should be allotted to defence? And once allotted, how is it to be divided between three ruinously expensive and hotly competing services? And for what type war, with what type of weapons, have they to prepare? In comparison with problems such as these, the political issues of the past seem almost nostalgically simple.[62]

The military analyst, Stefan T. Possony, has also noted the 'fundamental' importance of this matter: 'This is the fundamental question which divides the services in peace as well as in war. This is the problem on which hinges the efficiency of preparedness, tactics and strategy. This is the puzzle which no one has yet solved.'[63] Trenchard and the Air Ministry were thus faced with a particularly difficult phase of an eternal defence problem right at the start of their peacetime charge and it helped trigger a life or death struggle for the RAF.[64]

At the same time, there were new responsibilities placed upon the air service which helped it survive this struggle. First, the Air Ministry was given full authority over British civilian air activity, which gave legal authorisation to what the Ministry was already doing. This was accomplished by the Air Navigation Act (27 February 1919) which provided that the 'purposes of the Air Council shall include all matters connected with air navigation'.[65] This became interpreted in the broadest manner possible; the RAF licensed civilian pilots and gave them their medical examinations, it placed orders for and approved models of civilian aircraft, and, in the words of a usually sympathetic observer, 'one sees that from the very beginning the Air Ministry had a stranglehold on Civil Aviation and civil aeroplanes of all sorts'.[66]

Given the common opinion of the day that civil and military aerial developments were inextricably linked,[67] this enactment was thoroughly logical. Nevertheless, the limited finances available at that time resulted in the civilian side receiving short shrift from its military leadership. The first two postwar Controller-Generals of Civil Aviation within the Air Ministry both complained of this financial neglect.[68] But still, civil air had become an added responsibility for the RAF which thus enhanced its importance and prospects of survival.

Another important addition to the Air Ministry's authority was the transfer to it (from the Ministry of Munitions) of the responsibility for all aircraft design, supply and inspection. This resulted in the creation of the office within the Air Ministry of a Director-General of Aircraft Production and Research, which was fully operating by January 1920.[69] Of awesome future importance for the Ministry was the transfer to it of the responsibility for home air defence. The first major step in that process was at a War Office conference chaired by Churchill in February 1919 in which that transfer was established in principle. In reality, only the principle was to be involved for some years as the Government was then in the process of removing almost all of the wartime home defence establishment. By the end of 1920, except for what was in storage, there was neither gun nor searchlight left in London's defence system, there was not even a fighter squadron specifically assigned to that duty, the air defence control rooms had been dismantled, and all that was left (and this still under War Office control) was a small AA gunnery and searchlight school.[70]

The RAF entered into one other activity in 1919 which greatly helped sustain its value within the corridors of power when it started to help police the outposts of the Empire. The first instance of this was in that frequent problem area, the North-West Frontier between India and Afghanistan. In late spring 1919, the new Amir of Afghanistan gave in to the war party of his state and sent troops into India. An Anglo-

Indian force including RAF units was able to put down this thrust fairly easily (the Waziristan campaign). The air support proved of greatest value; in fact, the first appearance of a plane over the Amir's capital city, Kabul, was sufficient to force him into negotiations.[71] The RAF next helped in the final defeat of the forces of that long persisting menace, the 'Mad Mullah' (Mohomed bin Abdillah Hassan) in Somaliland. Costly and sometimes disastrous campaigns had been sent against this self-proclaimed Mahdi since 1899, but never with lasting results. The World War had interrupted these British efforts and thus the Mullah had been able to control large expanses of the country. Once the war was over, the army proposed a large – and expensive – expedition to put down the Mullah once and for all. The Government decided against the army's plan because of its expense and this gave the air force spokesmen their chance. They claimed that the RAF could defeat the native force with merely a few planes, and with the Government's approval, they went ahead and did so. (Some support was provided by ground troops, especially the Somaliland Camel Corps.) The whole issue was decided in three weeks – after so many years of frustration – and for a total cost of £77,000 ('the cheapest war in history').[72]

But Afghanistan and Somaliland were only preludes to the most important RAF success in its imperial activity, the aerial policing of Iraq. There, in the summer of 1920, a local uprising along the Euphrates River forced England into a four month, £100,000,000 operation which involved 60,000 troops (of which 2,000 suffered casualties) before the rebellion was suppressed. Furthermore, 'a large and costly military garrison' was needed thereafter to maintain order. One result was a strong press reaction in England against the occupation of Iraq. Editorials denounced the continuation of such a costly enterprise and demanded that the mandate should be returned to the League of Nations while they also reminded the English that they had promised independence to the Arabs.[73] Obviously, here was an imperial trouble spot which offered an excellent opportunity for some service to relieve the Government of these burdens.

In May 1920, when he was still Air Minister, Churchill had sent a suggestion to the Cabinet that military responsibility over Mesopotamia be transferred to the air force, a suggestion that Trenchard was often voicing at that time. Then, in January 1921, Churchill became Secretary of State for the Colonies and the following March, he chaired a conference (the Cairo Conference) to study the Iraq situation. After Churchill opened the conference with a statement that the Government had to cut its costs in the area, Trenchard jumped at the opportunity and presented his 'Scheme for the Control of Mesopotamia by the Royal Air Force'.[74]

Churchill, already on record for such control, naturally favoured Trenchard's scheme and so did the conference members in general (including T.E. Lawrence) and thus the principle of air force control was accepted. The conference justified their approval not just on economic grounds but also for the advantages of giving air force personnel more war experience thus providing an opportunity to test 'the potentialities of the Air Force'. (The realised absence of proof of the claims of the air power advocates is clearly indicated here.) Thus was authorised the first joint service operation wherein the RAF was given overall command.[75]

This new activity of the RAF received a great deal of criticism, the nature of which correlates with major themes of this study. First, the morality of aerial bombardment again became a question of controversy. A rare agreement was established between Labour backbenchers and senior officers of the older services as they mutually spoke of 'indiscriminate bombing'. CIGS Sir Henry Wilson moralised against 'the bomb that falls from God knows where and lands on God knows what'.[76] There were even instances of air officers showing moral qualms. Air-Commodore L.E.O. Charlton resigned over the policy of bombing native peoples, although he was offered (and accepted) another job at the Air Ministry as Trenchard desired the least possible public attention on that issue. Hugh Dowding (who was to become the famous leader of Fighter Command during the Battle of Britain) insisted that prior warning must be given to the natives before punitive raids were undertaken against them.[77]

Of course most air force authorities then and since argued against these charges of immorality and they had a strong case. The air force did adopt a stringent advance warning policy and attempted to minimise fatalities while they worked at (in terms of the RAF War Manual of that day) 'interrupting the normal life of the enemy people to such an extent that a continuance of hostilities becomes impossible'.[78] Furthermore, as various airmen remind us, it is quite illogical to assume that aerial bombardment of native villages is more indiscriminate than shelling them by long-range artillery.[79]

Other arguments have been made against the 'air control' imperial policy[80] which was initiated in Iraq (soon to be followed elsewhere, first in Transjordan). A recent one is that it furnished the wrong type of experience for an air service that was all too soon to face the challenge of the Second World War. The policy forced the always straitened air service to order most of its airplane models for this use with the result that insufficient progress was made in developing high altitude performance aircraft and heavy duty bombers with proper defensive armament and with more efficient bombsights, etc.[81] A recent military polemicist against air power has written that such activities against the tribesmen would foster a 'Jove' complex among the airmen as they wrought 'destruction on disobedient groundlings'.[82] There is also the contention that this use of air power is not in accord with sound doctrine in that the natives do not possess 'vital centres' which are, according to doctrine, the most vulnerable targets for aerial attack. But that charge has been answered in a manner which indicates that with imperial air policy one has not really entirely left the realm of strategic bombing concepts. The former CAS Sir John Slessor has written that natives do have their vital centres of supply and sustainment; 'there are almost always some essentials without which he [the tribesman] cannot maintain his livelihood; they differ greatly but they are usually there, and it is these things that Intelligence has got to know and tell the airman'.[83]

No contention at the time against the air control policy had a chance to hamper its continuance because it did the job it was supposed to do; it greatly reduced the costs of imperial policing while still being effective in maintaining order. In the very first year of its operation in Iraq, air control reduced England's expenses there from about £20.1 million to £6.6 million and the savings kept increasing. By the sixth year in Iraq, air control costs were down to £1.65 million.[84] Here indeed is the type of success story which impresses governments! It could not help but ease the RAF's problem of survival.

Nevertheless, that problem — which was always connected with the older services' rivalry — was, in one respect, compounded by the imperial air control policy. This was due to the fact that this policy was the most outstanding example of a continuing argument by air advocates for 'substitution', the transference to the air service of various military activities being performed by the army or navy. The reaction by senior army and naval officers to RAF incursions into their former prerogatives was naturally strongly and emotionally antagonistic. For one case in point connected with the air control policy subject, after the Cairo Conference decision was made, CIGS Sir Henry Wilson refused to supply some specific army units to the new command and the air force found itself having to staff armoured car units and the like.[85]

We have thus returned to the topic of interservice rivalry and it is time to show specifically how this rivalry and the parallel struggle by the RAF to maintain its independence helped prompt the development of its interwar aerial doctrine.

The end of war hostilities brought forth the start of interservice hostility directed at the RAF. Army and navy ranking officers saw the junior service as 'a wartime expedient' which should thus, with the advent of peace, be returned back to its former constituent parts, the Royal Naval Air Service and the Royal Flying Corps.[86] As already described, army opposition soon crystallised into attacking the air force policy of 'substitution', especially with respect to the responsibility of policing the Empire. The navy manifested its opposition earlier and on a variety of issues. As early as November 1918 the Admiralty reported dissatisfaction among its officers with respect to the air service not attending to the navy's aerial requirements. Specific complaints included charges that the planes assigned to the navy's use were obsolescent,[87] that the personnel in the fleet air arm were poorly trained and that personnel assigned to ship duty in the fleet air arm should be from the naval service. The underlying purpose behind all such complaints was to ease the path of the Admiralty to regain control over its own air arm. Trenchard, whose 1919 White Paper had foreseen such a reversion, nevertheless fiercely rejected the idea as he claimed that it would be years before the RAF became so strong that its losing a part would not endanger the whole. He did agree with the Admiralty that the fleet air arm was not as it should be but argued that he must be given time – he spoke in terms of two to three years – to make the necessary improvements.[88]

But the problems worsened with time. Because of postwar economic stringencies, the fleet air arm was soon reduced to almost a paper force. Another irritation appeared in June 1920 when British relations with Turkey had deteriorated to the point of potential war. The Admiral in charge of fleet operations in the troubled area, Admiral de Robeck, asked for additional air support due to the crisis and he received neither response nor action from the Air Ministry for four months.[89] And yet another exacerbating element in the naval/air controversy became ever more intense, the debate about the airplane versus the battleship.

Simply put, the debate was whether the airplane had or had not made the battleship obsolete. Both sides were prone to overstate their position and both argued from insufficient evidence. Still, it was a controversy that touched the navy supporters to the quick.

To most admirals the respective value of battleships and aircraft was not basically a technological issue, but more in the nature of a spiritual issue. They cherished the battlefleet with a religious fervour, as an article of belief defying all scientific examination... A battleship had long been to an admiral what a cathedral is to a bishop.[90]

For some years there had been indications that the navy's sacred battleship was going to be a target of the air power promoters. A pioneer American airplane designer, Glenn Curtis, had, as early as 1910, made practice bombing runs at simulated battleships over Lake Keuka in New York. About that same time, a British air pioneer, Claude Grahame-White, made similar test runs at painted shapes of battleships at Blackpool, England.[91]

By the end of the war, claims by air power spokesmen that the era of the capital ship was definitely over became quite outspoken, frequent, and, for the naval supporters, all too noticeable. Nor were such claims made just by airmen. The former First Sea Lord, Admiral Lord Fisher, for example, wrote to *The Times* (12 September 1919) that 'by sea the only way to avoid the air is to get under the water... That's why I keep on emphasising that the whole Navy has to be scrapped.'[92] Another former noteworthy of the navy, Admiral Sir Percy Scott, had also become a convert to air power and he asserted that ships could not be given sufficient armour plating to protect them against air attacks.[93] On this matter, Scott also used *The Times* as a rostrum wherein he asked 'What is the use of a battleship?'[94] Of course, air force spokesmen were no less unreserved. Air Vice-Marshal Sir Sefton Brancker was another who made use of *The Times* where (on 22 December 1920) he proposed that defence money should be spent on strengthening the RAF rather than on such *passé* weapons as capital ships.[95]

The navy retaliated by answers both in the popular press and in official recommendations. That venerable spokesman for the Admiralty's cause, Sir Archibald Hurd, repeatedly argued the case for the battleship's continued supremacy in *The Fortnightly Review*.[96] The navy's official position was naturally unyielding. In August 1919, the Admiralty established a Postwar Questions Committee which examined the air-battleship question among others and which decided (in its final report published in March 1920) that 'we do not consider that aircraft using any known form of weapon will render the capital ship obsolete... and that the battleship retains her old predominant position'. This remained the position of the Admiralty Board throughout the 1920s.[97]

The question of the status of the battleship in the age of air power is one that is commonly associated with the theories and activities of General William ('Billy') Mitchell. At least by March 1919 Mitchell was looking forward to putting the battleship to the test of aerial bombardment and he wrote a continuous barrage of commentary, especially in popular journals, advocating such tests while making general claims for air supremacy.[98] Mitchell was quite unrealistic in many of his claims but spokesmen for the United States Navy were at least equally imprudent; Secretary of the Navy, Josephus Daniel, for instance, is deservedly remembered for his statement that he would stand bareheaded on the bridge of a battleship and fearlessly await all attacks from Mitchell's bombers.[99]

Mitchell was finally allowed his tests and naturally they did not stop the argumentation, although Mitchell certainly had the immediate advantage in that his planes did sink the *Ostfriesland*.[100] But the important point for this study is that these tests (and others that soon followed) attracted much interest in England and fed the controversy over the battleship issue. Sholto Douglas is one who admits that he had intensely identified with Mitchell and his cause — seeing Mitchell as a fellow airman beset by naval villainy.[101] A second point of relevance is that in America as well as in England, the air-battleship controversy brought forth much unrealistic emotional reaction. One scholar has discussed this fact (while sharing the controversy's inflated rhetoric), maintaining that 'of all the homeric battles of history, that between air and naval theorists is the most difficult to analyse', and he described it as being beyond 'rational comprehension'.[102] Certainly in England this emotionalism helped make the naval service become an enemy of serious proportions for the RAF.

The mention of 'Billy' Mitchell unfolds another issue concerning the development of English aerial doctrine that should be briefly noted. Quite frequently, Trenchard's name is linked with Mitchell's and that of General Giulio Douhet of Italy as representing the three pioneer theorists in air power. Together they have been called 'the great triumvirate'.[103] The question arises as to how much cross-fertilisation of thought existed between the three as it is certainly true that Mitchell and Douhet presented very similar theories about strategic bombing to those of Trenchard and his fellow RAF officers.

Mitchell, for example, wrote:

> It is now realized that the hostile main army in the field is a false objective and the real objectives are the vital centers ... The result of warfare by air will be to bring about quick decisions. Superior

air power will cause such havoc, or the threat of such havoc, in the opposing country that a long-drawn-out campaign will be impossible.[104]

As with the battleship controversy, Mitchell also was anxious to convince the public of his strategic bombing theories and just after the *Ostfriesland* tests, he sent planes on mock raids over Manhattan, Philadelphia, Wilmington and Baltimore in that endeavour. The press reactions in the cities indicate that Mitchell's confidence in air power was shared by the fourth estate. The staid *New York Times* headlined its report on the Manhattan raid with a banner reading 'City "Wiped Out" in Big War Game' and then reported that seventeen planes dropping (theoretically) twenty-one tons of gas, incendiaries and high explosive fragmentation bombs were able to accomplish that fact. All of the lower Manhattan population were reported dead or fleeing the city in panic. Philadelphia was listed as the next target although Mitchell had 'deferred the destruction of that city until next week'.[105]

Even more importance is imputed to General Douhet in the formulation of air war strategy; he has long been called 'the father of air power doctrine'.[106] Many authorities have evaluated Douhet as the theorist *par excellence* in military air concepts: the 'one distinguished name' for Bernard Brodie,[107] 'the greatest military writer of the Long Armistice' for Theodore Ropp,[108] and various other writers also agree to his preeminence.[109] However, while it is true that General Douhet was a passionate and unrestrained proponent of the strategic bombing policy,[110] there is strong debate as to whether he really exerted much influence on the development of American and English air doctrines.

In England, certainly, the theory of strategic air power had many self-generated impulses. The very advent of the airplane, as we have seen, created a spontaneous and widespread trend of thought with respect to its war usage. Then came the World War with the Zeppelin and Gotha experiences which made the English extremely aware of their special vulnerability to such attacks and which created the strong English emphasis upon strategic bombing in its deterrent (that is, defensive) role. Neither the thinking of Douhet, nor Mitchell, nor even Trenchard was required for these early manifestations and Douhet's contributions in the interwar years — at the most — could only have been of a bolstering nature to beliefs already formed.[111] But even this much effect is in question. Sir John Slessor, for example, denies even *hearing* the name Douhet during the 1920s and states (as of 1956) that he had not yet read his works. Liddell Hart in his last work (published posthumously) claimed that Douhet not only had *no* influence in Europe generally during the early interwar years but that he was of small importance at that time in official Italian military circles.[112] But these negative judgements may be attempting too strong a revisionist view about Douhet. For instance, while Liddell Hart observed that the first published English translations of Douhet only appeared in America in 1942 and in Britain in 1943, this ignores the manuscript translations made for the United States Army starting in 1921.[113] Nor could official Italian interest in Douhet be as slight as Liddell Hart stated since it was the Italian Air Ministry which published the 1921 edition of his *Command of the Air* and it was the Italian Ministry of Culture who reissued that work in its 1927 edition.[114] Nor could Britain have been totally untouched by 'Douhetism'; it is impossible for this writer to believe that not one RAF air attaché in Rome participated, during the 1920s, in a discussion of Douhet's ideas. Moreover, the former Air Minister, Hoare, has recorded, 'in the twenties we in the Air Ministry were specially interested in the Italian Air Force . . . and we had always maintained a close and friendly contact from the start with Italian aviation'.[115] But the essential point is that the RAF hierarchy would not have been much affected by 'Douhetism' — they had already travelled down that road.

Some scholars see that Douhet-like strategic bombing policy of the RAF as the single most inflammatory factor of all those which divided the naval and air services.[116] While this writer holds that the strategic bombing policy was in part advocated as a reaction *to* the service rivalry, there is no doubt that it played an important role in worsening that rivalry. One obvious instance was the Air Ministry's choice of Air Vice-Marshal A. V. Vyvyan as the first Commander of Coastal Area Command (established in September 1919), the command which was in control of all shore based aircraft that were allocated to navy oriented operations. (This command was renamed Coastal Command in 1936.) Vyvyan's beliefs about air power were hardly of a nature to appeal to the naval people with whom he had to work in close association. He spoke against the use of airplanes for submarine patrols and convoy escort duty, stating that such use of planes was 'a purely defensive policy' and that 'the nature of warfare in the air is eminently Offensive and not Defensive', viewpoints thoroughly echoing those of his commander, CAS Trenchard.[117]

The First Sea Lord, Admiral the Earl Beatty, one of Trenchard's most fearsome opponents, showed early on that he held a strong distaste for Trenchard's strategic bombing views. He claimed that they were primarily an attempt to justify the autonomy of the air service and that Trenchard grossly overstated his case so as to alarm (on false evidence) the English political authorities as to the consequences if they did not support the RAF cause. Beatty made this charge against Trenchard for one instance in 1920 in reaction to the air chief's speech at the second anniversary dinner of the IAF in which Trenchard warned 'we may yet see governments living in dug-outs and holding Cabinet meetings in the bowels of the earth'.[118]

Naturally the Government could not stay aloof from such inter-service rivalry in its midst. Pressures from Parliament about the unsettled state of the air force (noted in the previous chapter) also demanded recognition. And above all responsible Government leaders could not ignore the fundamental issues concerning national defence that were involved. Thus, starting in 1921, a series of extremely important committee enquiries examined these questions and — noteworthy for the theme of this chapter — the air service defended itself at these hearings primarily on the basis of its strategic bombing function, especially with respect to its *defensive* importance.

Trenchard himself helped to initiate this series of committee enquiries when he took the offensive against the older services by sending an RAF policy paper to the Committee of Imperial Defence (CID) in March 1921. This paper included the argument for 'substitution', that giving to the RAF duties formerly assigned to the older services would make for financial savings and increased efficiency; it advised the readjustment of defence budgets so that the air force received more and outmoded war means (such as battleships) received much less; but above all it stressed that the major military menace to Great Britain was now that of strategic air attack and that the older services had nothing to contribute to this problem. Further, 'unless we can put up an adequate defence', Trenchard wrote, 'we must be prepared for a dislocation of national life to a degree unthought of in the past'. The air service should therefore be given full responsibility for that defence and such responsibility should include the maintenance of a skeleton home defence system such as existed in the last war. Nevertheless, the effectiveness of England's air defence would primarily depend upon the RAF's counterstrike potential.[119]

Neither the War Office nor the Admiralty let Trenchard's paper go unchallenged and the CID soon received opposing papers from both these sources. The committee pursued the matter by considering further testimony (both oral and written) from various service leaders. Trenchard thereupon increased his claims. After repeating his warning about the new menace threatening England's defences, he argued that the RAF should be assigned responsibility for the English home defences against both air and sea attack and that this assignment should include command over the army and navy units involved in such defences. (Trenchard received the potent support of Churchill, the Colonial Secretary, in that demand.) This new statement of course further increased the wrath among army and navy leaders and they retaliated by strong pleas for their right to regain control over their own air arms. The CID realised it faced a developing deadlock and turned the controversy over to its Standing Defence Subcommittee under the chairmanship of the Lord President of the Council, Lord Balfour.[120]

This subcommittee was given the responsibility of deciding upon a great many defence questions during the early postwar years and, in turn, it created controversial questions as many believed that Lloyd George gave responsibility to Balfour that he should have reserved for himself. Still, it was a quality committee. Lord Balfour had been concerned with defence problems for years – he was instrumental, for example, in the original establishment of the CID – and he had a general reputation for possessing a detached and penetrating intelligence.[121] His committee was also staffed with knowledgeable members including leaders from the Government and the services. When Balfour was temporarily unavailable, Winston Churchill replaced him as temporary chairman. It was this committee, then, which was first to judge upon the postwar fate of the air service.[122]

If one were to judge from Balfour's views during the war, the RAF's fate would seem gravely endangered; then, as First Lord of the Admiralty, Balfour had been a strong supporter of the navy's prerogatives over its own air branch.[123] But at the Versailles Conference, Balfour showed that he was one of the converts to the new fearsome view of future air war and he strenuously warned the Conference about German civil aviation and the 'alarming' military risk that it represented.[124]

Such fears as these about air power won the day for the RAF for Balfour strongly supported the cause of the air service in his decision paper issued in July 1921. He reasoned that the case for an independent air force depended upon the need for that service to fulfil a dominant military role and that such a role existed in its furnishing the defence against enemy strategic air raids. 'The Air Force assert', wrote Balfour, 'that if there is another great war, the first and most formidable danger which this island will run will take the form of a great air attack directed by the enemy against London and other vital spots'. Balfour accepted that premise and decided that only the air force could defend against such attacks.

> Here, then, we have a military operation which not only can be carried out independently by the Air Force, but which cannot be carried out by anything else. The Air Force does not act as an auxiliary; it requires no aid either from the Navy or the Army unless, indeed, the anti-aircraft guns were controlled by the latter, which would be contrary to that principle of a single command. In any case, since the Air Force would do most of the work, it is they who should be responsible for its direction.[125]

This commentary clearly shows that the strategic bombing philosophy was extremely useful, indeed, probably essential, for the RAF's struggle for postwar survival. In fact, Balfour made this yet more clear as he chastised the older services for their attempts to downgrade the threat of strategic air raids. In his decision paper he made a scathing comment on this:

> There is a tendency in some of the papers laid before the Standing Committee to minimize the military effect on this country of air-raids successfully carried out on a very great scale. In the memorandum prepared by the General Staff there is a picture drawn of Great Britain with its capital in ruins and the Admiralty and War Office carrying on their duties undismayed in the safe but obscure retreat supplied by some disused coal-mine. Even such a catastrophe as this, they say, would not force a decision; and perhaps they are right.

> I would, however, observe that as a matter of history, peace has usually been arranged between belligerents long before the worsted party was reduced to so pitiable a condition; and while the position of the General Staffs of the army and navy heroically carrying on their functions at the bottom of a coal pit might in some respects be less disastrous than it seems, seeing that in the contingency supposed they would have little to do, the enemy aeroplanes wandering at will over the country could carry out their work of destruction, however numerous and however heroic might be the armies and the navies of the country they were reducing to ruin.[126]

Trenchard could hardly have asked for more. The newly appointed Air Minister, Major Frederick Guest, was certainly triumphant when he wrote Trenchard after the decision that 'I think it's a great victory due to your persistent and clear arguments'.[127] There was just a touch of solace for the other services, however, in Balfour's paper. While he opposed any division in the RAF, no return of auxiliary air arms, he did advise that air force units taking part in tactical, cooperative operations should be under the command of the surface commander. He also recognised the service rivalries being generated and advised the services to be tactful and cooperative in the future, to give each other the courteous treatment proper between recognised equals. These recommendations by Balfour were quickly endorsed by the Cabinet and the Royal Air Force had survived its first major postwar crisis.[128]

But in the 1920s, the air service never had long to wait before it was subjected to another trial. While Admiral Beatty had been leading the attack against it through the period of Balfour's enquiry, now CIGS Sir Henry Wilson initiated a new period of trouble for the Air Ministry. Wilson had deeply resented Balfour's decisions and saw a tool for retaliation in the new Geddes Committee, recently created to effect governmental economic savings. Wilson therefore demanded that the finances of the RAF be examined and charged that the air service did not spend money with sufficient care. Lord Beatty quickly gave Admiralty support to Wilson's request. Trenchard then immediately countered by demanding that all three services be equally investigated and he claimed that such an enquiry would show that the air force provided the most efficient military service of the three on the basis of respective funding. 'Such is the background to the decision, provoked by Wilson but supported by Beatty, to link the future of the Royal Air Force with the investigation by the Geddes Committee into the whole field of national expenditure.'[129]

Presumably the Geddes Committee would have examined the armed services anyway but perhaps not in such a thorough manner if not driven to it by Wilson's and Trenchard's polemics. Such an investigation would certainly have been within the committee's purview. The committee had been established by Lloyd George in August 1921 as an answer to the economy campaign being pushed by banking figures (such as Montagu Norman, Governor of the Bank of England), the press, and others in reaction to the postwar depression. Lloyd George told Sir Eric Geddes to find ways to cut expenses 'with an axe' and the reputation ever since of the 'Geddes' Axe' indicates how successfully his committee (the Committee on National Expenditure) pursued that assignment.[130]

Geddes had been the First Lord during the period when the RAF was formed and he had fostered the navy's approval of the transfer of the RNAS to the new service. He was thus not remembered fondly by the navy's ruling circle.[131] Nevertheless, he opened his service investigations with the RAF which seemed to reflect Wilson's charge and he also mentioned his doubts about the need to maintain a home defence air force as well as other air commands. So here again was a real challenge for Trenchard and once again the CAS successfully defended his service. He argued that the RAF leadership was very economy minded, that greater savings would accrue with more RAF 'substitution', and that his service had to be retained because of the trends in modern military strategy. An interim report by the committee showed this last argument had a telling effect as it affirmed that 'aerial attack must be one of the principle considerations of the future . . .'. The final committee recommendations on the air service question represented another RAF victory; they stressed the need to maintain its independence, that the return of any of its components to other services' control would not produce substantial savings, and that more transfer of duties to the air service (i.e. 'substitution') should be sought. To top off the victory, the committee also commended the RAF for its thrifty ways.[132]

The older services were as unhappy with the Geddes Committee's decisions as they were with Balfour's. They both presented official rebuttals and again pressed the need for their own control over their own air arms. Lord Beatty also wrote a separate protest under his own signature. CIGS Wilson took time to deliver a speech fulminating against the RAF's bombing strategy and accusing it of being in essence just a method to kill women and children. The Cabinet was not deterred by any of these counterblasts and issued its recommendations (following those of the Geddes Committee) that the RAF should remain separate and unified. Austen Chamberlain so informed Parliament on 16 March 1922.[133]

The high committee perils for the RAF were not yet over, but, by this time, another factor had entered into play that eased its problems of survival and this was a growing fear among the English of the French air strength and its possible use against them. It is difficult now for the historian to accept seriously the premise of an Anglo-French air war in the 1920s. There were many who could not take it seriously at the time. One it seems was Trenchard who 'privately dismissed the threat from France as a political chimera of the first magnitude'. But Trenchard's biographer also admits that this 'chimera' helped solidify the air service's position.[134] In fact, the evidence strongly indicates that the French air 'scare' is another instance where air force proponents used existing fears of strategic bombing to further their own vested interests.

If fears of a French war were unrealistic, they did, at least, reflect deteriorating relations between the *Entente* powers. As is so well known, the coming of peace proved just how much the *Entente* had been a marriage of convenience. 'In the early post-war years their rivalry and dissensions were more in evidence than their cooperation'; this divergence above all being manifested in their respective Middle Eastern and German policies.[135] Those who took control of foreign affairs in England at this time were not apt to ameliorate the French tensions as both Lloyd George and his Foreign Minister Lord Curzon 'detested' France.[136] Within one year after Versailles during the period of the Geneva Conference (a conference sponsored by Lloyd George), there were widespread rumours that the Prime Minister had spoken to the French about cancelling the *Entente*; rumours that were exaggerated but with some substance.[137] The following year *Entente* relations showed themselves yet more vitriolic at the Washington Conference, especially over the French submarine programme.[138] The French were adamant in their refusal to limit that programme and Lloyd George reacted by saying that the French were endangering the peace of Europe. King George V also saw the same war clouds and he frequently warned the Government that England was gravely endangered because she was so vulnerable to French air raids on her homeland. The King's reasoning was predominantly based on the Air Ministry's statements of that same danger.[139]

Given that climate of opinion in high government circles, supported by some military authorities, the decision taken by the CID to investigate the extent of the threat to England of strategic air raids seems thoroughly sensible. The report from that enquiry again represented an air force victory. The subcommittee accepted the RAFs figures for the potential tonnage of bombs that the country could expect (hypothetically from France). The committee members accepted the RAF claim that the only effective way to counter air attacks was by the use of the air service (and they thereby implicitly rejected the Admiralty position that warship attacks against enemy ports would force the enemy to cease these air raids). The members also argued the need for an expanded air force, and here too they used the Air Ministry's suggestions for a home defence force of fourteen bomber and nine fighter squadrons. Trenchard had made that suggestion when he cited the imbalance between the French and English air strengths. Lord Balfour was chairman of the CID when this report was issued and he passed its recommendations on to the Cabinet with his strong endorsement. Lloyd George then persuaded the Cabinet to make the recommendations for an expanded air force into official policy. In sum, the *very* problematical threat to England represented by the French air force, in conjunction with the impassioned general view about the nature of future air war, reversed the defence spending policy recommended by the Geddes committee and turned around the general Government emphasis on reducing expenses.[140]

It is at this point that the articles by General P.R.C. Groves in the British press made the French air threat become a public issue.[141] Furthermore, the *Entente* tensions worsened in 1922. By January 1922 Lloyd George's close adviser Philip Kerr (to be Lord Lothian) believed that 'we are rapidly drifting into crisis with France . . .'.[142] One reason for this exacerbation was the election of Raymond Poincaré to the French premiership on 13 January. Both Lloyd George and Curzon strongly disliked Poincaré who, in his turn, was much more obstructionistic than was his predecessor, Aristide Briand, to British attempts at European appeasement.[143] It was in 1922 that the *Entente* became deeply divided over the Chanak crisis.[144] But the most important confrontation — and the one most relevant to our present concern — resulted from the French preparations for a Ruhr occupation as a response to German defaults on reparation payments.

Poincaré had shown his partiality for the use of the Ruhr as 'un gage productif' as early as the Versailles Conference.[145] Colonel Repington, in talks with French leaders in 1921, often heard them speak about occupying the Ruhr and (in a diary note of 2 May 1921) once reported that the French were 'simply *aching* to get hold of the Ruhr, and I don't think that they can keep off it'.[146] Nevertheless, it was in 1922 that the final French plans were made for the climacteric occupation of the Ruhr which occurred on 11 January 1923.[147]

Britain strove to prevent that event. Andrew Bonar Law, by then Prime Minister, travelled to Paris with vain protests; Prime Minister Smuts of South Africa even wired a suggestion for the Empire to take concerted action and break all French ties. But Poincaré was adamant — *Le Temps* seemed to sum up the French official position with its comment (on 26 November 1922): 'Enough. Since we shall only get what we grab, let us grab.'[148] Even after the *fait accompli*, Britain maintained its strong opposition. Curzon, the foreign minister, kept demonstrating a bitter attitude; the Labour Party called for an ending of the *Entente* and Stanley Baldwin warned France that the *Entente* might well break apart over the crisis.[149] Today scholars agree that the Ruhr issue represents, possibly, the nadir in modern Anglo-French relations since the formation of their alliance.[150]

But still it is difficult today to view this crisis in the context of a *casus belli*, even though some did feel war tremors at the time. As Lord Douglas recalls (he was then serving with the Air Ministry): 'Ridiculous as it may sound, that [the Ruhr occupation] forced us at the Air Ministry to start thinking in terms of a possible war, with France as the potential enemy.'[151] There is evidence from the time which shows just how widespread were such thoughts. Poincaré saw a need to talk to Japan in order to gain a new alliance partner. Gustav Stresemann, a man of much common sense, believed that an Anglo-French war had become a possibility. As a last example, that titan of thought, Oswald Spengler, wrote various people that he interpreted the French occupation as a preparatory step in their war plans against England (the occupied area provided the means of 'the construction of an operational basin for submarines and aircraft on the north coast against England').[152]

As a reaction to these increasing tensions of 1922-3, the first expansion scheme for the RAF of twenty-three home defence squadrons was revised upwards to a programme scheduling fifty-two squadrons. Once again, the political and popular climate of opinion and pressures combined with air force argumentation worked to advance the military aerial developments in England. The air service propaganda had not let up, very much due to the fact that the interservice rivalry and the threat to the RAF's very survival had not let up. The change in Prime Ministers in October 1922 did not appear to be an improvement because when the new one, Andrew Bonar Law, offered the Air Minister's post to Sir Samuel Hoare, he warned that it might be an existing office for only a few weeks' duration. Bonar Law repeated this threat when Hoare attended his first CID meeting in November 1922. Hoare countered with a request for a thorough investigation into the question before such action was taken. This started another flurry of interservice polemics. Lord Beatty and Trenchard each threatened to resign and Trenchard issued another outspoken memorandum on air power capability and the French menace. Bonar Law bent with the wind and established a CID subcommittee 'to enquire into the co-operation and co-relation between the Navy, Army and Air Force from the point of view of National and Imperial Defence generally . . .', while the committee's charge also included the questions of the tactical air arms and the advisability of an overriding single defence authority. It was also to recommend 'the standard to be aimed at for defining the strength of the Air Force for Home and Imperial Defence'.[153]

Thus was created 'the famed Salisbury Committee'[154] chaired by the Marquis of Salisbury, Lord President of the Council, and which included some of the most eminent figures in the Government such as Lords Curzon, Devonshire and Peel, the three service ministers, the Chancellor of the Exchequer (Stanley Baldwin) and also men of stature from outside the Cabinet, especially Lords Balfour and Weir. Weir, the former Air Minister, possibly exerted the most influence within the committee membership. Also important in what was to be another air force victory was the fact that Trenchard made a far more favourable impression in the hearings than did Beatty in what developed into a heroic struggle between the two service chiefs. 'During the enquiry Trenchard's dogged determinism and singlemindedness of purpose compared favourably with Beatty's dogmatism . . .'[155]

The hearings were unusually thorough. To simplify proceedings, the Salisbury Committee decided at its first meeting to establish its own subcommittee to survey the problem connected with naval-air cooperation and the question of the fleet air arm. This subcommittee became known as the Balfour committee. All told the two committees held thirty-two sessions and produced eighty-six memoranda. And once again the RAF arguments carried the day with the usual few sops extended to the navy and army. Trenchard won the right to maintain control over *all* air units. He received approval on two major innovations; the establishment of a unified Chiefs of Staff organisation and the establishment of an Imperial Defence College for senior officers from all three services (both proposals of course had the advantage to the RAF of giving it an equal standing with the older services). Finally, the Salisbury Committee recommended that the home defence air force be expanded over a five year period to a strength of fifty-two squadrons.[156]

The Board of Admiralty did not take kindly to this further setback; in fact, they caused a short-lived Government crisis by threatening a resignation *en masse*. Prime Minister Baldwin was able to overcome this *crise de nerfs* with the mediating help of Leo Amery, the First Lord, but it was clear that navy-air tensions remained bitterly unresolved.[157]

The decision to further expand the home defence squadrons was again, as with the first expansion scheme, a response to RAF reasoning based on the French bombing threat.[158] The Salisbury Committee hearings were held when the *Entente* relations were at their worst. It was a most opportune time for the air service to face such a review. The

RAF's warnings found a sympathetic response — for example, the prestigious Lord Haldane agreed that England needed added aerial protection against France — and once again the implications of the strategic air doctrine gained the day.[159]

The specific figure of fifty-two squadrons (which originally was planned to have a complement of some 600 firstline airplanes in the home based force) was derived from Trenchard's proposals to the committee as was its planned breakdown into thirty-five bombing and seventeen fighter squadrons.[160] Throughout his interwar years of service as CAS, Trenchard was wholly pessimistic about the fighter airplane. He recognised that popular pressure would make the employment of fighters a political necessity but he considered this to be their sole value, a sop to the demands of an ignorant populace and an accommodation to political reality. In a speech Trenchard gave at Cambridge in 1925, he justified the bomber preponderance (of two thirds) in the home *defence* air command in terms that clearly illustrated this attitude: 'Although it is necessary to have some defence to keep up the morale of your own people, it is infinitely more necessary to lower the morale of the people against you by *attacking* them wherever they may be'.[161] Trenchard's predilection for the bomber in Britain's air defence plans did not stand unchallenged. As was seen in the preceding chapter, there was much debate in Parliament over this issue; ranking officers in the army and navy expressed their belief in an air defence of just fighter planes; there were even proponents of the fighter over the bomber within the RAF leadership.[162] But Trenchard had the authority and persuasive power to force through his ideas. He established the bomber emphasis in the expansion plan in an RAF conference held on 19 July 1923 with such arguments as the following: 'I feel that although there would be an outcry, the French in a bombing duel would probably squeal before we did. That was really the final thing. The nation that would stand being bombed longest would win in the end'.[163]

It was thus, on the basis, in part, of such unscientific thinking as who was apt to 'squeal' first that RAF air doctrine was developed in the early interwar years. Air force apologists may argue that the two to one bomber-fighter ratio was admittedly arbitrary but that there was no exact way to calulate a proper ratio and that the fighter was far less dependable before the development of radar[164] but it is nevertheless obvious that Trenchard's priorities with respect to methods of air defence were not those which proved out in 1940 and that there were authorities in the 1920s who argued for a reversal of his priorities.

Noteworthy among these was England's foremost expert in AA ground defences, General E.B. Ashmore. In a work published in 1929, Ashmore deplored the bomber preponderance in the home defence command and sensibly pointed out that counterbombing did not deter the Germans in the Great War but that the fighter defence over England did prove increasingly effective. He reasoned that since England would never *start* an air war, the current reliance on a bomber-oriented defence would inescapably result in England suffering relatively unimpeded air strikes against her cities at the opening of the next major war. 'I submit', he wrote, 'that we are relying too much on the defensive power of offensive bombing ... and that over-reliance in bombing has been carried over to under-reliance on fighter aeroplane defence.'[165]

Although a philosophical base for England's air defences had been set, a base which was to last until after the mid-1930s, the actual physical expansion of the home defences did not proceed as scheduled. (The official designation of this defence force, on 1 January 1925, was the Air Defence of Great Britain Command.) In that the impulse for this expansion depended so much upon contemporary fear of France, when this fear subsided so too did the motivation to spend large sums on the ADGB programme. Therefore, in this regard too, the 'spirit of Locarno' had its assuaging effect.[166] Thus, a Cabinet Committee on Air Force Expansion for Home Defence, chaired by Lord Birkenhead, reported on 27 November 1925 that the 1923 expansion scheme need not be completed in its stipulated five years but that it could be phased over the much longer period terminating in the fiscal year 1935-6. Thereafter all attempts to hasten the rate of expansion were blocked by the Treasury's budget minded officials and by the Ten Year Rule. (The famous 'appeaser', Sir Samuel Hoare, has remembered with acid pleasure that Winston Churchill when Chancellor of the Exchequer was very instrumental in enforcing delays to the expansion programme which abetted the crucial English military weakness of the later 1930s.[167]) Finally, the Labour Cabinet of 1930 decided to further delay the programme in the interests of economy and postponed its completion date to 1938.[168]

The same events and fears which prompted the air expansion programme also prompted the revival of England's AA ground defences as well as the start of her interwar ARP activity. Plans had been made for AA ground defences to work in cooperation with the nine fighter squadrons stipulated in the original twenty-three home defence squadron programme. The integrated ground-air programme is known as the Steel-Bartholomew Plan (of February 1923), named after the two officers who headed a joint service committee established to devise such a plan (Air Commodore J.M. Steel and Colonel W.H. Bartholomew). This plan delineated an air defence arrangement for the greater London area which closely followed the London Air Defence Area scheme set up during World War I by Major-General E.B. Ashmore. It consisted, first, of an advanced observation area of sound locators on the southeastern coast backed by a network of manned observation posts. Further inland was an 'outer artillery zone' which was flanked and backed up by a long belt (about fifteen miles wide) for fighter defence operations. Over London itself was an 'inner artillery zone'. This Steel-Bartholomew Plan can be seen in retrospect as that which set the pattern followed continuously thereafter until its successful testing in 1940. The most important addition to the plan was in its early warning organisation in which radar was later to play such an essential role.[169]

The plan was of course amended many times, the first change coming almost immediately when the air expansion programme was changed from twenty-three to fifty-two squadrons. An important event in the process of activating the plan was the assignment of General Ashmore to command the ground forces within it, an obvious recognition of his precedent-setting role during the Great War. His previous experience immediately manifested its worth as he set up the observation network on the former wartime pattern, with subcontrol centres correlating the findings of some twenty-five observation posts each. Once again England had an air defence organisation which included the grid maps, coloured counters, aircraft movement plotters, etc. — all of which was to gain so much acclaim in the war to come. The observation posts were manned by volunteers who were appointed to the position of special constables and who were under the immediate control of the local chief constables. The entire ground defence organisation was made part of the ADGB command and thus the Air Ministry had overall command responsibility. The defences separate from the bombing squadrons were put under a subordinate command officially designated as Fighting Area. (The two RAF officers who first commanded ADGB and its subordinate Fighting Area were both to gain renown in the second war; Air Marshal Sir John Salmond and Air Vice-Marshal H.R.M. Brooke-Popham respectively.) The War Office still retained financial and immediate operational responsibility over the gun and searchlight units which were wholly manned by territorial army personnel. Thus the ADGB command was one of divided authority which in retrospect still seems cumbersome and which was much criticised at the time — especially by General Ashmore — but nevertheless which worked very effectively in the crucial test of the Battle of Britain.

However, as with the air squadrons, economic stringency kept the ground units far below paper strength during the 1920s, even though the War Office had accepted the principle that these forces should be kept at wartime readiness because air warfare did not allow time for preparation after hostilities started. Nevertheless the realities of the times were delay and economic deprivation for all the units of the ADGB command throughout the early interwar years.[170]

That pattern of early impulse and then delay is repeated in the history of interwar air raid precautions activity. The bogey of a war with France prompted the formation of an ARP committee and starting with its first meeting (in May 1924), the committee evidenced fears based upon Air Ministry prognostications about the awesome threat of the aerial knockout blow. Still, the ARP preparations throughout the remainder of the decade were hampered by financial stringencies imposed by the Treasury.[171] The committee's first report, issued in 1925, set the trend for ARP concepts in Britain for years to come. It offered no doubt that London would be the key target area for enemy bombers and that they would not pinpoint on specific targets as often as they would engage in area bombardment in order to undermine morale. These conclusions forced the committee into much pessimism: any concealment attempt was considered fairly hopeless because there was no way to camouflage the whole city; the maintenance of public morale appeared to them as another awesome problem, and they expressed the possible necessity of *forcing* needed workers to stay in London — even to the point of placing police cordons around the city; the committee also considered all their future efforts to be, at best, just palliatives as they accepted the Air Ministry's doubts about the effectiveness of any type of home air defence. Because the committee saw the problem of public morale as essential, they strongly advised education of the public in the best known methods of self-preservation but in this, too, defeatism was to prevail because the Government vetoed such public orientation in fear of spreading panicky apprehensions.[172]

The ARP committee took the tack of delegating responsibility to various functioning departments and committees for the task of formulating specific plans and techniques. In the 1920s such allocation of authority eventually involved twelve governmental departments plus six specialised committees. The most active of these, by 1930, in developing plans was the Ministry of Health whose responsibilities were basically concerned with the handling of casualties. (Of note is that this included research in mental facilities; the air war to come was seen as a major threat to mental stability.) All of the ARP connected agencies were hampered by financial restraints, for reasons already noted. The committee had stated the need for research in such matters as camouflage, lighting, bomb damage capability, anti-gas measures and air raid shelters, but all such operational research was delayed for lack of funds. Nor did all theoretical answers come easily. For example, anti-gas requirements appeared to conflict with the Office of Work's plans for shelters; there was observed 'a clear, and apparently irreconcilable, conflict between the need to send the public underground for protection against high explosives and the need to keep them above ground for protection against gas'. Surveys were also made, but again no great progress was achieved, in plans for an early warning system, evacuation requirements, manpower needs and, in general, the various activities that characterised ARP work in England during World War II. During the 1920s, the basic paths were being laid, but the Government pressure for secrecy and economy left much to do in the next decade.[173]

This chapter has discussed many aspects of Britain's activity in air war preparations at the service and Government levels. Naturally, this activity reflected a general set of assumptions about air power. The assumptions held by a great many service leaders coincided to a striking degree with those held in non-expert circles described in the previous chapter. The military airpower spokesmen were not reticent in expressing their views, although the publication used might determine a certain amount of prudence on their part. A survey of such military writing will now be presented of which RAF statements will naturally predominate. As with the previous chapter, a certain amount of repetition in ideas is not only inevitable but very much to the point. What is being depicted is a pervasive viewpoint rather than unique concepts by a few *avant garde* theorists.

The *Journal of the Royal United Service Institution* is an excellent source for official military opinion. The Institution's location, in the centre of Whitehall across from Horse Guards, symbolises its central position among British military journals. It is 'the major general Service magazine',[174] and of definite influence.[175] The journal not only prints independently written articles of military (and other) interests but also reprints papers delivered at meetings sponsored by the Institution, often on topics concerning vital areas of military policy. We will now examine some examples that deal with air power topics.

Air Commodore Brooke-Popham was asked by the Institution soon after the armistice to give a talk on the newly created air service.

Naturally, his comments mostly consisted of a review of the air force's work in the Great War. Still, he made some statements which pointed towards the future and which are relevant here. After admitting that bombing techniques had been quite primitive during the recent war, this air force authority then warned, 'we may one day have a very rude awakening if we take the very limited results obtained either by ourselves or by the Germans as a criterion of what will happen in the next war'. Even with respect to the last war, he added, strategic bombing raids were able to obtain a significant impact on public morale and he spoke of the value of forcing work stoppages and wasted expenditure in home defence protection in both men and *matériel*. Lastly, he prophesied that small incendiary bombs might prove to be an extremely dangerous weapon in future urban air attacks.[176]

Another talk delivered at the Institution shortly after the war was specifically directed to the topic of future war techniques. In this address, Major-General Sir Louis C. Jackson portrayed the airplane as the most potent new military weapon, especially with respect to its strategic bombing function. Jackson was one of those who believed that London was particularly vulnerable to this new war strategy, and he warned that the use of incendiaries might virtually burn down the city. He thus stressed the need to spend heavily on AA defence and cautioned that victory in the next conflict would go to the nation that was best prepared for aerial warfare. It is also of note that both Jackson and Viscount Peel, the Under-Secretary of State for War (who chaired the meeting for which Jackson spoke), agreed that chemical warfare should not be banned but, rather, that Britain should lead in this war technology so that it would continue being victorious in war.[177] (It may well be that a people's confidence in new weaponry would be at its highest just after a successful war effort.)

Just over a year after Brooke-Popham's appearance in the journal, there was another major piece by an RAF officer, this was an article by Group Captain Chamier about air war strategy, and like Brooke-Popham (and repeating the Trenchard doctrine) Chamier emphasised the immense impact that aerial bombardment had on morale. He too mentioned the wastage forced upon the enemy in home defence efforts, efforts that were the inevitable result of popular pressure for action urged upon political leaders. Chamier shared the common belief that the immediacy of the threat pertaining to air war precluded any postwar opportunity for preparation and mobilisation; that all must be readied in advance. Chamier broke from 'Trenchardism' — to coin a phrase — in his belief that the fighter plane had the advantage over the bomber and that adequate *home* air defences would prevent a breakdown in the nation's will-to-war under aerial attack. But he returned to normal RAF dogma in his affirmation that once aerial supremacy was gained, then the road was open to achieving such a breakdown of morale.[178]

Some six months later, another article by an RAF officer adhered even more closely to the Trenchard line. This piece, by Flight-Lieutenant Mackay, was concerned with the role of aircraft in imperial defence and its message was that the role of the strategic bomber was by far the most important. The air force should maintain the *minimum* number of fighter airplanes that the public would accept and its main expenditure should be on a retaliatory strategic bombing strike force. Mackay was yet another who emphasised public morale as a target 'by striking at the Government and the people themselves', while in agreement with Chamier that such strategic attacks should wait upon the attainment of aerial supremacy. He also temporised about fighter aircraft by stating that a balanced air defence system — fighters plus all other anti-aircraft devices including a balloon-steel network obstruction system — could seriously hinder enemy raiders. He assumed gas would be used in future air attacks and lastly (and typically in the year 1922) he was very fearful of current dangers, especially from the French air force.[179]

In the same issue, yet another RAF officer wrote of the immense moral impact of strategic bombing and the attendant advantage of diverting enemy effort and resources to home defence in a piece which described RAF action in the last year of the war.[180]

Also in 1922, the Institution's annual Gold Medal Prize Essay contest was dedicated to the subject of what changes science would bring about in the next continental war — the subject in itself indicating the growing climate of opinion which is the concern of this study. Even more indicative is the content in the winning essay, written by Major R Chenevix-Trench. The author's answer affirmed that air power constituted the most formidable change. He was somewhat equivocal about the potential of strategic bombing with respect to whether it could furnish the knockout blow all by itself when he claimed that the primary question in future war would be whether civilians could endure such attacks. Otherwise, Chenevix-Trench very closely paralleled Trenchardism. For air defence he believed the country should primarily depend upon a deterrent shield and not spend great amounts on home defence, money that would be wasted because such defence was ineffectual. The populace therefore must be forced to realise the basic priorities of military necessities: 'in short, the ordinary citizen will have to absorb a lesson in strategy that has been too hard in the past for most statesmen and not a few soldiers'. A public outbreak of panic was another problem that must be faced. In sum, Chenevix-Trench's prize essay showed that he considered science and war to be almost an equivalent way of describing strategic bombing.[181]

In the same issue there was an article which discussed the problem of home air defence and its author believed that only two alternatives 'in present circumstances' could offer security: sufficient distance from enemy air force bases or a deterrent striking force of such strength as to frighten the enemy into not starting upon this type of air warfare.[182]

If Chenevix-Trench had some question about the ability of air power to deliver the knockout blow, just one year later another paper presented no doubts at all. This was a reprinting of an address given to the Institution by an RAF spokesman, Wing Commander Edmonds. This authority in fact believed the English air service had the ability to deliver such a blow at that present moment. He was very specific about the nature of that victory blow: 'We must make him [the enemy] feel that life has become so impossible that he prefers to accept peace on our terms. That is our object — to destroy the enemy's *morale*'. And, to repeat, he stated the RAF bombing force could achieve that goal whenever called upon. (How empty was that assumption can best be judged by the next war, yet fifteen years away!) Edmonds was typical of prevailing RAF doctrine in deprecating all home air defence methods including the fighter plane but he nevertheless prudently stated that his opinions were not necessarily representing official air staff policy. This speech prompted various comments from the audience of which one made by a ranking naval officer bears notice *because* of that service affiliation; Vice-Admiral V.H.G. Bernard agreed with the speaker that the next war would presumably be an undeclared one starting with a surprise air attack on the English home front and that the outcome of such an attack was of crucial importance. To minimise its surprise value, Admiral Bernard stated the vital need to have a superior secret service organisation.[183]

A further dimension of the air threat was highlighted a year later in an article about chemical warfare by Major E.R. MacPherson. This expert saw no hope at all in attempts to ban gas warfare and believed that the next continental war would start with airborne poison gas attacks against urban centres, unless the opposing side had the stronger air arm so that the potential attacker would be too apprehensive to try such a strategy. Such a deterrent force was the major's only hopeful defensive suggestion.[184]

Yet another RAF speaker at the Institution was Group Captain MacNeece who, in 1925, spoke on the problems of air defence. He had very little comfort to offer on that topic. He affirmed that the urban centres would be the primary targets for future air war and then explained how Britain was preparing for that possibility with its ADGB system. He justified the bombing emphasis in that command upon the premise that air *defence* (as such) just did not work. 'It follows that victory will come to the nation which in an air war in the future has not only the strongest striking force, but which shows the greatest stoicism and ingenuity in meeting strange and devastating terrors'. During the floor discussion which followed this talk, there was no basic disagreement with that premise. There was comment on the necessity for public education if so much depended upon public morale. MacNeece agreed with that logic but added that such education had to wait upon governmental action.[185] (Which meant a long wait as the Government did not want to promulgate ARP information to the public in fear of further frightening them.[186] Thus by the mid-1920s, general concurrence in the current air war doctrines did not lead to effective action. Rather, the doctrines led to such a hopeless pessimism that *inaction* was the normal result.)

In 1926 Colonel Villiers-Stuart spoke to the Institution about civil interactions with military concerns, and he too focused on the problem of strategic bombing raids. This speaker had no doubt about the inevitability of such raids; they were 'not only a possibility but a certainty . . .'. They would probably include gas and no defence would prevent the bombers from reaching their targets. Public panic was, therefore, a major concern and public education was thus an obvious necessity. But then Stuart vacillated just as did the Government. The people must not be frightened by such briefings so it would be best to adopt a filtering down scheme wherein at first only local authorities would be informed of the nation's ARP planning. As to the nature of such planning, Stuart advised the preparation of gas-tight rooms in all threatened residences, the formation of a variety of civil defence organisations, and other such activities — which logically would require mass education as a prelude. And he then concluded with another piece of advice which added further to his logical inconsistency as he emphasised that action was needed *immediately*.[187]

According to an article the following year, the public had received an inordinate amount of misleading information with respect to 'the first annual tactical exercises of the Air Defence of Great Britain . . .'. Despite the term 'tactical', these air exercises, held during the last week of July 1927, presumed strategic air raids on London by the enemy state 'Eastland'. As a result of these raids, the defending state 'Westland' had to remove its governmental activities from London to Manchester. Still, the author argued, the local press had exaggerated the results by stating that the capital city had been 'wiped out' by the bombers. Otherwise, this report, written by another RAF officer, offered little that would give confidence. The 'lag' time factor between first observation of the enemy and the first opportunity to intercept was a great problem as it now took only twenty minutes for a bomber to reach London from the coast; the fighters were outclassed by the bombers when they did intercept; and the main lesson to be learned from the exercises was 'that the best possible system of ground and air defence cannot guarantee that no enemy aircraft will penetrate those defences'.[188]

These air exercises, starting in 1927, became an annual air drill in England, and their results were certainly disquieting for the English people. Those who were dedicated to promulgating the current air war fears could always use these exercises for bolstering evidence.[189] Still, a sensible analysis of these tests should not have led to hopeless conclusions but, rather, to the conclusion that the two primary factors in successful air defence were sufficient early detection techniques and adequate fighter aircraft, and that England needed much improvement in both cases. The fighter certainly must have the advantage over the bomber in speed and manoeuverability. London's proximity to the coast put a special burden on observation requirements. This burden would be ever greater as bomber development improved, especially in speed, but in the 1920s, the problem should not have been deemed insurmountable. In the first exercise, most of the raiders were detected and intercepted.[190]

A second appearance by Group Captain MacNeece Foster (the 'Foster' having been added in the interim) was made in the journal in 1928 with a reprint of his talk on 'air power and its application'. His forecasts now were wholly catastrophic in character. London could not hold out for more than weeks against air attacks which would leave the city 'flooded with gas, set alight with incendiary shells and bombed by high explosives'. The effect on morale would be incalculable but certainly bad, 'and in these days when politics and social discontent might not fail to turn to violence, it is possible that air attacks might be a prelude to an internal political upheaval'. He explicitly foresaw the extinction of all life throughout entire urban areas. As for protection against such horrors, this expert offered no hope but in the possession of the world's strongest air force. Not one of the military authorities in MacNeece Foster's audience voiced disagreement with his prophecies of utter urban ruin.[191]

Within weeks, the Institution's members were subjected to yet another voice of doom when Major Lefebure spoke on chemical warfare. This expert described gas as easily becoming the 'decisive weapon' of the future. Air raids which included gas would produce 'chaos and terror'. The only method that the speaker could suggest to avoid such a fate was an international ban on gas war, the approach that so many others believed impossible to achieve. As for the audience's reaction, one member, Major J. Davidson Pratt, believed the horrors of gas attack had been exaggerated, but the chairman of the session, General Sir Noel Birch, wholly supported the terrifying visions of Major Lefebure.[192]

Just one issue later, the journal contained a statement by Major C.C. Turner which rejected the possibility of banning gas warfare, in his article reporting on the 1928 air exercises. His conclusions about the tests were very grim; he saw little hope in defence measures and believed that the people must be prepared to stoically endure the opening raids before the prime necessity, a sufficiently strong retaliatory force, forced the enemy to discontinue its raids. Poison gas would be the greatest danger but proper public education and preparation would bring this danger into sustainable dimensions. Therefore, he strongly urged such 'civilian education and discipline. I suggest this must either be tackled now and tackled effectively, or the Government and the barracks and the factories had better be moved at least 100 miles further inland'.[193] Major Turner thus furnishes another good illustration of the impasse then being experienced in Britain's military preparations as neither of his two alternatives were within possibility of attainment at that time.

One of the few optimistic statements to appear in the journal about the possible effectiveness of air defences was by Colonel H.W. Hill. He believed that the fighters could adequately deal with the raiders if they received sufficient notice and thus all depended upon the advanced observation and warning techniques (which was always an awesome problem for England until the development of radar). But all was not hopeful in Hill's article. He foresaw some terrifying possibilities in gas and germ warfare and was another who believed that panic could be expected from the urban populations if they were not sufficiently prepared and the need was to start that public orientation immediately.[194]

The 1930 air exercises occurred shortly after Hill's statement of cautious optimism and, according to the report in the journal, their results offered no evidence to support Hill's confidence. Once again the air defence side had little success in stopping the enemy attacks.[195]

In the final year of this study's purview, the journal contained three articles pertinent to our interests. The first, by Squadron Leader Slessor (who later, from 1950-53, served as CAS), discussed the historical development of the RAF. When he came to the establishment of the ADGB command, he justified its two-thirds bomber preponderance in the defensive context so often seen in this and the preceding chapter. He first argued that 'purely passive self-protection . . . has never been the British conception of national defence . . .' He then added that air defence, strictly defined, did not afford effective protection anyway;

> So the policy is to provide the essential minimum of fighters for close defence in co-operation with the ground AA defences, and to concentrate the bulk of our resources on the maintenance of a formidable striking force of bombers, the *positive proportion of the defences*, to enable us to launch a counter-offensive if we are attacked.[196] (Italics added.)

Next, Major C.C. Turner again reported on the annual air exercises and this time his reaction to their results was not as negative as three years before.[197] Nevertheless, his updated reasoning was hardly comforting: 'It seems inevitable that in the event of war with a first-class Continental Power, one of the first measures to be taken would, of necessity, be the removal of the seat of government to Manchester, or to Liverpool'. Turner was also still convinced that a strong retaliatory force was an essential component of air defence.[198]

Finally, and notably, an article by Squadron Leader Andrews indicates that, by 1931, an officer of the RAF could publish doubts about the assumptions that dominated the 'Trenchardism' type of doctrine throughout the 1920s. (Trenchard himself had retired from the RAF in December 1929.) This officer analysed the strategic bombing doctrine and found it wanting as a policy for Great Britain. He realised that the policy had been adopted because air defence was considered so feeble a crutch, but he countered with the question of whether it was not very poor planning for the English to adopt a policy that put her at such disadvantage to her potential continental enemies. 'It would be difficult to name any enemy area that would be in any way comparable to London in respect to the importance, vulnerability and accessibility of the legitimate bombing objectives it contains.' He then questioned whether air defences were as inadequate as the doctrine presumed. He was one who believed that early observation techniques and fighter aircraft — with the help of ground defences — would be able to cope, if their quality of performance was kept at highest standards.[199] This was, of course, the revisionist trend of thought which would finally be adopted by the Government and air service and which would prove up to the test of 1940. Later advances in technology (especially in aircraft design and in radar) would allow such revisionist views to gain ascendancy.

Still, in 1931, Squadron Leader Andrews was very much the exception to the general view, even with respect to military opinion outside the RAF, as has been indicated by this review of the articles in the *Journal of the Royal United Service Institution*. This all-service medium presents a surprising manifestation of basic agreement with the RAF's strategic bombing assumptions (surprising when one remembers the interservice rivalries of the era). However, if one turns to a more specialised journal orientated to one military (non-air) service, one still might expect a basic disagreement with the air force position. Thus this study will now analyse *The Army Quarterly* through 1931 (it was first issued in October 1920) for its even more surprising evidence of how air force concepts received general acceptance.

A first, and striking example, is that Chief of the Air Staff Trenchard published an article in this quarterly just six months after its inauguration. Most of this article was concerned with the early postwar problems of the fledgling service such as organisation, recruitment and training. Nevertheless, Trenchard did not waste the opportunity to inform his army readers that the next war would start in the air, that the air service had to be at wartime strength during peacetime to a far greater degree than the other services, that the most important early campaign would be for air 'predominance' with overall final victory probably hanging in the balance, that air bombardment had an inordinate impact on morale compared to actual physical damage, and that the independent role of the air service would therefore include as its targets 'the morale of his [the enemy's] civil population and government'. Trenchard also prudently added — considering the journal — that the RAF would always have a cooperative role to play with the ground services.[200] Much more prudence was exhibited a year later by RAF Group Captain Chamier. This author, who had been outspoken about air power's independent role in an article in the RUSI journal just months before,[201] now devoted himself wholly to the cooperative role of air power in *The Army Quarterly*.[202]

Just one issue later, Major-General Bird returned the favour with an article which discussed the strategic bombing function of the air service.

He was not thoroughly convinced by the claims of the air theorists; he stated that new weapons tend to be exaggerated as to their value (although perhaps gas *could* do what its claimants say). Nevertheless, this general was mostly in agreement with RAF stances. 'The end of war is usually attained when one nation has been able to bring such pressure to bear on another that public opinion obliges the Government to sue for peace.' In the past, this result has usually followed victory over opposing armed forces. But now? While he expressed doubt that air raids on civilian capitals would alone be enough to accomplish this result 'so long as the power of effective retaliation is possessed . . .', he must leave the question open if one side gained air supremacy over the other. Then, yes, certainly 'terror raids' would be an awesome threat.[203]

A full exposure of the RAF positions was presented in *The Army Quarterly* just three months later in an article by Squadron Leader Walser. He argued that the next war would start in the air, that the first major military effort should be to gain air supremacy, and then to follow that with a primary programme of strategic bombing (and thus the need, he claimed, for an independent air service). He proposed that England must build a strong offensive bombing force in the immediate future as air war allowed no time for preparation once it started. Typically, he spoke of that force in terms of its deterrent value. Lastly, and of special note considering the journal, he argued for the need for England to adopt the policy of substitution, replacing the older services with the RAF in various responsibilities.[204]

Walser touched upon the interservice rivalry question, of course, as he pointedly justified the status of the independent air service. Shortly afterwards, Major-General Bird, in a second article, also approached that question and added his support to continued RAF independence — although he did believe the service should place more importance on its tactical supporting role.[205] Another vote of confidence was given six months later by Major Stoehr who returned to air power's independent role for his justification. He endorsed the arguments of the air warfare propagandists (they appeared, he said, 'to be quite incontrovertible'), such as the awesome nature of the bombing threat and its potentially decisive influence.[206] In sum, the RAF was winning the interservice debate over its status even in *The Army Quarterly*.

That journal's next article of relevance to our purpose was by an officer in the Royal Artillery who, significantly, considering his allegiance, wrote a very weak defence of AA gunnery effectiveness. While he protested the then current press trend of printing horrifying articles about air attacks and their reports of the increasing inadequacy of air defences, he nevertheless agreed that cities did represent a tremendous problem for those defences and that neither guns nor planes afforded sufficient protection at that time. His only hope in urban defence was that future answers might come from new scientific breakthroughs.[207]

Within six months another article appeared by an officer of the Royal Engineers which very explicitly deprecated the results of AA fire as it paraded without demur the most outspoken claims of the air power advocates. This army expert even included that terrifying concept of the interwar years, revolution via air raids. He opened his comments with the claim that he would base 'all deductions upon accomplished facts . . .'. Such deductions thereupon included poison gas aerial attacks and the following result: 'The severe bombing of a large town, say 300 tons a day for several days, can scarcely fail to produce complete panic, coupled possibly with food riots and revolution. Against such raids all defence methods would be unavailing except that of a retaliatory bomber offensive.'[208]

Air force thinking was again displayed in this journal less than a year later in an article by Captain Rupert de la Bere, then a professor at the RAF College at Cranwell. He analysed the impact of modern science upon war and found it of a *decisive* magnitude, especially in its application of aerial warfare including use of poison gas and possibly germ spreading missiles. By such war strategy, he believed that victory might well be achieved in days, perhaps hours. He was truly a 'victory through air power' proponent. Strategic bombing attacks on the enemy vitals would definitely be the opening phase of the next war, declared or not. One of the antagonists would thereby suffer such a breakdown of public morale that it would be forced to attempt 'peace at any price'. De la Bere had no protective answers to offer except the customary one of sufficient offensive action.[209]

The following issue of the journal had another doomsday prophecy about chemical warfare, this by a writer preferring the anonymity of the pseudonym 'OTAC'. He too expressed his belief that air power plus CBW would probably be the 'decisive' element in military practices.[210] Then Brigadier-General Hartley soon added his affirmation that gas warfare would surely be used in future wars, both against troops and urban populations, and that it would produce a grave impact on morale.[211]

The Army Quarterly also had a report on the 1927 air exercises and its writer, Major Oliver Stewart, presented a more positive verdict about the air defences than the one presented in the *JRUSI*. Although noting that the exercises involved the assumption that the London Government was forced to move inland, he nevertheless reported very favourably on the fighter defence efforts. 'The fighter is more than a match for the bomber, assuming an approximately equal standard of design, and can even take on superior odds in bombers with a great chance of success'. He believed, however, that England was letting a bad disparity of performance develop between the two types of aircraft. The fighters *just* had time to intercept the bombers during these trials. The production of faster fighter airplanes was an essential need for England's future defence. Major Stewart's final summation was that 'adequate anti-aircraft defences and increased performance in fighting aircraft are the two needs which have been vividly demonstrated by the air exercises'.[212]

Major Stewart also reported on the 1928 exercises and this time his impressions were more duplex. His first sentence indicates this quality: 'The widely published reports stating that the recent Air Exercises demonstrated that London could not be defended against air attack were in many instances pulled out of truth and coloured to make sensational reading.' The author believed, rather, that the tests showed that London could receive 'a sensible measure' of defence *if* the attacks occurred in good weather. (Londoners would receive little comfort from that conclusion!) He admitted that the fighter performance was poor, that anti-gas measures should have been disseminated to the populace, and that London would have gone through very hard times if the attacks had been for real — but not to the extent of 'the often-quoted effect of making the civil population bring pressure to bear on the Government to sue for peace'. Although the performance levels of the fighters had dropped even further behind those of the bombers since the year before, they still achieved a good percentage of interceptions. Still, new and faster designs were a major necessity.[213]

Stewart's confidence in the public's ability to endure was echoed by Captain C.B. Thorne who wrote an article on urban air defence problems. This officer argued that while some bombers (probably carrying poison gas) would always penetrate the defences, that this still was not the means that would break public morale. He contended that this would only enrage the people and steady them in their war purpose — even though some panic might also ensue. Therefore, the author's final conclusion was that strategic bombing would not be a major element in obtaining victory in future wars, a rare statement in the 1920s.[214]

The quarterly's final interest for this study (for the time span concerned) lies in two further reports on the English air exercises by Major Stewart. In both articles, the author strongly argued against the bomber emphasis then prevailing among the RAF leadership. He claimed that England's need was, rather, for more fighters, and that if the country achieved a sufficient quantity of these planes she would be able to repel enemy air attacks. He did admit that the public did not share his confidence and that its general response to the air problem was a hopeless dismay: 'Without doubt the lay public has been persuaded that, if war occurred, London would soon be dust and ashes'.[215] That opinion supports the evidence shown in the preceding chapter; that such defeatism reflected a great amount of military thinking as well has already been shown.

The air service started publishing its own quarterly in 1930, and, significantly, the Trenchard school of thought was solidly represented in its first issue. An article by Wing-Commander Garrod supported the strategy of aerial attacks against public morale, the presumption of future wars starting with such raids, the conclusion that aircraft were pre-eminently offensive weapons and that the air force should thus be predominantly composed of bombers – in sum, a strong restatement of Trenchardism. There was, however, one sign that revisionism was taking place: 'It must always be remembered that disorganisation rather than destruction is the aim of this kind of warfare, partly because it is more economical and partly because it is equally effective'. Garrod sensibly reasoned that the English should not follow a policy of obliterating potential postwar markets for its goods.[216]

The second issue of this new journal again had an article championing the strategic bombing policy, this by the controversial but highly regarded ('the most brilliant and unorthodox speculative mind then in the Army . . .'[217]) General J.F.C. Fuller.[218] In this article, Fuller presented the thesis that all war techniques except air operations were becoming ever more mired in immobility and thus the dominant weapon of the future would be the airplane, especially in its use against the 'civil will'.[219]

General Fuller was certainly not a typical member of the military caste but he does represent here a fairly common aspect of interest; when the officers of the older services were not directly involved in the battle between the services, they were often voicing similar conclusions about air war as the most dedicated air power enthusiasts. Examples of this have already been seen here in articles from the service journals. Examples can also be found in books written by such officers. What follows is just a brief review of this type of evidence.

Lieutenant-Colonel Etherton, im his autobiography, discussed the future character of war. In these prognostications, air warfare played the dominant role. Aerial fleets of hundreds of bombers, each carrying at least two tons of bombs containing gas, incendiaries and high explosives, would depart for enemy capitals at the very outset of war. 'Entire districts' would be left 'untenable'. Design improvements in bombing machines would make defence an ever greater challenge. The final answer might be the alternatives of the utopia of eternal world peace or 'total annihilation' for the world community.[220] Major General Maurice, in 1926, described himself as being 'amongst those who believe that in future wars the prime object of the contending nations will not be the destruction of the opposing forces, but what the Germans call the will to victory of the opposing peoples'. And, he added, air power now allowed an antagonist to attack at first hand that civil will-to-war with, possibly, the opportunity to avoid confronting the enemy armed forces at all. 'For a people may find the continuance of war to be intolerable.'[221] The onetime CIGS, Sir William Robertson, presented like views in his autobiography. Future warfare would start with attacks from the air (possible unannounced) upon the enemy homeland. He believed such war methods were barbarous but still, henceforth, inevitable. The goal of such raids was 'the breaking down of the country's morale'. This goal might not be achieved, but England would be the most vulnerable to such methods. As to what to do in that eventuality, 'our best form of defence will lie in the ability to attack'.[222] The onetime First Sea Lord Admiral Fisher furnishes a final example. His advocacy of air power in *The Times* has already been noted.[223] He also expressed such beliefs in book form as early as 1920.[224]

Throughout the period under present study, then, the English public and politicians were continually exposed to military air theory which blended with and buttressed their own fevered views. Even the university lecture platform served as a means of such communication. Trenchard's address at Cambridge has already been cited.[225] Air Vice-Marshal Brooke-Popham gave a lecture at the University of London which also stressed the strategic bombing function of aircraft and that its primary target must always be the public's will-to-war. Indeed, Brooke-Popham used that opportunity to expound almost all of Trenchard's assumptions including the arguments that attack was always the best defence, that the fighter plane was a very ineffective instrument, and that England must start preparing because the next war could well begin both undeclared and with its opening blows directed at London.[226]

The time has arrived for final conclusions. There is, certainly, considerable evidence that by the end of the first decade of the interwar years, a generalised concept about the future impact of air warfare had been accepted by a large share of the English populace. The idea that the airplane's primary war usage would be as a raider of urban centres had become almost a truism. The major English military concern had become the question of defence against such urban attacks and the common conclusion was that air defence depended mostly upon the possession of an offensive deterrent bomber force.

The evidence also indicates that the gradual English adoption of this air war doctrine must be understood in the wide context of an interaction of military, political and public responses and that these responses started in connection with events that occurred before World War I, that they were strengthened by events during that war, and that they were solidly entrenched by events and trends in the immediate postwar years.

These conclusions have a corollary premise that the thinking uncovered by this study affords a helpful understanding of the appeasement era in English history. It is certainly beyond the purpose here to describe *that* era in depth. But to illustrate the point, I will quote from three comments about the appeasement period wherein all become more explicable within the context furnished by this study.

Firstly, Thomas Jones, in his diary note of 7 January 1936, recorded comments by Prime Minister Baldwin on why England was unable to consider a war with Italy during the Ethiopian crisis. Baldwin's words were (in part):

> I had repeatedly told Sam Hoare 'Keep us out of war, we are not ready for it . . .'. Italian bombers could get to London. I had also Germany in mind. Had we gone to war our anti-aircraft munitions would have been exhausted in a week.[227]

Secondly, another diary note, this by General Sir Edmund Ironside (soon to be CIGS) during the Munich crisis: 'I am now told that all the authorities have insisted upon the parcelling out of troops all over London during air raids . . . They want the sign of uniforms to quieten the people.'[228]

Lastly, the following passage is from a 1954 talk delivered at the Royal United Service Institution by Sir John Slessor in which he attempted to justify the established emphasis of the RAF upon strategic bombing. While arguing for his current policy, he admitted some errors connected with it during the appeasement era (when he was Director of

Plans for the Chiefs of Staff).

> Air power did indeed have a decisive effect in 1938. It was, above all, the 'thug with the bomber force' as someone called him at the time, that caused our bloodless defeat at Munich . . . I do not believe anyone — not even Mr Churchill — could have taken the responsibility of leading the nation into war in the appalling condition of air weakness in which we still were in September, 1938 . . . It is true that we did alarm ourselves and our political chiefs unduly in the years immediately before the war by visions of the 'knockout blow' — a phrase which made a pretty frequent appearance in the documents and discussions. But you must remember how it looked to us, in our inexperienced ignorance at the time, in 1938 . . .[229]

English appeasement was truly a 'coat of many colours' and all its threads have not been completely unravelled.

EPILOGUE

As everyone knows, England eventually did experience strategic bombing attacks but the result was not the holocaust that had been prophesied and England's air defence system by World War II depended primarily on the sensible emphases of fighter aircraft and ground defences. For England to have continued to depend upon a deterrent threat when the country recognised that, among the major European states, it was the most vulnerable to strategic bombing would have been increasingly 'curiouser and curiouser'. But the changeover to the more realistic air defence approach had to wait upon a justifiable confidence in an early warning system. The development of radar was therefore a major watershed in this later history. Great Britain's effort in this regard is a fascinating story somewhat analogous to the American undertaking to land men on the moon — a splendid uniting of science and industry in intense and successful endeavour.[1] Once radar offered reasonable hope that home defences could be effectively rallied before enemy raiders were already overhead and unloading, then 'slide-rule mentality' could gain influence within military and political planning.

This would affect not only the readjustment of air defence priorities of course, but also the overall British defensive strategy. A good example of this more realistic perspective is the 'European Appreciation 1939-40' prepared for the Committee of Imperial Defence early in 1939.[2] Therein, the air threat to England is, naturally, still recognised but only as one aspect within (potentially) a worldwide array of problems and it certainly is not overemphasised. The maintenance of public order in the aftermath of bombing raids is still a concern but not a predominant one. The deterrent is now recognised as potentially counterproductive and thus the restraints that were put upon Bomber Command in the early phase of World War II are foreshadowed: 'We assume that it would be no part of our national policy, and it would certainly not be in our own interests, to initiate air attacks against objectives which must involve casualties to the enemy civil population'.

But if the military planning shows a more realistic basis in the 1930s, this does not mean that the service leaders were not still prone to 'gusts of emotion' and were suddenly ready to have all their preconceptions and efforts be subjected to objective testing and analysis. The eminent 'boffin', P.M.S. Blackett, among others, has written about British mili-

Epilogue

tary resistance to operational research on through the World War II era.[3] Likewise, the aerial fears and delusions of the era we have studied were not eliminated by the more practical efforts of the 1930s. Air defence had to *prove* itself first. Until such proof, assumptions about air power would remain inflated in a great many preconditioned minds, not least among those in government. An incident remembered by Vincent Massey (then Canadian High Commissioner in London) during the Munich crisis offers an intriguing example; he was surprised to find thirty-seven territorial army troops assigned to duty on the roof of Canada House, not with anti-aircraft weapons but with twelve machine guns.[4] Clearly, the Government was still anxious about the public response to potential air attack.

Nevertheless, studies concerning British air policy in the 1930s emphasise the practical rather than the delusive — naturally so since the success story of the Battle of Britain awaits as the (relatively) happy ending.[5] (Even so, the delusive aspects offer more aid to understand the appeasement era.) The finale *does* represent a spectacular example of a successful change of direction and recovery of initiative. Certainly here, the British 'Establishment' did not just muddle through!

NOTES

PREFACE

1. For representative examples consult Walter Millis, *Arms and Men: A Study in American Military History* (New York: Mentor, 1956), pp. 231-2; E.L.M. Burns, *Megamurder* (New York: Pantheon Books, 1966), p. 13; and Robert C. Batchelder, *The Irreversible Decision, 1939-1950* (Boston: Houghton Mifflin Company, 1962), pp. 170-89, 210 and 221-2.
2. In an address given at Columbia University, the eminent Lord Harlech presented this prospect for an atomic war: 'It is entirely possible that all semblance of human civilization on this planet would be obliterated. Whether life of any sort could long be supported is questionable, and even if it could, we can be sure that the living would envy the dead.' David Ormsby-Gore, Lord Harlech, *Must the West Decline?* (New York: Columbia University Press, 1966), p. 44.

CHAPTER 1

1. Major Raymond H. Fredette, *The Sky on Fire: The First Battle of Britain 1917-1918 and the Birth of the Royal Air Force* (New York: Holt, Rinehart and Winston, 1966), pp. 34-6. (Hereinafter referred to as *The Sky on Fire*.) Also cf. Major Fredette's earlier statement on these early German aerial plans in 'Bombers of the Black Cross: German Bombardment Aviation in World War I', *The Airpower Historian*, VII (July, 1960). (Hereinafter referred to as 'Bombers of the Black Cross'.)
2. Fredette, *The Sky on Fire*, p. 35.
3. *Ibid.*, p. 36.
4. Fredette, 'Bombers of the Black Cross', p. 170.
5. As seen, for example, in the authoritative work by Theodore Ropp, *War in the Modern World*, rev. ed. (New York: Collier Books, 1962), p. 269.
6. Cap-Commandant aviateur Desmet, 'Rôle des aviations belge et française: sur le front occidental pendant la grande guerre', Part III, *Bulletin Belge des Sciences Militaires*, III (March, 1922), 348.
7. Arch Whitehouse, *The Years of the Sky Kings* (Garden City, New York: Doubleday & Co., Inc., 1959), p. 27. (Hereinafter referred to as *Sky Kings*.)
8. Major-General Sir F.H. Sykes, *Aviation in Peace and War* (London: Edward Arnold & Co., 1922), p. 35.
9. As in so many other areas, the Soviet Bloc has recently taken credit for the invention of the Zeppelin. A recent claim from Budapest states that Count von Zeppelin bought the original Zeppelin designs from the widow of the true inventor, the Hungarian David Schwarz. Associated Press bulletin published in the *Courier-Journal* (Louisville, Kentucky), 1 February 1968.
10. Hugo Eckener, *Count Zeppelin: The Man and his Work*, trans. Leigh Farnell (London: Massie Publishing Company, Ltd., 1938), p. 156. A very good short review of prewar interests and developments in airships in general may be found in Henri Azeau, Gilbert Caseneuve and Louis Saurel, *L'Enfance du Siècle (1900-1912)*, Vol. II: *Histoire Vivante du XXe*

Siècle (Paris: Robert Laffont, 1966), pp. 184-5.
11. Sykes, *op. cit.*, pp. 15-16.
12. *Ibid.*, and Eckener, *op. cit.*, pp. 159-60.
13. Douglas H. Robinson, 'The Zeppelin Bomber: High Policy Guided by Wishful Thinking', *The Airpower Historian*, VIII (July, 1961), 132. By the time the war started, von Zeppelin was preaching the need for ruthless aerial bombardment, believing that national security justified all means to preserve it. Eckener, *op. cit.*, p. 272.
14. Eckener, *ibid.*, p. 273. According to his memoirs, von Tirpitz was doubtful about the Zeppelins primarily because they would be too subject to the whims of nature. Grand Admiral von Tirpitz, *My Memoirs*, 1 (New York: Dodd, Mead and Company, 1919), p. 181.
15. Robinson, *op. cit.*, p. 133. As will be seen in Chapter Three, Bethmann-Hollweg was still in opposition to the bombing of civil target areas in 1917 when Germany had turned to the Gotha (long-range bomber) raids.
16. *Ibid.*, see also Fredette, *The Sky on Fire*, pp. 30-31.
17. As quoted in Robinson, *loc. cit.*
18. As quoted in George H. Quester, *Deterrence Before Hiroshima: The Airpower Background of Modern Strategy* (New York: John Wiley & Sons, Inc., 1966), p. 24.
19. *Ibid.*, p. 23.
20. As cited in Robinson, *op. cit.*, p. 137.
21. Edward Hallett Carr, *The Twenty Years' Crisis, 1919-1939: An Introduction to the Study of International Relations* 2nd. ed. (New York: Harper Torchbooks, 1964), p. 94.
22. The term 'knockout blow' was in very common use during World War I and originally meant a military victory which could bring an enemy to final defeat — an obvious adoption from its pugilistic connotation. The famous Memorandum by Lord Lansdowne to the Cabinet of 13 November 1916, furnishes a useful, typical example of this usage. 'Let our naval, military and economic advisers tell us frankly whether they are satisfied that the knock-out blow can and will be delivered'. As quoted in Paul Guinn, *British Strategy and Politics: 1914 to 1918* (Oxford: The Clarendon Press, 1965), p. 175. As will be seen, this term gradually becomes most commonly associated with the results hoped for by strategic bombing enthusiasts.
23. Cyril Falls, *A Hundred Years of War, 1850-1950* (New York: Collier Books, 1962), pp. 186-7. See also Raymond Aron, *The Century of Total War* (Boston: The Beacon Press, 1965). E.g. in this work's opening paragraph appears this statement: 'In the twentieth century, the soldier and citizen have become interchangeable . . .', p. 9.
24. Roger Ashley Leonard (ed.), *A Short Guide to Clausewitz on War* (London: Weidenfeld and Nicolson, 1967), p. 9. The quote is given in the editor's 'Introduction'.
25 *Ibid.*, p. 209. (This is from Vol. III, Book VIII, Chap. IV of Clausewitz' *Vom Kriege*.)
26. Stefan T. Possony, *Strategic Air Power: The Pattern of Dynamic Security* (Washington: Infantry Journal Press, 1949), p. 147.
27. Quester, *op. cit.*, p. 18.
28. Tirpitz, *op. cit.*, II (New York: Dodd, Mead and Company, 1919), pp. 294-5.
29. *Ibid.*, pp. 271-2.
30. Robinson, *op. cit.*, p. 138. May 31st is the correct date, not 1 May as stated in Hanson W. Baldwin, *World War I: An Outline History* (New York:

Harper and Row, (1962), p. 57. Baldwin's conclusion about this raid and earlier ones elsewhere is appropriate, however. 'Both combatants [France and England] were learning that war in the twentieth century meant no holds barred.' *Ibid.* With respect to casualty reports, authorities will vary of course. Casualty counting is an imprecise activity always. Thus another source, C.W. Glover's *Civil Defence: A Practical Manual Presenting with Working Drawings the Methods Required for Adequate Protection Against Aerial Attack* (London: Chapman and Hall, Ltd., 1938), p. 9, agrees with the figure of seven dead but lists only thirty-two injured. Glover also sets a figure on the property damage at £18,396.

31. Sykes, *op. cit.*, p. 45. General Sykes was the first commander of the RFC and his memory (in 1922) should be accurate. 'We considered it essential to dispatch at once to France every available machine and pilot, because both political and military authorities were of the opinion that for economic and financial reasons a war with a great European power could not last more than a few months.' *Ibid.* See also Norman Macmillan, *Tales of Two Air Wars* (London: G. Bell and Sons, Ltd., 1963), p. 46. Captain Macmillan, a prolific recorder of British aerial history, is also one of the most reliably informed in that history.
32. Air Chief Marshal Sir Philip Joubert de la Ferté, *The Third Service: The Story Behind the Royal Air Force* (London: Thames and Hudson, 1955), p. 24.
33. Winston S. Churchill, *The World Crisis*, I (New York: Charles Scribner's Sons, 1923), pp. 336-7. Also see the documents furnished in Captain S.W. Roskill (ed.), *Documents Relating to the Naval Air Service*, Vol. I, *1908-1918* (London: The Navy Records Society, 1969), pp. 167-79.
34. Major-General E.B. Ashmore, *Air Defence* (London: Longmans, Green & Co., 1929), p. 3.
35. Churchill, *op. cit.*, p. 348.
36. Violet Bonham Carter, *Winston Churchill As I Knew Him* (London: Eyre & Spottiswoode and Collins, 1965), pp. 327, 331. Lloyd George also bears witness to Churchill's quick readiness to assume this responsibility. David Lloyd George, *War Memoirs*, Vol. IV, *1917* (Boston: Little, Brown and Company, 1934), p. 105.
37. 'The lengthy career of one of the greatest of Englishmen [Churchill] was to be intimately wedded to the evolution of air power.' Eugene M. Emme (ed.), *The Impact of Air Power: National Security and World Politics* (Princeton, New Jersey: D. Van Nostrand Company, Inc., 1959), p. 6. The comment is by the editor.
38. Macmillan, *op. cit.*, p. 66.
39. Churchill, *op. cit.*, p. 340. The transfer of air defence responsibility officially made on 3 September was obviously an understood arrangement at least two days earlier, the date of this instruction.
40. *Ibid.*, pp. 340-41. The quotation is from Churchill's note of 5 September 1914. Four of these instructions by Churchill to Captain Sueter are reprinted in Emme, *op. cit.*, pp. 27-9. This work is the best anthology of writings on air power available.
41. Prime Minister Stanley Baldwin, in a famous speech in the House of Commons on 10 November 1932, said that 'the bomber will always get through . . .' and thus concluded that 'the only defence is in offence . . .'. His phrase thereafter was repeatedly used in both the popular press and in military circles. The speech is quoted at length in Emme, *op. cit.*, pp. 51-2.
42. Dr Eugene M. Emme, the distinguished historian of the National Aeronautics and Space Administration (NASA), universalises this conclusion from the

same set of evidence. 'It was thus that the indivisible relationship between offensive and defensive operations in air warfare was clearly documented in the first weeks of World War I.' *Ibid.*, p. 6.
43. Authorities differ regarding the distinctions which should be made between active and passive defence. In general, those who advocate offensive measures would make almost all forms of 'pure' defence be passive. However, it would seem that the most sensible distinction between the terms can be related to the moment of bomb release. Any efforts to *prevent* that moment, be they air strikes on enemy aircraft factories or fighter defence over the targets or anti-aircraft gunnery operations, would seem to be properly termed an *active* form of defence. Then, in counterdistinction, any measures to *alleviate* the effects of bombs after their release can be suitably described as passive defence. Obviously, Air Raid Precautions activities would properly fit into this latter category.
44. Churchill, *op. cit.*, p. 340.
45. Ashmore, *op. cit.*, is an excellent source for the history of this development. Alfred Rawlinson, *The Defence of London: 1915-1918*, 2nd. ed. (London: Andrew Melrose, Ltd., 1923) is not the specialised historical monograph that its title suggests but more of a personal reminiscence about the author's experiences as an officer over an anti-aircraft gunnery unit. As such, it has some historical value. The official history commissioned by the Committee of Imperial Defence, Sir Walter Alexander Raleigh and H.A. Jones, *The War in the Air: Being the Story of the Part Played in the Great War by the Royal Air Force* (6 vols.; Oxford: The Clarendon Press, 1922-37), touches on home defence matters at times and is a useful source. (The first volume was written by Raleigh, the latter five by Jones.)
46. Captain J.H. Boyd, 'AA Development in the Royal Engineers', *The Royal Engineers Journal*, XLIX (September, 1935), 372.
47. General Sir Frederick Pile, *Ack-Ack: Britain's Defence Against Air Attack during the Second World War* (London: George C. Harrap & Co., Ltd., 1949), p. 45.
48. Churchill, *op. cit.*, pp. 340-41. In the directive of 5 September Churchill also stressed the forward defence policy as the best answer.
49. Admiral Sir Percy Scott, *Fifty Years in the Royal Navy* (New York: George H. Doran Company, 1919), p. 289.
50. Rawlinson, *op. cit.*, p. 44. Two pictures of the French auto-canon are presented in this book facing p. 16.
51. Ashmore, *op. cit.*, pp. 4-5.
52. Rawlinson, *op. cit.*, pp. 84-5, 121. (He believed that his men had hit one Zeppelin by then but it certainly had not come down in Allied territory.)
53. *Ibid.*, pp. 84-5.
54. Scott, *op. cit.*, p. 290.
55. Rawlinson, *op. cit.*, pp. 38-40.
56. Air Commodore L.E.O. Charlton, *War Over England* (London: Longmans, Green and Co., 1936), p. 15.
57. Pile, *op. cit.*, pp. 46-7.
58. Ashmore, *op. cit.*, p. 5.
59. Ibid., pp. 5-6. Cf. General Ashmore's lecture to the Royal United Service Institution of earlier date. E.B. Ashmore, 'Anti-Aircraft Defence', *Journal of the Royal United Service Institution*, LXXII (February, 1927).
60. Churchill, *op. cit.*, p. 341.
61. Boyd, *op. cit.*, p. 373.
62. Commander R.H. Keate, 'Searchlights', *Journal of the Royal United Service*

Institution, LXI (November, 1916), 926. The 90-cm. model increased the range by some 1,100 yards. *Ibid.*
63. Boyd, *loc. cit.*
64. Keate, *op. cit.*, p. 927. Keate's article is basically a report on an article by Captain di Tondo in the *Revista di Artiglieria e Genio.*
65. Pile, *op. cit.*, p. 45.
66. Jourbert de la Ferté, *op. cit.*, p. 30.
67. Norman Macmillan, *Sir Sefton Brancker* (London: William Heinemann, Ltd., 1935), p. 61. This work on the life of one of the most important army air officers in the early stages of British aerial development is not accurately described by its proper bibliographical form. It is divided into two major sections, Book I covering Brancker's life story from 1877 to 1918 and Book II covering the years 1919-30. Book I is really Brancker's own *memoirs* with some few footnotes and additional material by Macmillan who thus is only the *editor* for this period. Macmillan wrote the whole of Book II. Thus, information from this book covering events mentioned in this chapter is almost all by Brancker and not Macmillan.
68. Jourbert de la Ferté, *op. cit.*, pp. 30-31.
69. Baldwin, *op. cit.*, p. 97.
70. Present mindedness is always a problem with respect to understanding this early stage in air warfare. One military historian makes another observation in this regard: 'We think now entirely in terms of aeroplanes, but the German dirigibles of W.W. I could carry more bombs further than any heavier-than-air machine of that era.' Jac Weller, *Weapons and Tactics: Hastings to Berlin* (London: Nicholas Vane, 1966), p. 91.
71. As quoted in Robinson, *op. cit.*, p. 142.
72. Winston S. Churchill, *The World Crisis*, II (New York: Charles Scribner's Sons, 1923), 51-2.
73. Arthur J. Marder (ed.), *Fear God and Dread Nought: The Correspondence of Admiral of the Fleet Lord Fisher of Kilverstone*, Vol. III: *Restoration, Abdication, and Last Years, 1914-1920* (London: Jonathan Cape, 1959), pp. 116-7. The italics are Lord Fisher's.
74. As quoted in Lloyd George, *op. cit.*, p. 104.
75. Lord Fisher, *Memories and Records*, Vol. I: *Memories* (New York: George H. Doran Company, 1920), pp. 129-30. Comments and correspondence about this proposal and resignation threat can also be found in Churchill, *The World Crisis*, II, pp. 52-3 and in Marder, *op. cit.*, p. 124. In that Fisher's account differs somewhat from Churchill's (I have presented Churchill's interpretation), it is also important to note the jotting that Prime Minister Asquith made on 5 January 1915. 'Old Fisher seriously proposed, by way of reprisals for the Zeppelin raids, to shoot all the German prisoners here, and when Winston refused to embrace this statesman-like suggestion sent in a formal resignation of his office [*sic*].' The Earl of Oxford and Asquith, *Memories and Reflections, 1852-1927*, II (London: Cassell and Company, Ltd., 1928), p. 51. This basically substantiates, of course, the Churchill version of the incident, albeit with an exaggerated statement.
76. As quoted in John Evelyn Wrench, *Struggle: 1914-1920* (London: Ivor Nicholson & Watson Limited, 1935), p. 161.
77. Including Queen Alexandra, who wrote Lord Fisher, 'This is *too* bad, those beasts actually went straight to Sandringham, I suppose in the hopes of exterminating us with their Zeppelin bombs – though, thank God, they failed this time! . . .' As quoted in Marder, *op. cit.*, p. 143.
78. Charlton, *op. cit.*, p. 8. Philip Littell used the term 'Zeppelinitis' in an

October 1915 piece in the *New Republic*, as a much more benign form of this fixation among certain social circles in the United States. This article is reprinted in Philip Littell, *Books and Things* (New York: Harcourt, Brace and Howe, 1919), pp. 57-63.
79. B.H. Liddell Hart, *The Memoirs of Captain Liddell Hart*, I (London: Cassell, 1965), pp. 17-18.
80. Terence H. O'Brien, *Civil Defence* (London: HMSO, 1955), p. 7. This work is one of the Civil Series volumes in the *History of the Second World War* series edited by Sir Keith Hancock. Further evidence of the panic reactions by the Hull populace can be found in Charlton, *op. cit.*, pp. 130-32. The Air Ministry compiled evidence of adverse English crowd reaction to World War I air attacks (including the Hull story). See Great Britain, Public Record Office, AIR 9/69, Folio 22.
81. David Mitchell, *Monstrous Regiment: The Story of the Women of the First World War* (New York: The Macmillan Company, 1965), p. 217.
82. B.H. Liddell Hart, *Paris, or the Future of War* (New York: E.P. Dutton & Company, 1925), p. 40.
83. Air Vice-Marshal E.J. Kingston-McCloughry, *Defence Policy and Strategy* (New York: Frederick A. Praeger, 1960), p. 209.
84. O'Brien, *op. cit.*, p. 8.
85. Wrench, *op. cit.*, p. 144.
86. Pilots were even assaulted on the streets because of the lack of air protection they furnished. Quester, *op. cit.*, p. 28.
87. The ever recurring problem of whether to hyphenate or not is complicated in this case by the gentleman's own inconsistency in this regard. Usually, he did not use a hyphen – nor is his name hyphenated in *Hansard*. However, in his first published work, the hyphen is used throughout. N. Pemberton Billing, *Air War: How to Wage It* (London: Gale and Polden, Ltd., 1916). There are no monograph studies concerning Pemberton Billing. The following account is based primarily upon contemporary newspaper reports and upon his statements in *Hansard*.
88. One notable exception is the recent work by Major Raymond H. Fredette already cited, *The Sky on Fire*.
89. A.J.P. Taylor, *English History, 1914-1945* (New York: Oxford University Press, 1965), p. 44. This is Vol. XV of *The Oxford History of England* series, edited by Sir George Clark. With respect to this quotation, this writer has some qualms about Taylor's use of the word 'initiating', but otherwise feels that his summary is valid.
90. Pemberton Billing, *op. cit.*, p. 5.
91. Brigadier General F.G. Stone, 'Aeroplanes', *The Nineteenth Century and After*, LXXXVIII (July, 1920), p. 151.
92. Fredette, *op. cit.*, p. 26. In his first book, Pemberton Billing wrote that the only way to defeat the airships was 'to trap them in their nests'. *Op. cit.*, p. 15.
93. Pemberton Billing is one among many who illustrates the tendency (mentioned in this study's 'Preface') to see defensive and offensive air action in one total context. Therefore, to say that his platform stressed air *defence* is not to restrict his aerial doctrine *only* to defence, narrowly defined.
94. Lloyd George made this conclusion about Pemberton Billing's victory. *Op. cit.*, p. 106.
95. The comments in this paragraph are taken from news reports by the London correspondent of *The New York Times*, who was fascinated by Pemberton Billing and (as he viewed it) his urban American style of

campaigning. Cf. *The New York Times*, 24, 26 January and 15 March 1916. The cover picture of Pemberton Billing's *Air War: How to Wage It* is a good example of the serious, yet dashing, image of the man.
96. *The New York Times*, 24 January 1916. The returns are given in the 26 January edition.
97. Frank Owen, *Tempestuous Journey: Lloyd George, His Life and Times* (London: Hutchinson, 1954), p. 485.
98. *The Times*, 19 February 1916.
99. *The Times*, 7 March 1916. For other information in this paragraph see also the editions of 23 and 28 February and 11 March 1916.
100. *The Times* was obviously using the term 'passive' as a bitter complaint and not in its more modern, technical sense. Also, this editorial is just one more example where defence concerns quickly blend with offensive doctrine.
101. *The Times*, 11 March 1916. In a separate column that day entitled 'Political Notes', *The Times* declared that the election was a great upset; that no political pundit had thought a coalition candidate could be defeated by an independent without strong local ties.
102. Fredette, *loc. cit.*
103. 80 H.C. Deb. 5s., columns 1965-1969.
104. Macmillan, *Sir Sefton Brancker*, p. 123.
105. Churchill described these experiences in the section 'In the Air' in Winston S. Churchill, *Amid These Storms: Thoughts and Adventures* (New York: Charles Scribner's Sons, 1932) pp. 181-98.
106. H.A. Taylor, *Jix, Viscount Brentford: Being the Authoritative and Official Biography of the Rt. Hon. William Joynson-Hicks, First Viscount Brentford of Newick* (London: Stanley Paul & Co., Ltd., 1933), pp. 124, 135.
107. W. Joynson-Hicks, *The Command of the Air or Prophecies Fulfilled: Being Speeches Delivered in the House of Commons* (London: Nisbet & Co. Ltd., 1916), pp. 112-13, 132, 135.
108. The first two volumes of *Hansard* covering the sessions immediately after Billing's victory sufficiently illustrate these points. Between 22 March and 1 June 1916, he spoke on nineteen different days. (81 H.C. Deb. 5s. and 82 H.C. Deb. 5s., *passim*). Usually he presented more than one question during the questioning period — on 6 April for an extreme example, he presented ten questions. (81 H.C. Deb. 5s., columns 1344-1347, 1369, and 1376-1377.) For just one example of the increasing irritation evidenced by the House, see 82 H.C. Deb. 5s., column 1567. And Pemberton Billing recognised the fact of this irritation in comments seen in 82 H.C. Deb. 5s., columns 1612 and 3049.
109. The Northcliffe press was the most outspoken among the dailies in advocating increased air power. Lord Northcliffe was also an active supporter of Pemberton Billing, both in his candidacies and afterwards, in his viewpoints. Cf. Pemberton Billing, *op. cit.*, p. 3 and Wrench, *op. cit.*, pp. 160-61.
110. This speech and meeting is reported at length in *The Times*, 28 March 1916.
111. A description of General Douhet's theories will be presented in Chapter Six. His writings started to appear in the early postwar era.
112. General Mitchell's theories will also be discussed in Chapter Six.
113. *The Times*, *loc. cit.* The only divergence between this resolution and the views of Pemberton Billing was the felt need to express these views 'respectfully'.

Notes 217

114. Wrench, *op. cit.,* pp. 141-2.
115. These articles are reprinted in R.D. Blumenfeld, *All in a Lifetime* (London: Ernest Benn Limited, 1931), *passim.*
116. Viscount Norwich, *Old Men Forget: The Autobiography of Duff Cooper* (London: Rupert Hart-Davis, 1953), pp. 49-50.

CHAPTER 2

1. This descriptive term is used by the English air historian Norman Macmillan in his *Tales of Two Air Wars* (London: G. Bell and Sons, Ltd., 1963), p. 66.
2. N. Pemberton-Billing, *Air War: How to Wage It* (London: Gale & Polden, Ltd., 1916), p. 15.
3. Director of the Air Division Sueter later gave a thorough account of his activities in aerial defence in World War I. See Rear-Admiral Murray F. Sueter, *Airmen or Noahs: Fair Play for our Airmen; the Great 'Neon' Air Myth Exposed* (London: Sir Issac Pitman & Sons, Ltd., 1928), *passim.* For his positive thinking about the forward defence policy, see *ibid.,* p. 14.
4. Major Raymond H. Fredette, *The Sky on Fire: The First Battle of Britain 1917-1918 and the Birth of the Royal Air Force* (New York: Holt, Rinehart and Winston, 1966), p. 11 and Captain S.W. Roskill (ed.), *Documents Relating to the Naval Air Service* Vol. I, *1908-1918* (London: The Navy Records Society, 1969), p. 180.
5. Arch Whitehouse, *The Years of the Sky Kings* (Garden City, New York: Doubleday & Co., Inc., 1959), p. 81.
6. A brief but useful chronology of the Zeppelin era is in George H. Quester, *Deterrence before Hiroshima: The Airpower Background of Modern Strategy* (New York: John Wiley & Sons, Inc., 1966), pp. 18-19.
7. As quoted in Sir Philip Joubert de la Ferté, *The Third Service: The Story Behind the Royal Air Force* (London: Thames and Hudson, 1955), pp. 32-3.
8. Winston S. Churchill, *The World Crisis,* I (New York: Charles Scribner's Sons, 1923), p. 345.
9. Whitehouse, *op. cit.,* p. 35.
10. Churchill, *op. cit.,* pp. 344-5.
11. One brief and knowledgeable account of the development of the tank is in B.H. Liddell Hart, *The Real War, 1914-1918* (Boston: Little, Brown & Co., 1930), pp. 250-5. Sueter, *op. cit.,* pp. 179-220, takes note of other claims to the parentage of the tank but makes a strong argument for the Naval Air Service's role. He also furnishes many photographs of the naval prototypes of the tank.
12. Churchill, *op. cit.,* p. 340.
13. The story of Linnarz's challenge and the Dunkirk unit's success can be found in Whitehouse, *op. cit.,* pp. 84-8. Mr Whitehouse dates that first raid on London ten days too early, however. 31 May is the correct date.
14. Macmillan, *op. cit.,* p. 67.
15. As was noted in Chapter 1, Noel Pemberton Billing was just such a spokesman. For one more reference where he clearly states this position that offensive air strikes against Zeppelin bases were 'the only really satisfactory method' of defence, see *The Times,* 13 April 1916.
16. See Chapter 1, pp. 17-18. Yet one more authoritative view which reaches this same conclusion about relative dangers to friend and foe is in Joubert de la

Ferté, *op. cit.*, p. 42.
17. Churchill, *op. cit.*, pp. 341, 343-4.
18. As quoted in W. Joynson-Hicks, *The Command of the Air or Prophecies Fulfilled: Being Speeches Delivered in the House of Commons* (London: Nisbet & Co., Ltd., 1916), p. 148.
19. Captain J.H. Boyd, 'AA Development in the Royal Engineers', *The Royal Engineers Journal*, LXIX (September, 1935), p. 373.
20. Ronald W. Clark, *The Rise of the Boffins* (London: Phoenix House, Ltd., 1962), p. 8. A 'boffin' is a scientist who works in conjunction with the armed forces. The term became commonly used in the late 1930s primarily with respect to the scientists associated with RAF work. *Ibid.*, pp. vii-viii.
21. These speeches are reprinted in A.V. Hill, *The Ethical Dilemma of Science and Other Writings* (London: The Scientific Book Guild, 1962), see especially pp. 92, 175, 265-6, 307-8.
22. *Ibid.*, pp. 92, 175, 308.
23. This is the phrase used by another famous 'boffin', Professor P.M.S. Blackett, in a paper written December 1941, to describe the unscientific reasoning common among so many military leaders. Cf. Ronald W. Clark, *op. cit.*, p. 215. The phrase became 'one of the more famous remarks of the war'. *Ibid.* If Mr. Clark overstates this 'fame', it nevertheless indicates how meaningful was the charge of falsely grounded emotionalism within military dogma.
24. One specific example is found in Sir Frederick Pile, *Ack-Ack: Britain's Defence against Air Attack during the Second World War* (London: George C. Harrap & Co., Ltd., 1949), p. 49. Clark makes the point into a blanket indictment against all British armed services during World War I. *Op. cit.*, p. 10.
25. Alfred Rawlinson, *The Defence of London: 1915-1918* 2nd. ed. (London: Andrew Melrose, Ltd., 1923), pp. 38-41.
26. Radar devices for directional gun-laying were not generally effective until 1942. They proved 'a godsend' for Britain in the winter of 1943, in this case the G.L.3 of Canadian development. Cf. Pile, *op. cit.*, pp. 1, 197.
27. Air Commodore L.E.O. Charlton, *War Over England* (London: Longmans, Green and Co., 1936), p. 45. A picture of the L.12 being towed into Ostend is included in Charlton's account. The usually reliable General Sir Frederick Pile is thus wrong in stating that the Dover AA fire had not succeeded in bringing the Zeppelin down. Pile, *op. cit.*, p. 46.
28. Major-General E. B. Ashmore, *Air Defence* (London: Longmans, Green & Co., 1929), pp. 21-2.
29. A rare statement of such confidence is in Claude Grahame-White and Harry Harper, 'Zeppelin Airships: Their Record in the War', *The Fortnightly Review* XCVIII, No. 585 (1 September, 1915), 542-56. These two early British aerial authorities were certainly being sensible in this article by attempting to downgrade the danger of the Zeppelins. But their confidence in the vulnerability of the airships to AA fire was based upon very little positive evidence. They do mention three probable 'kills' by Allied AA fire along the Western front. Far more reflective of the actual results of anti-aircraft fire up to that time was the conclusion of another authority in the same journal three months later. A.J. Liversedge, 'Possibilities of the Large Airship', *The Fortnightly Review*, XCVIII, No. 588 (1 December, 1915), 1079-1092. In this article the author prophesied that London could not be adequately protected by AA fire for years to come (p. 1089).

30. Pile, *op. cit.*, p. 47.
31. For technical data regarding World War I searchlights, see Commander R.H. Keate, 'Searchlights', *Journal of the Royal United Service Institution*, LXI (November 1916), 925-933. The quality of brightness lessened, of course, at higher altitudes. In the specific case of the 90-cm models, they produced fifty-three candlepower brilliance at 1,111 yards, only twelve candlepower strength at 2,222 yards and just five candlepower at their maximum height of 3,333 yards – and these brightnesses held true only under maximum weather conditions. For the English use of these lights in the Zeppelin era, see Pile, *op. cit.*, pp. 46-7, Ashmore, *op. cit.*, p. 15, and Boyd, *op. cit.*, p. 373.
32. Major-General E.B. Ashmore, 'Anti-Aircraft Defence', *Journal of the Royal United Service Institution*, LXXII (February, 1927), 8. This is a transcript of a speech given by General Ashmore who was then in charge of the London aerial defences.
33. Rawlinson, *op. cit.*, p. 55.
34. Sueter, *op. cit.*, pp. 56-7, 175. Rear-Admiral Sueter is one of those who stresses this advantage even though he earlier admits the gross inefficiency in bombsight techniques.
35. Ernest Wilhelm von Hoeppner, *Germany's War in the Air: A Restrospect on the Development and the Work of our Military Aviation Forces in the World War*, trans. J. Hawley Larnev (Leipsig: A.F. Koehler, 1921), pp. 40-41. This bibliographic information accurately restates the title page information but nevertheless the copy of this work consulted was obviously a typescript copy produced by the Air University, Maxwell Air Force Base, Montgomery, Alabama, where it is presently shelved.
36. 'To the infant invention of aviation, the War proved to be a forcing house of tropical intensity.' David Lloyd George, *War Memoirs*, Vol. IV, *1917* (Boston: Little, Brown, and Co., 1934), p. 103.
37. Churchill, *op. cit.*, p. 339.
38. Norman Macmillan, *Sir Sefton Brancker* (London: William Heinemann Ltd., 1935), p. 116, n. 1, and Ashmore, *Air Defence*, p. 16.
39. Sueter, *op. cit.*, p. 22. B.E. stands for Blériot Experimental, named after Louis Blériot, the first man who flew across the English Channel.
40. Macmillan, *Sir Sefton Brancker*, p. 28. Later knighted, Sir Geoffrey de Havilland owned his own aircraft works after World War I, and he and his company were to have a crucial role in developing and producing airplanes for British use in World War II. Cf. M.M. Postan, D. Hay, and J.D. Scott, *Design and Development of Weapons: Studies in Government and Industrial Organisation* (London: HMSO, 1964), Part I 'Aircraft', *passim*. This work is one of the Civil Series volumes in the *History of the Second World War* series edited by Sir Keith Hancock.
41. Macmillan, *Sir Sefton Brancker*, p. 47, n.1.
42. This specific specification requirement came from Sir Sefton Brancker, Deputy Director of Military Aeronautics in the War Office, who 'considered that an aeroplane should be able to fly straight and level without the intervention of the pilot . . .', Joubert de la Ferté, *op. cit.*, p. 21.
43. When the postwar era brought a renewed interest in the advantages of stability rather than those of manoeuvrability for purposes of dog-fighting, this particular feature of the B.E. 2C design was returned to in various peacetime models. *Ibid.*
44. These rockets were fitted into hollow tubes and were triggered electronically by an ignition switch in the cockpit. Whitehouse, *op. cit.*, p. 77. Other sources used for this description of the B.E.2C are Joubert de la Ferté,

op. cit., p. 54, Ashmore, *Air Defence*, pp. 16-17, and Captain Raymond H. Fredette, 'First Gothas over London: The Story of a Daring Raid and its Aftermath', *The Airpower Historian*, VIII (October, 1961), p. 202, and Fredette, *The Sky on Fire*, p. 64.

45. Macmillan, *Sir Sefton Brancker*, pp. 73 and 110.
46. *Ibid.*, p. 61.
47. 59 H.C. Deb. 5s., Column 1913. Winston Churchill first used this 'hornet' metaphor in a speech at the Mansion House banquet of 5 May 1913. Robin Higham, *The British Rigid Airship, 1908-1931: A Study in Weapons Policy* (London: G.T. Foulis & Co., Ltd., 1961), p. 72, n.1.
48. The 'often-quoted simile of the hornets' (Churchill, *loc. cit.*) was not always cordially remembered. In a speech to the House of Commons on 11 November 1915, William Joynson-Hicks (later Viscount Brentford of Newick) attacked Churchill by reiterating that phrase 'swarm of hornets' while questioning their existence. He pointed out that the Zeppelins were attacking with virtually no effective opposition. This speech is reprinted in Joynson-Hicks, *op. cit.*, pp. 118-35.
49. Whitehouse, *op. cit.*, pp. 88-92. Warneford died in a freak accident just a few days after his triumph. See also Sueter, *op. cit.*, pp. 13-15 for other details on Flight Sub-Lieutenant Warneford including the text of the Royal message conferring the Victoria Cross. Warneford's official report of this event is reprinted in Roskill, *op. cit.*, pp. 206-7.
50. Churchill, *op. cit.*, p. 344.
51. Ashmore, *Air Defence*, p. 16.
52. Admiral Sir Percy Scott, *Fifty Years in the Royal Navy* (New York: George H. Doran Company, 1919), p. 292.
53. A good study of the development of the various incendiary bullets is in Arch Whitehouse, *Decisive Air Battles of the First World War* (New York: Duell, Sloan and Pearce, 1963), pp. 44-7.
54. Douglas H. Robinson, 'The Zeppelin Bomber: High Policy Guided by Wishful Thinking', *The Airpower Historian*, VIII (July, 1961), p. 142.
55. *Ibid.*
56. For an excellent summary of the principles of active air defence, see Major General Frederic H. Smith, Jr., 'Current Practice in Air Defense: Part I; Principles and Problems', *Air University Quarterly Review*, VI (Spring, 1953), 2-13.
57. Ashmore, *Air Defence*, p. 5.
58. Pile, *op. cit.*, p. 46.
59. Hoeppner, *op. cit.*, pp. 26, 47, 49.
60. See Chapter 1, pp. 00-00 for a more detailed description of this early system.
61. Ashmore, *Air Defence*, pp. 18-19 and for a complementary account, also consult his 'Anti-Aircraft Defence', p. 9.
62. Ashmore, *Air Defence*, p. 19. Another good brief account is in Terence H. O'Brien, *Civil Defence* (London: HMSO, 1955), pp. 8-9. O'Brien claims there were eight such divisions but one may presume Ashmore is correct in that he took over command of this whole system in 1917.
63. Rawlinson, *op. cit.*, pp. 173-6.
64. An excellent account of the main operations centre of World War II, the Cabinet War Room, is in James Leasor, *The Clock With Four Hands: Based on the Experiences of General Sir Leslie Hollis* (New York: Raynal & Co., 1959), *passim*.
65. The first Defence of the Realm Act is printed in full in Arthur Marwick, *The Deluge: British Society and the First World War* (Boston: Little, Brown and Company, 1965), pp. 36-7.

Notes 221

66. As well as for other purposes of course. My comments here are limited to the concerns of this study.
67. Rees Jenkins, 'Civil Aspects of Air Defence', *Journal of the Royal United Service Institution*, LXXII (August, 1927), 533-4. Mr Jenkins was an authority with established governmental connections. He was, for example, the British appointee on the Allied committee to oversee the destruction of the *Luftwaffe* equipment according to the Versailles Treaty specifications (from 1918 to 1921). *Ibid.*, p. 521.
68. Ashmore, *Air Defence*, p. 6. A good source for the official provisions and their intentions is O'Brien, *op. cit.*, pp. 7-8.
69. Charlton, *op. cit.*, pp. 7-10.
70. *Ibid.*, p. 10 and Ashmore, *Air Defence*, p. 6. There was, also, an attempt to dim sound as well as light. Clock chimes were banned early in 1916. Marwick, *op. cit.*, p. 137. One may well assume that clock chimes would furnish little help to Zeppelin commanders flying at their high altitudes, but this ban stayed in force until the war was over. R.D. Blumenfeld, *All in a Lifetime* (London: Ernest Benn Limited, 1931), p. 116.
71. Marwick, *loc. cit.*
72. Report of the Voluntary Patrollers as quoted in David Mitchell, *Monstrous Regiment: The Story of the Women of the First World War* (New York: The Macmillan Company, 1965), p. 213. The Voluntary Patrollers were women recruited to aid police in street patrol duty, primarily to minimise acts of public immorality. *Ibid.*
73. O'Brien, *op. cit.*, p. 8.
74. Air Vice-Marshal E.J. Kingston-McCloughry, *Defence Policy and Strategy* (New York: Frederick A. Praeger, 1960), p. 208. This comment is an excellent example of the tendency among British air authorities, oriented as they usually are to an offensive view of aerial defence, to confuse *home* defence with *passive* defence. That is, by 1915 (when the decision to transfer the entire dim-out task to the Home Office was made), many lessons had been learned with respect to *active* home defence, but this was an early lesson – if not the first – in *passive*, civil defence.
75. Pile, *op. cit.*, pp. 46-7.
76. O'Brien, *op. cit.*, pp. 8-9, Marwick, *op.cit.*, p. 137 and Charlton, *op. cit.*, p. 11.
77. Marwick, *op. cit.*, p. 163, states the date to be 18 June 1915. He obviously confused the date of the newspaper report with the date of the issuance of the public notice. Marwick also gives no details about these instructions, thereby ignoring some interesting sociomilitary history.
78. As quoted in *The New York Times*, 18 June 1915. The details about these instructions by Scotland Yard are given in this news report.
79. Liddell Hart, *op. cit.*, pp. 129-30.
80. Augustin M. Prentiss, *Chemicals in War: A Treatise on Chemical Warfare* (New York: McGraw-Hill Book Company, Inc., 1937), p. 515.
81. O'Brien, *op. cit.*, p. 232. For the interwar years of concern, *ibid.*, *passim*.
82. Fredette, *The Sky on Fire*, p. 160 states that 70 per cent of the bombs that year were incendiaries. The very first use ever of aerial incendiary bombs was during the first scheduled Zeppelin raid on London (31 May 1915). Prentiss, *op. cit.*, p. 251.
83. Fredette, *The Sky on Fire*, p. 161.
84. C.W. Glover, *Civil Defence: A Practical Manual Presenting with Working Drawings the Methods Required for Adequate Protection Against Aerial Attack* (London: Chapman & Hall, Ltd., 1938), p. 6. This author, a former captain in the Royal Artillery Corps, received official support

from the Government in the preparation of this study (p. ix).
85. Fredette, *The Sky on Fire, loc. cit.*, and Glover, *op. cit.*, p. 6.
86. O'Brien, *op. cit.*, p. 276.
87. Glover, *loc. cit.*,
88. Cmd. 8250, *Reports from Commissioners*, II (1916).
89. Sir Norman Hill et al., *War and Insurance* (London: Oxford University Press, 1927), p. 57. This work, one among the British group of the *Economic and Social History of the World War* series sponsored by the Carnegie Endowment for International Peace, is an excellent source for this subject (pp. 55-97). All comments here are from this source or from the following official governmental papers. Cmd. 7997 (1915), 'Government War Risks Insurance Scheme' (Report of the Aircraft Insurance Committee), *Reports from Commissioners, 1914-1916*, Vol. XXXVII and 'War Risks (Insurance by Trustees) Act, 1916', *Bills Public, 1916*, Vol. II, Bill 13.
90. No doubt the decision to follow this World War I policy was aided by the fact it turned out to be profitable. Even though the first established rates were later halved (in March 1917) and though a further concession allowed free coverage for losses up to £500, the government received over £11,000,000 profit during the war from this insurance scheme. Fredette, *The Sky on Fire*, p. 169. The man chosen to chair the government's Aircraft Insurance Committee was the brilliant financier and later governor of the Bank of England, Montagu Norman. Andrew Boyle, *Montagu Norman: A Biography* (London: Cassell, 1967), p. 103. Presumably much credit for these profits can be attributed to his management.
91. Lloyd George, *op. cit.*, p. 106. Although this volume of Lloyd George's *War Memoirs* is entitled 1917, the author is here discussing the earliest reorganisation of aerial defence as it faced the Zeppelin threat.
92. Chapter 1, n.77 *et passim*.
93. Rawlinson, *op. cit.*, p. 9.
94. An excellent brief account of the maneuverings behind this change is in A.J.P. Taylor, *English History, 1914-1945* (New York: Oxford University Press, 1965), pp. 29-31.
95. Sueter, *op. cit.*, p. 174.
96. Joubert de la Ferté, *op. cit.*, p. 36.
97. Scott, *op. cit.*, pp. 292-3.
98. Sueter, *op. cit.*, pp. 163-4.
99. Macmillan, *Tales of Two Air Wars*, p. 46. War Committee discussions which led to this change in command are reprinted in Roskill, *op. cit.*, pp. 294-5.
100. Rawlinson, *op. cit.*, p. 57.
101. The emotions and history of this changeover are well told in Roy Jenkins, *Asquith* (London: Collins, 1964), pp. 382-8.
102. Scott, *op. cit.*, p. 299. The naval gunnery units set up by Scott remained as part of the defences. These units were thereafter under operational orders of the Army but were still supplied and paid by the Navy. Rawlinson, *loc. cit.*
103. This speech is quoted in full in Joynson-Hicks, *op. cit.*, pp. 138-59. The *Hansard* citation is 80 H.C. Deb. 5s., columns 81-95. For another account of Joynson-Hicks' views at this time, see H.A. Taylor, *Jix: Viscount Brentford; Being the Authoritative and Official Biography of the Rt. Hon. William Joynson-Hicks, first Viscount Brentford of Newick* (London: Stanley Paul & Co., Ltd., 1933), pp. 135-6.
104. 80 H.C. Deb. 5s., column 126.
105. H.J. Tennant, Under-Secretary of State for War, said 'the complete defence

Notes 223

of every part of the United Kingdom against the attack of long-range aircraft by any system of guns or purely defensive aircraft can never be complete [sic], and if carried out even partially would impose an intolerable strain on our resources', 80 H.C. Deb. 5s., column 104.

106. For the whole air defence debate of that day, see 80 H.C. Deb. 5s., columns 81-156.
107. Macmillan, *Sir Sefton Brancker*, pp. 113-15. The comments are by Sir Sefton himself as this work is an autobiography for the years through 1918.
108. *Ibid.*, pp. 78-9.
109. Besides the debate already cited which illustrates all these points, see e.g. *The Times*' editorials of 7 March and 10 March 1916.
110. Lloyd George, *op. cit.*, p. 106.
111. John Evelyn Wrench, *Struggle: 1914-1920* (London: Ivor Nicholson & Watson, 1935), p. 153. This on 5 February 1916.
112. Fredette, *The Sky on Fire*, p. 112. This in late 1915.
113. Lloyd George, *loc. cit.* For Lord Montagu's strong statements on air power, see Fredette, *The Sky on Fire*, p. 68. Even in 1911, he was predicting that air attacks 'would be more nerve-shattering and would do more to shake the confidence of a people than a definite threat on sea or land'. *Ibid.*
114. Lords Derby and Montagu both spoke in an air debate in the House of Lords on 23 May 1916, and were in agreement about the deficiencies of their charge in the Derby Committee. Cf. 22 H.C. Deb. 5s., columns 101-102 and 115-116. Other useful accounts of this committee are in Macmillan, *Sir Sefton Brancker*, pp. 116-117, and Lloyd George, *op. cit.*, pp. 106-7. The brief for the Derby Committee and correspondence showing the disenchantment soon felt by Lords Derby and Montagu are in Roskill, *op. cit.*, pp. 306-9, 315-6, 325-8.
115. Macmillan, *Sir Sefton Brancker*, p. 116, and Lloyd George, *op. cit.*, p. 106.
116. Macmillan, *Sir Sefton Brancker*, *loc. cit.*; the expression is General Brancker's who worked closely with this committee.
117. Lord Derby as quoted in Fredette, *The Sky on Fire*, p. 112. In a contemporary letter (23 June 1916) to his friend Lord Charles Montagu, Lord Derby said 'I only resigned after A.J.B. had told me he would resist any further powers given to me – after that I had no alternative.' As quoted in Randolph S. Churchill, *Lord Derby, 'King of Lancashire': The Official Life of Edward, Seventeenth Earl of Derby; 1865-1948*, (London: Heinemann, 1959), p. 212. A.J. Balfour was then First Lord of the Admiralty and he fully supported the naval obstinacy in aerial matters.
118. The present Earl of Derby (Edward Stanley) was seventeenth in line to hold that venerable title.
119. 'I accepted that post with a full knowledge that I was absolutely unacquainted with anything connected with the air . . .' 22 H.L. Deb. 5s., column 114.
120. 81 H.C. Deb. 5s., column 240. See also columns 368-9.
121. *The Times*, 13 April 1916. Foreshadowing Chamberlain's later uncertainties about aerial matters, he was somewhat ambiguous in this early stance. A letter from the Lord Mayor appeared in *The Times* two days later (5 April 1916) which stated that although he did believe that air supremacy was essential, he was not advocating just one solution (an air ministry) to obtain it.
122. A.J.P. Taylor, *op. cit.*, pp. 19, 204, 224. Interestingly, Curzon also was one of the political figures most disliked by Lord Derby whom he succeeded. Randolph S. Churchill, *op. cit.*, pp. 259-60.
123. Macmillan, *Sir Sefton Brancker*, pp. 125-6.

124. 22 H.L. Deb. 5s., columns 151-66. The trend is clearly seen here of the fusion of offensive and defensive air doctrines and *their* connection to the mission of an independent air service. This trend will culminate, of course, in the creation of the Royal Air Force.
125. Macmillan, *Sir Sefton Brancker*, p. 125.
126. A.J.P. Taylor, *op. cit.*, p. 204.
127. Macmillan, *Sir Sefton Brancker*, p. 126.
128. Andrew Boyle, *Trenchard* (London: Collins, 1962), pp. 176-7.
129. Not surprisingly, since the Board's decisions usually went against the stands taken by the naval representatives. Macmillan, *Sir Sefton Brancker*, p. 128. Curzon's first report of the Air Board's activities included bitter comments on the lack of Admiralty cooperation; Roskill, *op.cit.*, pp. 395-7.
130. As quoted in Lloyd George, *op. cit.*, p. 109.
131. Blanche E.C. Dugdale, *Arthur James Balfour: First Earl of Balfour, K.G., O.M., F.R.S., etc.*, Vol. II, *1906-1930* (New York: G.P. Putnam's Sons, 1937), p. 112.
132. Lloyd George, *op. cit.*, p. 108.
133. As quoted in Dugdale, *loc. cit.*
134. These are graphically described by Sir Sefton in Macmillan, *Sir Sefton Brancker*, pp. 128-37.
135. David Lloyd George, *War Memoirs*, Vol. II, *1915-1916* (Boston: Little, Brown, and Company, 1933), pp. 381-4. Naturally, Lloyd George would wish to emphasise the impotence prevalent at that time.
136. Boyle, *op. cit.*, p. 205.
137. Macmillan, *Sir Sefton Brancker*, p. 138.
138. Jenkins, *op. cit.*, pp. 421-63.
139. *Ibid.*, pp. 453-4.
140. Lord Riddell, *War Diary, 1914-1918* (London: Ivor Nicholson & Watson Ltd., 1933), p. 254.
141. Fredette, *The Sky on Fire*, p. 114.
142. Ashmore, *Air Defence*, p. 30.
143. Fredette, *The Sky on Fire*, *loc. cit.*
144. Ashmore, *Air Defence*, pp. 14-15.
145. Fredette, 'First Gothas over London', p. 202. By then there were over seventeen thousand people involved in home aerial defence in all its phases. Robinson, *op. cit.*, p. 147.
146. Both the Zeppelin and the Schütte-Lanz were rigid airships, but the Zeppelin's framework was of metal while the latter's was of wood. Georg W. Feuchter, *Geschichte des Luftkriegs: Entwicklung und Zukunft* (Bonn: Athenäum-Verlag, 1954), p. 18.
147. Whitehouse, *The Years of the Sky Kings*, p. 96.
148. Lieutenant W. Leefe-Robinson's report (the pilot concerned) is quoted in full in *ibid.*, pp. 94-5, plus an accompanying thorough account, pp. 93-6.
149. Sueter, *op. cit.*, pp. 18-21.
150. Wrench, *op. cit.*, p. 170.
151. As quoted in George Macaulay Trevelyan, *Grey of Fallodon: The Life and Letters of Sir Edward Grey, afterwards Viscount Grey of Fallodon* (Boston: Houghton Mifflin Company, 1937), p. 366.
152. Hoeppner, *op. cit.*, p. 66. This disillusionment in the Zeppelins was not evident earlier in the year. For example, the Kaiser was informed that a raid of 31 March 1916, 'was very successful'. This is from the diary note of Admiral von Müller. Walter Görlitz, (ed.), *The Kaiser and His Court: The Diaries, Note Books and Letters of Admiral Georg Alexander von Müller, Chief of the Naval Cabinet, 1914-1918* (New York: Harcourt, Brace &

World, Inc., 1959), p. 150.
153. Hoeppner, *loc. cit.*
154. Feuchter, *op. cit.*, p. 20.
155. Fredette, *The Sky on Fire*, pp. 161-2.
156. Winston S. Churchill, *The World Crisis*, II (New York: Charles Scribner's Sons, 1923), p. 51.
157. A recent example is in Arthur Marwick, *Britain in the Century of Total War: War, Peace and Social Change, 1900-1967* (Boston: Little, Brown and Company, 1968), p. 68.
158. *The Times*, 13 January 1919; Robinson, *op. cit.*, p. 147; and Hilton P. Goss, *Civilian Morale under Aerial Bombardment, 1914-1939* (Maxwell Air Force Base, Montgomery, Alabama: United States Air University Documentary Research Study No. 14, 1948), p. 11. Robinson and Goss differ widely in the estimated property loss from airship attack. Goss lists this figure at £1,527,985 while Robinson has it as £7,332,408 of which £4,080,523 was suffered in London alone.
159. W.O. Horsnaill, 'Air Navies of the Future', *The Fortnightly Review*, XCIX (June 1, 1916), 1056-1063.

CHAPTER 3

1. An important recent monograph of this phase of World War I history is Major Raymond H. Fredette, *The Sky on Fire: The First Battle of Britain 1917-1918 and the Birth of the Royal Air Force* (New York: Holt, Rinehart and Winston, 1966). The official British history is also an excellent source of information. Sir Walter Alexander Raleigh and H.A. Jones, *The War in the Air: Being the Story of the Part Played in the Great War by the Royal Air Force* (6 vols.; Oxford: The Clarendon Press, 1922-37).
2. Fredette, *op. cit.*, pp. 53-61, and Jones, *op. cit.*, Vol. V (1935), pp. 26-7. Figures on war losses are always open to question and sources almost always can be found which disagree. With one exception I have used the statistics given in the above two works which are carefully researched studies. But, for example, Sir Sefton Brancker, a leading administrative figure in the British Air Service, remembered the casualty figure for this raid as being only between four and five hundred while Sir Basil Thomson, of Scotland Yard, recorded in his diary on that day a death toll of ninety-seven. Norman Macmillan, *Sir Sefton Brancker* (London: William Heinemann Ltd., 1935), p. 153, and Sir Basil Thomson, *The Scene Changes* (Garden City, New York: Doubleday, Doran & Company, Inc., 1937), p. 375. The one exception mentioned was with respect to the casualty count in the school tragedy where I have used the totals stated by Mr Herbert Fisher, then President of the Board of Education, in his official report to the House of Commons. 95 H.C. Deb. 5s., Cols. 15-16.
3. See Chapter 1, p. 11.
4. Fredette, *op. cit.*, pp. 36-7.
5. Marcel Poëte, 'La Physionomie de Paris pendant la Guerre' in Henri Sellier, A. Bruggeman and Marcel Poëte, *Paris pendant la Guerre* (Paris: Les Presses Universitaires, 1926), p. 82. The Gothas appeared over Paris in 1918 (p. 81). The descriptions of the G.IVs are taken from Fredette, *op. cit.*, pp. 38-9, Jones, *op. cit.*, pp. 19-20, and Major-General E.B. Ashmore, *Air Defence* (London: Longmans, Green & Co., 1929), p. 32.
6. Sir Llewellyn Woodward, *Great Britain and the War of 1914-1918* (London:

Methuen & Co. Ltd., 1967), p. 372.
7. *The Diary of Lord Bertie of Thame, 1914-1918*, Lady Algernon Gordon Lennox (ed.) (London: Hodder and Stoughton, Ltd., 1924) II, p. 72.
8. Thomson, *op. cit.*, p. 336. The target was, in fact, not Buckingham Palace but the Admiralty. Ashmore, *op. cit.*, p. 29.
9. *The Times* expressed editorial concern about the Londoners' passive reaction, even though, admittedly, the damage had been slight. 'We would only add a warning to our own public that, like all fresh portents of the kind, this isolated visit is by no means to be ignored . . . it is wise to regard it as the prelude to further visits of the kind on an extensive scale, and to lay our plans accordingly. We have always believed that the method of raiding by aeroplanes, which are relatively cheap and elusive, has far more dangerous possibilities than the large and costly Zeppelin.' *The Times*, 29 November 1916.
10. Fredette, *op. cit.*, pp. 19-24, 39, 44-8.
11. See Chapter 1, pp. 12-13.
12. As quoted in General Ludendorff, *The General Staff and Its Problems: The History of the Relations between the High Command and the German Imperial Government as Revealed by Official Documents*, trans. F. A. Holt (New York: E.P. Dutton and Company, n.d.), II, p. 452. Fritz Fischer quotes the last sentences of this same passage in quite a variant form of wording but the meaning remains unchanged. Fritz Fischer, *Germany's Aims in the First World War* (New York: W.W. Norton & Company, 1967), p. 393.
13. As quoted in Ludendorff, *op. cit.*, pp. 456-7.
14. Fredette, *op. cit.*, pp. 39-40.
15. Fischer, *op. cit.*, pp. 399-401.
16. *Ibid.*, p. 394.
17. As early as 1915 Rudyard Kipling published these lines in his poem 'The Beginnings'.
    It was not part of their blood,
        It came to them very late
    With long arrears to make good,
        When the English began to hate.
    Quoted in Bernard Bergonzi, *Heroes' Twilight: A Study of the Literature of the Great War* (New York: Coward-McCann, Inc., 1966), p. 138.
18. *The Times*, 14 June 1917.
19. A picture of such crowds is in Fredette, *op. cit.*, facing p. 72.
20. See, for example, the *Daily Mail*, 15 June and 16 June 1917. These pictures are still able to move the viewer; pictures of kindergarten-aged war casualties are particularly pathetic.
21. Cf. Chapter 2, p 39.
22. Sir Charles Webster and Noble Frankland, *The Strategic Air Offensive Against Germany, 1939-1945*, Vol. I: *Preparation* (London: HMSO, 1961), p. 35.
23. See 94 H.C. Deb. 5s., Cols. 1135-1136, 1283, 1286-1288, 1415, 1457-1458, 1609, 1787 and 1957-1958 and 95 H.C. Deb. 5s., Cols. 14, 25, 191-194, 373 and 513.
24. 94 H.C. Deb. 5s., Cols. 1135-1136.
25. 25 H.L. Deb. 5s., Col. 439.
26. Rear-Admiral Murray F. Sueter, *Airmen or Noahs: Fair Play for our Airmen; the Great 'Neon' Air Myth Exposed* (London: Sir Isaac Pitman & Sons, Ltd., 1928), pp. 237, 243.
27. 25 H.L. Deb. 5s., Col. 622.
28. Jones, *op. cit.*, p. 46.

Notes

29. Major Oliver Stewart, 'The Doctrine of Strategical Bombing', *Journal of the Royal United Service Institution*, LXXXI (February, 1936), p. 99.
30. 94 H.C. Deb. 5s., Cols. 1287-1288.
31. 95 H.C. Deb. 5s., Cols. 1627-1647.
32. 95 H.C. Deb. 5s., Cols. 1627-1628.
33. 95 H.C. Deb. 5s., Col. 1629.
34. 95 H.C. Deb. 5s., Cols. 1633-1634.
35. 95 H.C. Deb. 5s., Cols. 1640-1643.
36. 95 H.C. Deb. 5s., Col. 2102.
37. Terence H. O'Brien, *Civil Defence* (London: HMSO, 1955), p. 10.
38. *Ibid.*, and Jones, *op. cit.*, pp. 46-51.
39. Lady Cynthia Asquith, *Diaries, 1915-1918* (London: Hutchinson, 1968), p. 349. Lady Cynthia was the former Prime Minister's daughter-in-law.
40. Jones, *op. cit.*, p. 114.
41. 101 H.C. Deb. 5s., Col. 1436.
42. Jones, *op. cit.*, pp. 89-90, 135-6 and Fredette, *op. cit.*, pp. 143-4, 177.
43. Air Commodore L.E.O. Charlton, *War Over England* (London: Longmans, Green and Co., 1936), p. 12.
44. Frank Owen, *Tempestuous Journey: Lloyd George, His Life and Times* (London: Hutchinson, 1954), p. 409.
45. See Chapter 1, pp. 22-6.
46. 103 H.C. Deb. 5s., Cols. 908-909.
47. 103 H.C. Deb. 5s., Cols. 909-910.
48. Fredette, *op. cit.*, p. 144.
49. Lieut.-Col. C. à Court Repington, *The First World War, 1914-1918: Personal Experiences* (London: Constable & Co., Ltd., 1920) II, p. 233.
50. Fredette, *loc. cit.*
51. Arthur Marwick, *The Deluge: British Society and the First World War* (Boston: Little, Brown & Co., 1965), p. 198.
52. Squadron Leader B.E. Sutton, 'Some Aspects of the Work of the Royal Air Force with the BEF in 1918', *Journal of the Royal United Service Institution*, LXVII (May, 1922), 346.
53. A.J.P. Taylor, *English History, 1914-1945* (New York: Oxford University Press, 1965), p. 73, n. 4. Sir Basil Thomson of Scotland Yard, after telling another anecdote about Lloyd George's faint-heartedness, summed up with a like conclusion: 'The fact was that among his many conspicuous qualities was an exaggerated solicitude for the safety of his own skin.' Thomson, *op. cit.*, p. 430.
54. Claude Grahame-White and Harry Harper, *Air Power: Naval, Military, Commercial* (London: Chapman & Hall, Ltd., 1917), p. 40.
55. They were. A great many shelters for endangered factory workers had already been provided in Germany by mid-1916. Ernest Wilhelm von Hoeppner, *Germany's War in the Air: A Restrospect on the Development and the Work of our Military Aviation Forces in the World War*, trans. J. Hawley Larner (Leipsig: A.F. Koehler, 1921), p. 48.
56. Winston S. Churchill, *The World Crisis*, IV (New York: Charles Scribner's Sons, 1927), Appendix IV, 294-5.
57. See Chapter 1, pp. 21-2.
58. *Supra*, p. 58.
59. W.T. Wells, 'Air Raid Precautions and the Public', *Fortnightly Review*, CXLIII (February, 1938), p. 149.
60. David Lloyd George, *War Memoirs*, Vol. IV, *1917* (Boston: Little, Brown, & Co., 1934), p. 116.
61. Jones, *op. cit.*, p. 89.

62. Webster and Frankland, *op. cit.*, p. 45. This idea of the need for troops to maintain home morale in England during aerial attack is still being advocated in official circles. In 1960, for example, an Air Vice-Marshal wrote, 'the Army will play their greatest part in home defence in the event of severe air or rocket attack, whether nuclear or otherwise, by maintaining morale as well as supporting local authorities and essential public services and utilities'. E. J. Kingston-McCloughry, *Defense Policy and Strategy* (New York: Frederick A. Praeger, 1960), p. 90. This military authority displays another continuing trend in the interpretations of World War I aerial history when he writes, 'the chief lesson to be learned from the First World War was that direct air attack on the people caused great loss of morale, out of all proportion to the casualties and material damage' (p. 209).
63. As described in Chapter 1, Professor Stefan Possony has questioned this connection between morale and the will to carry on a war which is made by strategic bombing advocates. Stefan T. Possony, *Strategic Air Power: The Pattern of Dynamic Security* (Washington: Infantry Journal Press, 1949), p. 147 and see the discussion on this subject in Chapter 1, p. 14.
64. Herbert Henry Asquith, *Letters of the Earl of Oxford and Asquith to a Friend; First Series, 1915-1922,* Desmond MacCarthy (ed.) (London: Geoffrey Bles, 1933), p. 38.
65. *Supra,* pp. 60-1.
66. Field-Marshal Sir William Robertson, *From Private to Field-Marshal* (Boston: Houghton Mifflin Co., 1921), p. 349.
67. As quoted in Raymond Postgate and Aylmer Vallance, *England Goes to Press: The English People's Opinion on Foreign Affairs as Reflected in Their Newspapers Since Waterloo. (1815-1937)* (New York: The Bobbs-Merrill Company, 1937), p. 276.
68. Woodward, *op. cit.*, p. 377. Professor Woodward, however, represents these two trends as being in opposition to each other whereas they are not mutually exclusive categories. The scared man may well be a more angry man after his panic subsides.
69. Frank P. Chambers, *The War behind the War, 1914-1918: A History of the Political and Civilian Fronts* (London: Faber and Faber Ltd., 1939), p. 226.
70. Maurice Rickards, *Posters of the First World War* (New York: Walker and Co., 1968), plate 15.
71. 103 H.C. Deb. 5s., Col. 1016. The fear of possible reactions by the London poor was, of course, heightened by the recent events in Russia.
72. *The Times,* 14 June 1917.
73. 94 H.C. Deb. 5s., Cols. 1135 and 1283-1285.
74. This letter is quoted in Field-Marshal Sir William Robertson, *Soldiers and Statesmen. 1914-1918* (New York: Charles Scribner's Sons, 1926), pp. 11 and 16-17.
75. Lloyd George, *op. cit.*, p. 115.
76. Fredette, *op. cit.*, p. 66.
77. Cf. Chapter 2, pp. 32-3.
78. Fredette, *op. cit.*, p. 64.
79. 25 H.L. Deb. 5s., Col. 618.
80. As quoted in Jones, *op. cit.*, p. 25.
81. Lord Riddell, *War Diary, 1914-1918* (London: Ivor Nicholson & Watson Ltd., 1933), p. 254.
82. Woodward, *op. cit.*, p. 360.
83. Fredette, *op. cit.*, p. 64.

Notes

84. See Chapter 2, pp. 28-30.
85. Jones, *op. cit.*, p. 30.
86. *Ibid.*, p. 31. Haig was in London primarily for discussions on the Flanders campaign then in progress. Cf. Victor Bonham-Carter, *The Strategy of Victory, 1914-1918: The Life and Times of the Master Strategist of World War I; Field Marshal Sir William Robertson* (New York: Holt, Rinehart and Winston, 1964), pp. 259-63.
87. Andrew Boyle, *Trenchard* (London: Collins, 1962), pp. 221-3.
88. Jones, *loc. cit.*, and Fredette, *op. cit.*, p. 65.
89. Cyril Falls, *A Hundred Years of War, 1850-1950* (New York: Collier Books, 1962), pp. 221-2.
90. Cecil Lewis, *Sagittarius Rising* (New York: Harcourt, Brace and Co., 1936), pp. 165, 167.
91. Boyle, *op. cit.*, p. 223 and Jones, *op. cit.*, p. 32.
92. Fredette, *op. cit.*, p. 81. The raid is described in detail in this source, pp. 75-9.
93. Robertson, *Soldiers and Statesmen*, p. 17.
94. Woodward, *op. cit.*, p. 373.
95. Robertson, *Soldiers and Statesmen*, p. 18.
96. L. E.O. Charlton, 'The Great Unprepared', *The Fortnightly Review*, CXLIV (October, 1938), 426. Historians, of course, have even longer memories. George H. Quester, in a work published in 1966, also makes the point that home front morale was the cause for this removal of needed planes from the Western front. George H. Quester, *Deterrence before Hiroshima: The Airpower Background of Modern Strategy* (New York: John Wiley & Sons, Inc., 1966), p. 36.
97. Air Commodore H.R. Brooke-Popham, 'The Air Force', *Journal of the Royal United Service Institution*, LXV (February, 1920), 55.
98. *Supra*, p. 54.
99. *Supra*, p. 54-5.
100. Hoeppner, *op. cit.*, p. 78.
101. Georg W. Feuchter, *Geschichte des Luftkriegs: Entwicklung und Zukunft* (Bonn: Athenaüm-Verlag, 1954), pp. 20-21.
102. Stewart, *loc. cit.*, lists most of the specific figures in these regards.
103. Lieutenant-Colonel K.M. Loch, 'Anti-Aircraft Defence', *Journal of the Royal United Service Institution*, LXXXII (May, 1937), 311.
104. Lewis, *op. cit.*, pp. 180, 199 and Fredette, *op. cit.*, pp. 65, 90, 179.
105. *Ibid.*, pp. 47, 100, 108, 177, 182.
106. Air Chief Marshal Sir Philip Joubert de la Ferté, *The Third Service: The Story Behind the Royal Air Force* (London: Thames and Hudson, 1955), p. 47.
107. Ashmore, *op. cit.*, pp. 60, 88.
108. Lloyd George, *op. cit.*, p. 115.
109. Macmillan, *op. cit.*, pp. 112, n. 1, 142.
110. Major-General Sir F.H. Sykes, *Aviation in Peace and War* (London: Edward Arnold & Co., 1922), p. 83.
111. Ashmore, *op. cit.*, pp. 37-8.
112. Jones, *op. cit.*, pp. 79, 87.
113. Fredette, *op. cit.*, pp. 24-5, 212. The first full-scale test of the system was not made until mid-April 1918. Ashmore, *op. cit.*, p. 84.
114. *Ibid.*, p. 42.
115. Earl of Swinton (in collaboration with James D. Margach), *Sixty Years of Power: Some Memories of the Men Who Wielded It* (New York: James H. Heineman, Inc., 1966), p. 45.
116. W.K. Hancock, *Smuts*, Vol. 1: *The Sanguine Years, 1870-1919* (Cambridge:

117. Taylor, *op. cit.*, p. 82.
118. Jones, *op. cit.*, pp. 41-2.
119. Lloyd George, *op. cit.*, pp. 118-9.
120. As quoted in Woodward, *op. cit.*, p. 373.
121. Jones, *op. cit.*, p. 43 and Fredette, *op. cit.*, p. 89.
122. *Ibid.*
123. Alfred Rawlinson, *The Defence of London: 1915-1918* (2nd ed.; London: Andrew Melrose, Ltd., 1923), p. 225.
124. Joubert de la Ferté, *op. cit.*, p. 53.
125. General Sir Frederick Pile, *Ack-Ack: Britain's Defence against Air Attack during the Second World War* (London: George G. Harrap & Co., Ltd., 1949), pp. 48-9.
126. General Ashmore was not a man, it seems, to share credits very generously. In his descriptions of the organisation and changes involved in the LADA command, he gives no recognition to the guidelines set up by the Smuts report. Ashmore, *op. cit., passim.* On the other hand, if one depended upon the account of one eulogiser of Smuts, the impression gained would be that Smuts *personally* accomplished all of his suggestions, rather than Ashmore and his organisation. F.S. Crafford, *Jan Smuts: A Biography* (2nd ed.; London: George Allen & Unwin Ltd., 1946), p. 148.
127. Ashmore, *op. cit.*, p. 41. The idea of a zone defence had been proposed long before, in the earliest days of the war. Commodore (later, Admiral) Sueter had presented Churchill with such a scheme in September 1914, about which the First Lord commented, 'admirable, but beyond our resources'. Sueter, *op. cit.*, p. 165.
128. Jones, *op. cit.*, pp. 64, 89.
129. See Chapter I, p. 18.
130. Jones, *op. cit.*, pp. 84, 91, 104.
131. Ashmore, *op. cit.*, p. 54.
132. Pile, *op. cit.*, p. 49. General Pile was speaking of the Great War's barrage fire system in this passage.
133. Ashmore, *op. cit.*, p. 73.
134. Alfred F. Havighurst, *Twentieth-Century Britain* (Evanston, Ill.; Row, Peterson & Co., 1962), pp. 65-6.
135. As quoted in Hancock, *op. cit.*, p. 442.
136. Jones, *op. cit.*, p. 66
137. Hoeppner, *op. cit.*, p. 93.
138. Jones, *op. cit.*, p. 67.
139. Fredette, *op. cit.*, p. 126 and Ashmore, *op. cit.*, p. 55. A picture of this apron effect is in *ibid.*, frontispiece.
140. *Ibid.*, p. 56.
141. Jones, *op. cit.*, pp. 68-70.
142. Ashmore, *loc. cit.*
143. Flight-Lieut. P. Worthington, 'Work of the Kite Balloon on Land and Sea', *Journal of the Royal United Service Institution*, LXV (August, 1920), 488-489 and Air Commodore J.G. Hearson, 'Balloon Barrages', *Journal of the Royal United Service Institution*, LXXXIII (February, 1938), 90.
144. Fredette, *op. cit.*, pp. 182-3.
145. *Ibid.*, pp. 132-6 and Arch Whitehouse, *The Years of the Sky Kings* (Garden City, New York: Doubleday & Co., Inc., 1959), pp. 183-4. A popular but accurate and *well illustrated* account is in Anon., 'It's a Bird, It's a Plane, It's the Riesenflugzeug of W. W. I', *Esquire*, LXVI (October, 1966).
146. Captain J.H. Boyd, 'AA Development in the Royal Engineers', *The Royal*

Engineers Journal, XLIX (September, 1935), 378.
147. Ashmore, op. cit., pp. 84-5. This is another example of how the threat of strategic air bombardment can mean a displacement of war personnel to home front duties.
148. Ibid., p. 75 and Boyd, op. cit., pp. 378-9.
149. As quoted in Fredette, op. cit., p. 173.
150. Infra, p. 74.
151. Jones, op. cit., pp. 73-6. Fredette, op. cit., pp. 127-8 and Boyd, op. cit., p. 378. The school established by the Royal Engineers for instruction in searchlight techniques also included training in the operation of sound locators. Ibid.
152. Jones, op. cit., pp. 71-3 and Pile, op. cit., p. 49.
153. Cf. Chapter 2, pp. 36-7.
154. Jones, op. cit., p. 23.
155. Fredette, op. cit., p. 107.
156. Ashmore, op. cit., pp. 92-5 and 136. As cited in Chapter 2, an excellent description of the Cabinet War Room operations in World War II is given in James Leasor, *The Clock with Four Hands: Based on the Experiences of General Sir Leslie Hollis* (New York: Reynal & Co., 1959), *passim.*
157. Sir Almeric Fitzroy, *Memoirs*, 11 (5th ed.; London: Hutchinson & Co., n.d), pp. 661-2.
158. Fredette, op. cit., p. 190.
159. Ashmore, op. cit., p. 125.
160. Ibid., p. 129.
161. Rees Jenkins, 'Civil Aspects of Air Defence', *Journal of the Royal United Service Institution*, LXXII (August, 1927), 527-528.
162. Charlton, *War Over England*, pp. 17-18.
163. Lady Cynthia Asquith, op. cit., p. 351.
164. Poëte, op. cit., p. 83.
165. Fredette, op. cit., p. 266.
166. Major-General E.B. Ashmore, 'Anti-Aircraft Defence', *Journal of the Royal United Service Institution*, LXXII (February, 1927), 2.
167. Fredette, op. cit., pp. 184, 215, 218.
168. Ibid., p. 8.

NOTES TO CHAPTER 4

1. See Chapter 3, p. 55.
2. As quoted in Charles Andler, *'Frightfulness' in Theory and Practice as Compared with Franco-British War Usages*, trans. Bernard Miall (London: T. Fisher Unwin, Ltd., 1916), pp. 173-4.
3. This felicitous phrase is attributed to George Wyndham in Winston S. Churchill, *The World Crisis*, I (New York: Charles Scribner's Sons, 1923), p. 508.
4. Ian Colvin, *The Life of Lord Carson*, III (New York: The Macmillan Company, 1937), p. 144. For Pemberton Billing, See Chapter 1, pp. 22-7.
5. *The Times*, 19 February 1916.
6. N. Pemberton Billing, *Air War: How to Wage It* (London: Gale & Polden, Ltd., 1916), p. 17. He further wrote that he was not advocating killing women and children as such but that Englishmen must kill those who sought to kill them. He said that this might well involve the death of German women and children but it matters not', the purpose for the raids would be much more important. (p. 18).

7. 80 H.C. Deb. 5s., Col. 1967.
8. H.A. Taylor, *Jix, Viscount Brentford: Being the Authoritative and Official Biography of the Rt. Hon. William Joynson-Hicks, First Viscount Brentford of Newick* (London: Stanley Paul & Co., Ltd., 1933), p. 135.
9. The speeches are reprinted in W. Joynson-Hicks, *The Command of the Air or Prophecies Fulfilled: Being Speeches Delivered in the House of Commons* (London: Nisbet & Co., Ltd., 1916), pp. 101-35. The quoted passage is from p. 135.
10. An excellent, although mordant, study of Joynson-Hicks is 'The Salutary Tale of Jix' by Ronald Blythe in his *The Age of Illusion: England in the Twenties and Thirties, 1919-1940* (London: Penguin Books, 1964), pp. 24-54.
11. The article, 'Aeroplanes in War', originally appeared in *English Review* and is reprinted in Joynson-Hicks, *op. cit.*, pp. 161-84. The quoted passage is from p. 184.
12. Andler, *op. cit., passim.* This charge against the Germans is still being repeated; in a recent study of the air war waged against England in World War I, the author states, 'the Prussian military mind was nurtured on the Clausewitzian dictum of "unlimited war". The bombing of cities was clearly in keeping with the Teutonic idea of Schrecklichkeit [*sic*] which held that acts of "frightfulness" served to paralyse an enemy's will to resist.' Major Raymond H. Fredette, *The Sky on Fire: The First Battle of Britain 1917-1918 and the Birth of the Royal Air Force* (New York: Holt, Rinehart and Winston, 1966), p. 29.
13. 80 H.C. Deb. 5s., Col. 131. Mr Cecil Harmsworth (later Lord Harmsworth) was the brother of the press lord Alfred Harmsworth, Lord Northcliffe.
14. H.F. Wyatt, 'Air Raids and the New War', *The Nineteenth Century and After*, LXXXII (July, 1917), 33-34.
15. The importance of Lanchester in the early formulation of aerial doctrine is described by Robin Higham in his *The Military Intellectuals in Britain: 1918-1939* (New Brunswick, New Jersey: Rutgers University Press, 1966), pp. 126-32.
16. As quoted in *ibid.*, p. 131.
17. One of the few English churchmen who actively fought the policy of strategic bombing in World War II, George Bell, Bishop of Chichester, recalled the efforts of Archbishop Davidson in a message to his communicants in 1939. Ronald C.D. Jasper, *George Bell, Bishop of Chichester* (London: Oxford University Press, 1967), p. 261.
18. 24 H.L. Deb. 5s., Cols. 1013-1015. The entire debate on the motion is an illuminating illustration of the emotions and reasoning regarding reprisals as of that date. *Ibid.*, Cols. 1013-1038.
19. As reprinted in G.A. Kertesz, (ed.), *Documents in the Political History of the European Continent, 1815-1939* (Oxford: Clarendon Press, 1968), p. 216.
20. Sir Llewellyn Woodward, *Great Britain and the War of 1914-1918* (London: Methuen & Co., Ltd., 1967), p. 370.
21. As quoted in Sir Thomas Barclay, 'Aircraft Bombs and International Law', *The Nineteenth Century and After*, LXXVI (November, 1914), 1033.
22. Woodward, *loc. cit.*
23. Barclay, *loc. cit.*
24. Geoffrey Bruun and Victor S. Mamatey, *The World in the Twentieth Century* (5th ed.; Boston: D.C. Heath and Company, 1967), p. 217.
25. Barclay, *op. cit.*, p. 1038.
26. Walter Görlitz (ed.), *The Kaiser and His Court: The Diaries, Note Books and Letters of Admiral Georg Alexander von Müller, Chief of the Naval*

Cabinet, 1914-1918 (New York: Harcourt, Brace & World, Inc., 1959), p. 83.
27. *Ibid.*, p. 106.
28. Ernest Wilhelm von Hoeppner, *Germany's War in the Air: A Retrospect on the Development and the Work of our Military Aviation Forces in the World War*, trans. J. Hawley Larnev (Leipsig: A.F. Koehler, 1921), pp. 77-8, 103-4.
29. Georg W. Feuchter, *Geschichte des Luftkriegs: Entwicklung und Zukunft* (Bonn: Athenäum-Verlag, 1954), p. 20.
30. The reprisal question was debated in all major participating states. The statesman Jules Cambon of France is cited in this regard in a diary note (5 February 1916), of Lord Bertie of Thame. 'Jules Cambon is not in favour of reprisals on the Germans for their airship and submarine atrocities: he is for keeping our hands clean in order that the Neutrals may make comparisons beneficial to us between German conduct and ours. He would not object to a slight bombardment of a German town just in order to remind the Germans what *could* be done, but no systematic reprisals.' Lord Bertie of Thame, *Diary, 1914 1918*, Lady Algernon Gordon Lennox, (ed.) (London: Hodder and Stoughton, Ltd., 1924), I, p. 298. Lord Bertie was the English ambassador to France.
31. The issue is discussed for example in Brig. Gen. Dale O. Smith, 'The Morality of Retaliation', *Air University Quarterly Review*, VII (Winter, 1954-1955), 55-59. General Smith argues that retaliatory attacks are as moral and natural as attacking 'an armed maniac' who is ravaging one's family (p. 57.)
32. Norman Macmillan, *Sir Sefton Brancker* (London: William Heinemann, Ltd., 1935), p. 67 and Sir George Arthur, *Life of Lord Kitchener*, III (New York: The Macmillan Company, 1920), p. 74, n. 1.
33. The phrase is derived, of course, from the title of one of the most famous books — it was even a Book-of-the-Month Club selection — ever written on air power. Major Alexander P. de Seversky, *Victory Through Air Power* (New York: Simon and Schuster, 1942).
34. Joynson-Hicks, *op. cit.*, p. 135.
35. A.J. Liversedge, 'Possibilities of the Large Airship', *The Fortnightly Review*, XCVIII (December 1, 1915), 1089.
36. His speech of 16 February, 1916. offers one example. 80 H.C. Deb. 5s., Col. 153,
37. Liversedge, *op. cit.*, p. 1087, Joynson-Hicks, *op. cit.*, pp. 114, 135, 80 H.C. Deb. 5s., Col. 92 and T.F. Farnum, 'Aeroplanes In War', *Blackwood's Magazine*, CC (December, 1916), 788-795.
38. Peter Lewis, *The British Bomber Since 1914: Fifty Years of Design and Development* (London: Putnam, 1967), p. 41. Regarding that 'Forward Policy', see Chapter 2, pp. 28-30.
39. Macmillan, *op. cit.*, pp. 103-4.
40. Churchill, *op. cit.*, II (New York: Charles Scribner's Sons, 1923), pp. 562-3, 559.
41. Lewis, *op. cit.*, pp. 52-3, 59.
42. Macmillan, *op. cit.*, pp. 104, 128, and Andrew Boyle, *Trenchard* (London: Collins, 1962), p. 208. Relevant documents are in Captain S.W. Roskill (ed.), *Documents Relating to the Naval Air Service, Vol. I: 1908-1918* (London: The Navy Records Society, 1969), pp. 332-6.
43. One of these raids on Karlsruhe resulted in one of the worst civilian tragedies suffered by Germany from air attack during the war. On Corpus Christi day 1916, bombs fell upon a crowd watching a circus and left 110 dead,

mostly children, and 123 injured. Concerning that raid, the commanding officer of the German air forces later wrote, 'it was more and more evident that our enemies were trying through aerial bombardment to reach not only their military objectives but the peaceful civilian populations as well – France gave the example for such disgraceful acts.' Hoeppner, *op. cit.*, p. 49. It should be pointed out that Germany – some time before – had already attacked Paris and London by air.

44. Cap-Commandant aviateur Desmet, 'Rôle des aviations belge et française: sur le front occidental pendant la grande guerre', *Bulletin Belge des Sciences Militaires*, III (January, 1922), 349-350 and George H. Quester, *Deterrence before Hiroshima: The Airpower Background of Modern Strategy* (New York: John Wiley & Sons, Inc., 1966), p. 29.
45. A glance at a map will show the reader that Belfort, Luxeuil, and Nancy are all in the same general area of France.
46. Woodward, *op. cit.*, p. 375 Macmillan, *op. cit.*, p. 135, Air Chief Marshal Sir Philip Joubert de la Ferté, *The Third Service: The Story behind the Royal Air Force* (London: Thames and Hudson, 1955), p. 38 and Roskill, *op. cit.*, pp. 386-7, 408-9, 411-412.
47. Cf. Chapter 2, pp. 46-8.
48. Boyle, *op. cit.*, pp. 204-6 and see my discussion on this, viewed in a broader context, in Chapter 2, pp. 47-8.
49. Cf. Chapter 2, p. 48.
50. Boyle, *op. cit.*, p. 208 and Higham *op. cit.*, p. 145.
51. Macmillan, *op. cit.*, p. 135 and Roskill, *op. cit.*, p. 476.
52. Air Marshal Sir George H. Mills, 'Bomber Command of the Royal Air Force', *Air University Quarterly Review*, VII, (Spring, 1955), 38.
53. Edward Mead Earle, 'The Influence of Air Power upon History', *The Yale Review*, XXV (June, 1946), 582.
54. *Daily Mail*, 15 June 1917.
55. For representative examples of these articles, see *ibid.*, June 14, 15, 16, 19, 20, July 4, 5, and 9, 1917.
56. Editorials in *The Times* of 14 June, 9 July, 26, 27 September, and 1 October 19
57. *Ibid.*, 15 June 1917. Note just how powerful Joynson-Hicks presumed air power to be; that it could 'blot out' a town. Another report on his aerial views is in *ibid.*, 4 October 1917.
58. An important aristocratic spokesman on aerial matters; cf. Chapter 2, p. 44.
59. *The Times*, 15 June, 3, 9, 24 and 27 July 1917.
60. See e.g. *ibid.*, 15 and 16 June 1917 and the *Daily Mail*, 16 June 1917.
61. *The Times*, 15 October 1917. On the following day, *The Times* published a letter from W.B. Selbie of Mansfield College who stated that the Free Churches joined in the sentiments expressed by Bishops Oxon and Ely and were 'strongly against any policy of making war on noncombatants for purposes of vengeance or terrorism'.
62. *Ibid.*, 15 and 18 June 1917 and the *Daily Mail*, 18 June 1917.
63. *The Times*, 19 and 22 June, 1917.
64. 94 H.C. Deb. 5s., Cols. 127-136, 1321-1324 and 1457-1458 and 95 H.C. Deb. 5s., Cols. 15 and 365-366 offer typical examples.
65. 94 H.C. Deb. 5s., Col. 1285.
66. For Lowther, see e.g. 94 H.C. Deb. 5s., Cols. 1787-1788 and 1995 and 95 H.C. Deb. 5s., Col. 15.
67. Arthur J. Marder (ed.), *Fear God and Dread Nought: The Correspondence of Admiral of the Fleet Lord Fisher of Kilverstone*, Vol. III: *Restoration, Abdication, and Last Years, 1914-1920* (London: Jonathan Cape, 1959), p. 464. Included in the 58th Psalm, of course, is the following thought:

'The righteous shall rejoice when he seeth the vengeance: he shall wash his feet in the blood of the wicked.'
68. Wyatt, *op. cit.*, followed by his 'Convocation versus the Church and the Bible', *The Nineteenth Century and After*, LXXXII (August, 1917), and 'The Air War and the Bishops' Religion', *The Nineteenth Century and After*, LXXXII (November, 1917).
69. Wyatt, 'Air Raids and the New War', p. 32.
70. See Chapter 1, p. 27.
71. R.D. Blumenfeld, *All in a Lifetime* (London: Ernest Benn Limited, 1931), pp. 85-6.
72. Field-Marshal Sir William Robertson, *Soldiers and Statesmen, 1914-1918*, II (New York: Charles Scribner's Sons, 1926), p. 16. Trenchard, the R F C commander in France, wrote this paper at the request of Haig. Cf. Chapter 3, p. 64.
73. Students of the RAF history are apt to go even further with regards to Trenchard's influence. E.g. Derek Wood and Derek Dempster, *The Battle of Britain and the Rise of Air Power 1930-40*, 2d ed. (London: Hutchinson, 1963), p. 66 wherein it is reported that 'Trenchard ... provided the entire basis on which the Air Force was to develop ... '. This study will indicate how much broader that 'basis' was.
74. Boyle, *op. cit.*, p. 222.
75. *Ibid.* Haig and Trenchard, along with Trenchard's aide (and famous litterateur) Maurice Baring, were in London for these talks between 17 June and 26 June 1917. Maurice Baring, *Flying Corps Headquarters, 1914-1918* (rev. ed.; London: William Heinemann, Ltd., 1930), pp. 228-9.
76. H.A. Jones, *The War in the Air: Being the Story of the Part Played in the Great War by the Royal Air Force*, IV (Oxford: The Clarendon Press, 1934), p. 153 n. 1, and V (Oxford: The Clarendon Press, 1935), p. 29.
77. *Ibid.*
78. See Chapter 3, p. 63.
79. Macmillan, *op. cit.*, p. 150 and Air Commodore J.A. Chamier, *The Birth of the Royal Air Force: The Early History and Experiences of the Flying Services* (London: Sir Isaac Pitman & Sons, Ltd., 1943), p. 172.
80. In a memorandum to the War Cabinet, reprinted in Churchill, *op. cit.*, IV (New York: Charles Scribner's Sons, 1927), pp. 293-4.
81. The phrase itself, of course, comes from a much later date, being used in President Roosevelt's 'fireside' address of 29 December 1940. Cordell Hull, *Memoirs* (New York: The Macmillan Company, 1948), I, p. 885.
82. James J. Hudson, *Hostile Skies: A Combat History of the American Air Service in World War I* (Syracuse, New York: Syracuse University Press, 1968), p. 4.
83. *Ibid.*, pp. 4-5 and I.B. Holley, Jr., *Ideas and Weapons: Exploitation of the Aerial Weapon by the United States during World War I; A Study in the Relationship of Technological Advance, Military Doctrine, and the Development of Weapons* (New Haven, Connecticut: Yale University Press, 1953), pp. 41-5.
84. The *Daily Mail*, 16 June 1917, T.F. Farman, 'The American Aviation Program', *The Contemporary Review*, CXII (September, 1917), 328-329, and 27 H.L. Deb. 5s., Col. 3.
85. Daniel R. Beaver, *Newton D. Baker and the American War Effort: 1917-1919* (Lincoln, Nebraska: University of Nebraska Press, 1966), p. 57.
86. *The New York Times*, 21 May and 10 and 15 June 1917.
87. As cited in Benjamin D. Foulois and Col. C.V. Glines, *From the Wright Brothers to the Astronauts: The Memoirs of Major General Benjamin D.*

Foulois (New York: McGraw-Hill Book Company, 1968), p. 150.
88. As quoted in Charles J. Kelly, Jr., *The Sky's the Limit: The History of the Airlines* (New York: Coward-McCann, Inc., 1963), p. 27.
89. *The New York Times*, 14 and 16 June and 7, 14, 22 and 25 July 1917.
90. Hoeppner, *op. cit.*, pp. 95-6.
91. Holley, *op. cit.*, pp. 118-9.
92. *Ibid.*, p. 96. Dr Holley adds various other contributing factors in his account, see pp. 68-123.
93. Kelly, *op. cit.*, p. 29 and Samuel Taylor Moore, *U.S. Airpower: The Story of American Fighting Planes and Missiles from Hydrogen Bags to Hydrogen War-heads* (New York: Greenberg, 1958), p. 49.
94. Hudson, *op. cit.*, p. 9.
95. Reginald Pound and Geoffrey Harmsworth, *Northcliffe* (London: Cassell, 1959), pp. 556-7, 562.
96. Macmillan, *op. cit.*, p. 152.
97. Jones, *op. cit.*, V, pp. 38-9 and David Lloyd George, *War Memoirs*, Vol. IV: *1917* (Boston: Little, Brown, and Company, 1934), pp. 116-7.
98. As quoted in Mills, *op. cit.*, p. 40. Since strategic bombing is not specifically after a *city's* surrender, this answer is certainly less than a completely satisfactory ruling. But it does usefully indicate a tie between war against civilians in its twentieth century phase and such earlier parallels as medieval siege warfare when the surrender of towns *was* a specific goal.
99. See Chapter 3, pp. 68-9.
100. See Chapter 3, p. 69.
101. Fredette, *op. cit.*, p. 111.
102. Eugene M. Emme (ed.), *The Impact of Air Power: National Security and World Politics* (Princeton, New Jersey: D. van Nostrand Company, Inc., 1959), p. 33.
103. The report is reprinted in full as Appendix I in Jones, *op. cit.*, VI (Oxford, The Clarendon Press, 1937), pp. 8-14. The following analysis of its content as well as all quoted passages is derived from this source.
104. This report is also available, with a few elisions, in Emme, *op. cit.*, pp. 33-7, a more easily attainable work than Jones' *The War in the Air*. A shorter analysis of the report is found in David Divine, *The Broken Wing: A Study in the British Exercise of Air Power* (London: Hutchinson, 1966), pp. 117-19, but some passages quoted there fill in gaps in Emme's rendition. Another complete reprinting is in Roskill, *op. cit.*, pp. 510-17.
105. Sir Arthur Harris, *Bomber Offensive* (New York: The Macmillan Company, 1947), p. 17.
106. Lord Tedder, *Air Power in War* (London: Hodder and Stoughton, 1948), pp. 21-2.
107. Sir Robert Saundby, *Air Bombardment: The Story of its Development* (London: Chatto & Windus, 1961), pp. 23-4.
108. Lord Trenchard, *Air Power: Three Papers by Marshal of the R.A.F. the Viscount Trenchard* (London: Air Ministry Pamphlet #229, 1946), p. 13.
109. Hans Rumpf, *The Bombing of Germany*, trans. Edward Fitzgerald (London: Frederick Muller Limited, 1961), p. 227.
110. Lieutenant-General E.L.M. Burns, *Megamurder* (New York: Pantheon Books, 1966), p. 19.
111. Sir Charles Webster and Noble Frankland, *The Strategic Air Offensive Against Germany, 1939-1945*, Vol. 1: *Preparation* (London: HMSO, 1961), p. 10.
112. Noble Frankland, *The Bombing Offensive Against Germany: Outlines and Perspectives* (London: Faber and Faber, 1965), p. 34. Frankland follows this comment with another presentation of that oft-quoted passage from

the report.
113. Divine, *op. cit.*, pp. 119-20.
114. W.K. Hancock, *Smuts*, Vol. I: *The Sanguine Years, 1870-1919* (Cambridge: The University Press, 1962), p. 438.
115. W.K. Hancock, *Four Studies of War and Peace in this Century* (Cambridge: The University Press, 1961), p. 48.
116. For just three examples among many that could be cited, see E. Colston Shepherd, *The Air Force of Today*, rev. by H.T. Winter (rev. ed.; London: Blackie & Son, Ltd., 1942), pp. 18-19, Chamier, *op. cit.*, p. 172 and Frankland, *op. cit.*, p.16.
117. Divine, *op. cit.*, pp. 32-4.
118. *Ibid.*, pp. 88-9 and also cf. Chapter 2, pp. 44-5.
119. Cf. Chapter 2, pp. 45-7.
120. Fredette, *op. cit.*, p. 114 and Christopher Addison, *Politics from Within, 1911-1918: Including Some Records of a Great National Effort* (London: Herbert Jenkins Ltd., 1924), pp. 11, 90, 168-9.
121. Woodward, *op. cit.*, p. 360 and Divine, *op. cit.*, p. 124.
122. *Supra*, p. 81.
123. Boyle, *op. cit.*, p. 208. Regarding Weir – who would later be the political chief of the RAF – also see Chapter 2, pp. 33-4 and Chapter 3, p. 67.
124. 26 H.L. Deb. 5s., Col. 1095.
125. 'It was the successful air raid of July 7, 1917, that crystallized the rather scattered discussions which had been proceeding for some time over the higher direction of our air forces.' Joubert de la Ferté, *op. cit.*, p. 48.
126. R.J. Minney, *No. 10 Downing Street: A House in History* (Boston: Little, Brown and Company, 1963), p. 358.
127. Webster and Frankland, *op. cit.*, p. 38.
128. Jones, *op. cit.*, VI, 13.
129. David Divine claims the Cabinet members were more panicked than the general public by the problem of the Gothas. Divine, *op. cit.*, pp. 105, 123.
130. Jones, *op. cit.*, VI, 1, and Frankland, *op. cit.*, p. 36. Captain B.H. Liddell Hart, *The Real War, 1914-1918* (Boston: Little, Brown & Co., 1930), pp. 388-9 gives a good account of the distrust between Haig and Lloyd George. C. à Court Repington, *The First World War, 1914-1918: Personal Experiences of Lieut.-Col. C. à Court Repington* (London: Constable and Co., Ltd., 1920), II, pp. 135-6 offers a contemporary account of the growing manpower crisis.
131. Jones, *op. cit.*, VI, p. 11.
132. Boyle, *op. cit.*, p. 230. The only Cabinet members actively pushing Smuts' programme by then were Curzon and Churchill. *Ibid.*
133. Jones, *op. cit.*, VI, pp. 13-19.
134. Robert Blake (ed.), *The Private Papers of Douglas Haig, 1914-1919: Being Selections from the Private Diary and Correspondence of Field-Marshal the Earl Haig of Bemersyde, K.T., G.C.B., O.M., etc.* (London: Eyre & Spottiswoode, 1952), p. 252.
135. Robertson, *op. cit.*, p. 17.
136. As evidenced in a diary note of 3 October 1917 in Lord Riddell, *War Diary, 1914-1918* (London: Ivor Nicholson & Watson, Ltd., 1933), p. 278.
137. Fredette, *op. cit.*, pp. 117-8. Note that Geddes was not conceding much with that position. At the same time, he was being advised by Lord Balfour, a former First Lord, not to let the Navy lose any of its prerogatives. Boyle, *op. cit.*, p. 231.
138. Mark Kerr, *Land, Sea, and Air: Reminiscences* (London: Longmans, Green

and Co., 1927), pp.289-91.
139. *Ibid.*, p. 291 and 98 H.C. Deb. 5s., Cols. 27-28.
140. See e.g. the *Daily Mail*, 6, 7, 8, 25 and 28 September 1917.
141. *The Times*, 26 and 27 September 1917.
142. *Ibid.*, 1 October 1917.
143. *Ibid.*, 4 and 5 October 1917.
144. Woodward, *op. cit.*, p. 364 and Boyle, *op. cit.*, p. 245.
145. *Parliamentary Papers (Accounts and Papers)*, XXXIII (1919), Cmd. 100, pp. 4-5.
146. 214 H.C. Deb. 5s., Col. 1543.
147. Anon., 'Les forces de l'air en Angleterre', *Bulletin Belge des Sciences Militaires*, 11 (September, 1921), 456-457.
148. *The Times*, 10 November 1917.
149. Noble Frankland is one historian who realised the significance of this point regarding the beginnings of the RAF. Frankland, *op. cit.*, p. 27.
150. Fredette, *op. cit.*, p. 202 and Blake, *op. cit.*, p. 273.
151. Fredette, *op. cit.*, p. 200.
152. This comment is from Fitzroy's diary note of 17 November 1917. Sir Almeric Fitzroy, *Memoirs*, 11 (5th ed.; London: Hutchinson & Co., n.d.), p. 664.
153. A.J.P. Taylor, *English History, 1914-1945* (Oxford: Oxford University Press, 1965), p. 27.
154. Lloyd George, *op. cit.*, p. 124.
155. Jones, *op. cit.*, VI, p. 21.
156. Divine, *op. cit.*, p. 125.
157. Fredette, *loc. cit.*
158. Addison, *op. cit.*, p. 168.
159. Good contrasting versions of this contretemps are in Lloyd George, *op. cit.*, pp. 124-8 and Pound and Harmsworth, *op. cit.*, pp. 592-4. Relevant correspondence by the three leading figures is in Roskill, *op. cit.*, pp. 582-5.
160. This second approach to a press lord indicates that Lloyd George was at least in part motivated by the wish to gain press support for the government's air policy. Cf. Colvin, *op. cit.*, p. 350.
161. Lord Northcliffe later claimed that the appointment resulted from his prompting. Bertie, *op. cit.*, 11, p. 269.
162. John Evelyn Wrench, *Struggle, 1914-1920* (London: Ivor Nicholson & Watson, Ltd., 1935), p. 283. The letter was dated 5 January 1918.
163. Blake, *op. cit.*, p. 280 and Boyle, *op. cit.*, pp. 250-53.
164. *Ibid.*, pp. 254, 261-2.
165. As quoted in Wrench, *op. cit.*, p. 278.
166. The passage comes from a letter from Rothermere to Bonar Law dated 3 May 1918, and quoted in Lord Beaverbrook, *Men and Power, 1916-1918* (New York: Duell, Sloan and Pearce, 1956), p. 236.
167. Divine, *op. cit.*, pp. 129-31.
168. Prince Albert (George VI to be) was in the RAF at the time. The exchange of comments between his father and himself on the matter are available in John W. Wheeler-Bennett, *King George VI: His Life and Reign* (New York: St. Martin's Press, 1958), pp. 110-12. Prince Albert's great admiration is also clearly stated in another letter of his quoted in *ibid.*, pp. 116-7.
169. Baring, *op. cit.*, p. 279.
170. Repington, *op. cit.*, p. 272.
171. 105 H.C. Deb. 5s., Cols. 1303-1370.

172. The sympathetic interpretation of Rothermere's resignation was that it was made necessary by ill health. Cf. Pound and Harmsworth, *op. cit.*, pp. 636-7. While it is true that the recent loss of his son, Vyvyan, certainly affected Rothermere's health, nevertheless, political pressure was predominantly the reason behind the resignation. Cf. Wrench, *op. cit.*, pp. 287, 295-6 and Repington, *op. cit.*, p. 283. Repington also notes that Air Force personnel waved newspapers cheering the announcement of Rothermere's departure (*ibid.*). As the Lloyd George Liberal, Christopher Addison, admits, Rothermere had been 'a conspicuous failure' as air minister. Addison, *op. cit.*, p. 169.
173. One who does say so is Viscount Simon in his memoirs. This famous 'appeaser' of later years was then an MP and also a member of the RAF serving as one of Trenchard's staff aides. Describing the emotions of the time and especially of the sympathy towards Trenchard shown in the Commons debate, Simon concluded that 'the pressure of opinion was such that the announcement was made that it had been decided to give him [Trenchard] a fresh and adequate post in the Air Service'. Viscount Simon, *Retrospect* (London: Hutchinson, 1952), pp. 115-9.
174. Baring, *op. cit.*, pp. 251-5, Fredette, *op. cit.*, pp. 151-2, 155-9, and Alexander McKee, *The Friendless Sky: The Story of Air Combat in World War I* (New York: William Morrow and Company, 1964), pp. 169-70. The 41st Bombing Wing was the first official designation of the Ochey unit.
175. Boyle, *op. cit.*, pp. 235-6, 239, Jones, *op. cit.*, V, p. 91, and Baring, *op. cit.*, p. 262.
176. Fredette, *op. cit.*, p. 222. Woodward, *op. cit.*, pp. 374-5, n. 1, and Jones, *op. cit.*, VI, p. 101. Of related interest, Emperor Charles of Austria had believed 'that he might promote the cause of peace by forbidding his troops to resort to offensive aerial bombardment ...'. Gordon A. Craig, 'The Military Cohesion of the Austro-German Alliance, 1914-1918', in his *War, Politics and Diplomacy: Selected Essays* (New York: Frederick A. Praeger, Publishers, 1966), p. 56.
177. Jones, *op. cit.*, VI, *pp. 102-3*.
178. Boyle, *op. cit.*, pp. 288-9, Jones, *op. cit.*, VI, pp. 103-4, and Fredette, *op. cit.*, pp. 222-3.
179. Air Marshal Sir Robert Saundby, 'The Air Battle', *Journal of the Royal United Service Institution*, XCVIII (February, 1953), 31.
180. Boyle, *op. cit.*, pp. 289, 295.
181. Major-General Sir F.H. Sykes, *Aviation in Peace and War* (London: Edward Arnold & Co., 1922), p. 88. Sykes replaced Trenchard as CAS.
182. The wording of the agreement signed at the Doullens Conference as cited in Cyril Falls, *The Great War* (New York: Capricorn Books, 1961), p. 335.
183. Jones, *op. cit.*, VI, pp. 106-11 and Fredette, *op. cit.*, pp. 224-5.
184. Jones, *op. cit.*, VI, p. 111.
185. Beaver, *op. cit.*, p. 169 and Alfred Francis Hurley, 'The Aeronautical Ideas of General William Mitchell' (unpublished Ph. D. dissertation, Dept. of History, Princeton University, 1961), p. 62. The reader will note that Baker used the wording of the Inter-Allied Force's official mandate when he described its unacceptable objectives.
186. Falls, *op. cit.*, p. 367. This capsule summary of the weather conditions is borne out in Maurice Baring's eyewitness account; Baring, *op. cit.*, p. 275.
187. Frankland, *op. cit.*, pp. 105-6, Sykes, *op. cit.*, p. 90. and Fredette, *op. cit.*, p. 226. The strong critic of the RAF, David Divine, points out that the cost of the IAF aircraft losses was more expensive than the damage inflicted on Germany. Divine, *op. cit.*, p. 143.

188. Shepherd, *op. cit.*, p. 64.
189. Jones, *op. cit.*, VI, p. 171 and Mills, *op. cit.*, p. 41.
190. Lewis, *op. cit.*, p. 96.
191. Holley, *op. cit.*, p. 138 and Saundby, *Air Bombardment*, p. 21.
192. Hilton P. Goss, *Civilian Morale under Aerial Bombardment, 1914-1939* (Maxwell Air Force Base, Alabama: Documentary Research Division, Air University Libraries, 1948), p. 38. n. 3 (this is the United States Air University Documentary Research Study No. 14), Quester, *op. cit.*, pp. 44-5, Seymour M. Hersh, *Chemical and Biological Warfare: America's Hidden Arsenal* (Indianapolis: The Bobbs-Merrill Company, 1968), p. 6 and Boyle, *op. cit.*, p. 312. Warnings about these raids to come were already being announced to the Germans in the closing days of the war by air-dropped propaganda leaflets. Pound and Harmsworth, *op. cit.*, p. 658.
193. Lloyd George, *op. cit.*, p. 131.

NOTES TO CHAPTER 5

1. The pertinent passage from Lana is quoted in Air Marshal Sir Robert Saundby, *Air Bombardment: The Story of its Development* (London: Chatto & Windus, 1961), p. 6. Other works which cite the prophecies of Lana include Peter Lewis, *The British Bomber Since 1914: Fifty Years of Design and Development* (London: Putnam, 1967), p. 12 and Eugene M. Emme (ed.), *The Impact of Air Power: National Security and World Politics* (Princeton, New Jersey: D. Van Nostrand Company, Inc., 1959), pp. 3-4.
2. As quoted in Bernard Brodie, *Sea Power in the Machine Age* (2d ed.; Princeton, New Jersey: Princeton University Press, 1943), p. 387.
3. Marshal of the Royal Air Force Sir John Slessor, *The Central Blue* (London: Cassell, 1956).
4. Alfred, Lord Tennyson, *Works: With Notes by the Author*, Hallam, Lord Tennyson (ed.) (New York: The Macmillan Company, 1935), p. 98.
5. Major-General Sir F.H. Sykes, *Aviation in Peace and War* (London: Edward Arnold & Co., 1922), p. 19.
6. The historian Edward Mead Earle has written of the pre-eminence of Wells' influence and has urged historians to discover the importance of research in the Wellsian *oeuvre*. Edward Mead Earle, 'H.G. Wells, British Patriot in Search of a World State', in Edward Mead Earle (ed.), *Nationalism and Internationalism: Essays Inscribed to Carlton J.H. Hayes* (New York: Columbia University Press, 1950), pp. 79, 83.
7. H.G. Wells, *Anticipations of the Reaction of Mechanical and Scientific Progress upon Human Life and Thought* (New York: Harper & Brothers, 1902). The article containing the air war concepts appeared in *The Fortnightly Review*, CCCCXVII (September 1, 1901), 545-548.
8. James Playsted Wood, *I Told You So!: A Life of H.G. Wells* (New York: Pantheon Books, 1969), p. 65. Another Wellsian scholar, W. Warren Wagar, makes a similar assertion in H.G. Wells, *Journalism and Prophecy, 1893-1946*, W. Warren Wagar (ed.) (Boston: Houghton Mifflin, 1964), p. 10.
9. Wells, *Anticipations*, pp. 208-13. One recent work has quite overstated these passages in *Anticipations* (e.g. 'cities razed to the ground . . . '), an obvious confusion with Wells' later work. Lovat Dickson, *H.G. Wells: His Turbulent Life and Times* (New York: Atheneum, 1969), p. 88.
10. H.G. Wells, *First and Last Things: A Confession of Faith and a Rule of Life*

(New York: G.P. Putnam's Sons, 1908), p. 224.
11. See Chapter 3, pp. 61-2.
12. See *infra*, pp. 124-6.
13. This aspect is worth special note, as Professor Edward Warner points out, because the question of air power versus sea power was to become a major issue in the interwar years. Edward Warner, 'Douhet, Mitchell, Seversky: Theories of Air Warfare', in Edward Mead Earle (ed.), *Makers of Modern Strategy: Military Thought from Machiavelli to Hitler* (Princeton, New Jersey: Princeton University Press, 1943), p. 486.
14. H.G. Wells, *The War in the Air: And Particularly how Mr Bert Smallways Fared While it Lasted* (London: George Ball and Sons, 1908), *passim*. See also the account in I.F. Clarke, *Voices Prophesying War, 1763-1984* (London: Oxford University Press, 1966), pp. 100-1. This book is a major study of literary prophecies about future wars and the author makes the sound point that Wells is an outstanding participant within a crowded literary genre during the last three decades before World War I, a genre given impulse by the spate of new inventions of the time and by the ever increasing sense of national competition. *Ibid.*, pp. 46-7, 66, 68-9, 79.
15. H.G. Wells, *Washington and the Riddle of Peace* (New York: The Macmillan Company, 1922), p. 221.
16. General Alberto Salinas Carranza gives a good account of this flight and its impact in his 'The Historic Flight Across the English Channel – July 25, 1909', *The Air Power Historian*, VI (April, 1959), 94-102. Also useful is Henri Azeau, Gilbert Caseneuve and Louis Saurel, *L'Enfance du Siècle (1900-1912)*, (Paris: Robert Laffont, 1966), pp. 177-8. This is Vol. II of *Histoire Vivante du XXe Siècle*, Henri Azeau (ed.).
17. *Ibid.*, p. 178 and Carranza, *op. cit.*, p. 101. The paradoxical mixture of jubilation and fear is noted in Georg W. Feuchter, *Geschichte des Luftkriegs: Entwicklung und Zukunft* (Bonn: Athenäum-Verlag, 1954), p. 25.
18. Louis P. Mouillard, *L'Empire de l'Air* (1881) as reprinted in Emme, *op. cit.*, p. 19. The translation is by the Smithsonian Institute in its *Annual Report* of 1892.
19. The article is reprinted in Wells, *Journalism and Prophecy*, pp. 64-7.
20. 99 H.C. Deb. 5s. (November 12, 1917), Col. 137.
21. Viscount Templewood, *Empire of the Air: The Advent of the Air Age, 1922-1929* (London: Collins, 1957), p. 12.
22. As cited in Franklyn Arthur Johnson, *Defence by Committee: The British Committee of Imperial Defence, 1885-1959* (London: Oxford University Press, 1960), p. 181.
23. Alfred Rawlinson, *The Defence of London: 1915-1918* (2d ed.; London: Andrew Melrose, Ltd., 1923), pp. 249-51. It is of interest that these are comments by a former *naval* gunnery expert.
24. Sir Walter Raleigh, *The War in the Air: Being the Story of the Part Played in the Great War by the Royal Air Force*, I (Oxford: The Clarendon Press, 1922), pp. 70-71. Major-General Sykes, an early leader in the British air service, has noted this interpretation of Raleigh's and argues, rather, that the resistance was primarily the result of 'blind prejudice'. But this does not really conflict with Raleigh's position. The 'prejudice' which, according to Sykes, 'amounted almost to mania', had to be based upon something and that might well have been because warplanes threatened the older services' customary role and techniques. Sir Frederick Sykes, *From Many Angles: An Autobiography* (London: George G. Harrap & Company, Ltd., 1942), pp. 90-91. Another, and later, RAF authority, Air

Vice-Marshal Kingston McCloughry, is one who certainly agrees with the 'ostrich' interpretation of Raleigh's. E.J. Kingston McCloughry, *The Direction of War: A Critique of the Political Direction and High Command in War* (New York: Frederick A. Praeger, 1955), p. 59.
25. Eugene M. Emme, 'The Impact of Air Power upon History', *Air University Quarterly Review*, II (Winter, 1948), 7.
26. Robin Higham, *The British Rigid Airship, 1908-1931: A Study in Weapons Policy* (London: G.T. Foulis & Co., Ltd., 1961), p. 9 and Sir Llewellyn Woodward, *Great Britain and the War of 1914-1918* (London: Methuen & Co., Ltd., 1967), p. 356. The question of motivation is explored in David Divine, *The Broken Wing: A Study in the British Exercise of Air Power* (London: Hutchinson, 1966), p. 22. Lord Haldane has taken credit for inspiring this governmental action and writes with pride as to how the committee's work paid off with considerable progress in aeronautical technology. Richard Burdon Haldane, *An Autobiography* (Garden City, New York: Doubleday, Doran & Company, Inc., 1929), pp. 248-9.
27. J.M. Spaight, *The Beginnings of Organised Air Power: A Historical Study* (London: Longmans, Green and Co., Ltd., 1927), p. 122 and Higham, *op. cit.*, pp. 26-8.
28. *Ibid.*, p. 26.
29. Reginald Pound and Geoffrey Harmsworth, *Northcliffe* (London: Cassell, 1959), pp. 462, 467, 523, 625. Northcliffe's early air enthusiasms receive excessive praise in Hamilton Fyfe, *Northcliffe: An Intimate Biography* (New York: The Macmillan Company, 1930), pp. 134, 140-1, 143-8.
30. The articles are quoted in Sykes, *From Many Angles*, pp. 113-18.
31. Saundby, *op. cit.*, p. 7 and Lewis, *op. cit.*, p. 19. The usually reliable Cyril Falls is therefore in error by stating the first instance of the use of airplanes in war was during the first Balkan War. Cyril Falls, *A Hundred Years of War, 1850-1950* (New York: Collier Books, 1962), pp. 333-4.
32. David Lloyd George, *War Memoirs*, Vol. IV: *1917* (Boston: Little, Brown and Company, 1934), p. 103.
33. Claude Grahame-White, *The Story of the Aeroplane* (Boston: Small, Maynard and Company, 1911), pp. 236-47 and 266-7.
34. Claude Grahame-White and Harry Harper, *The Aeroplane in War* (Philadelphia: J.B. Lippincott Company, 1912), p. viii.
35. Claude Grahame-White, 'Aircraft in the War', *The Fortnightly Review*, XCVI (October 1, 1914), 667-677.
36. Claude Grahame-White and Harry Harper, 'Two Years of Aerial War', *The Fortnightly Review*, C (August 1, 1916), 206-210.
37. Claude Grahame-White and Harry Harper, 'The Dawn of the Air Age', *The Contemporary Review*, CXII (July, 1917), 79.
38. Claude Grahame-White and Harry Harper, *Air Power: Naval, Military, Commercial* (London: Chapman & Hall, Ltd., 1917), pp. 1-46, 119-35, 197-237.
39. CBW (the common form, now, of denoting chemical and biological warfare) is not a wholly new terror of our century. Seymour Hersh has written a short but enlightening account which shows that its history extends back to the ancient world. Seymour M. Hersh, *Chemical and Biological Warfare: America's Hidden Arsenal* (Indianapolis: The Bobbs-Merrill Company, 1968), pp. 3-4.
40. Wells, *The War in the Air*, p. 355.
41. See *supra*, pp. 112-14.
42. Alexander Graham Bell, 'Preparedness for Aerial Defence', as reprinted in Emme (ed.), *The Impact of Air Power*, p. 32. Professor Brian Bond

reminds me that the 'decline of faith in [Admiral] Mahan's doctrine of sea power as essential to great power status' was a contributing factor at this time.
43. Cf. *infra*, pp. 117-18.
44. Norman Macmillan, *Sir Sefton Brancker* (London: William Heinemann Ltd., 1935), pp. 207-8, and Sykes, *From Many Angles*, p. 270.
45. Sykes' 'Review of Air Situation and Strategy for the Information of the Imperial War Cabinet' is reprinted in full as Appendix V in *ibid.*, pp. 544-54. The quoted passages are from p. 544.
46. Major-General E.B. Ashmore, *Air Defence* (London: Longmans, Green & Co., 1929), p. 108.
47. Sykes, *From Many Angles*, p. 257. Brigadier-General P.R.C. Groves was Sykes' deputy in the British Air Section at Versailles and he also was a strong believer in the Jekyll-Hyde character of commercial aircraft. Robin Higham, *The Military Intellectuals in Britain: 1918-1939* (New Brunswick, New Jersey: Rutgers University Press, 1966), pp. 170-1.
48. David Carlton, 'The Problem of Civil Aviation in British Air Disarmament Policy, 1919-1934', *Journal of the Royal United Service Institution*, CXI (November, 1966), 307. It was typical of the pattern of Versailles that France took a more stringent position with respect to German air development than did the English; the French search for security there almost inevitably meant that her demands were stronger than the English felt justified. Arnold Wolfers, *Britain and France between Two Wars: Conflicting Strategies of Peace from Versailles to World War II* (New York: The Norton Library edition, 1966), pp. 11-28.
49. E.L. Woodward and Rohan Butler (eds.), *Documents on British Foreign Policy, 1919-1939*, Vol. 1 (First Series): *1919* (London: HMSO, 1947), pp. 484, 672-3, 678.
50. Rohan Butler and J.P.T. Bury (eds.), *Documents on British Foreign Policy, 1919-1939*, Vol. VII (First Series): *1920* (London: HMSO, 1958), p. 579.
51. Sykes, *Aviation in Peace and War*, p. 103.
52. Carlton, *op. cit.*, pp. 307-16 presents a useful study of this subject. All comments hereafter on this problem are taken from this source unless otherwise cited.
53. See, for example, the account in Arthur James May, *Europe and Two World Wars* (New York: Charles Scribner's Sons, 1947), pp. 417-20 wherein only the naval and Far Eastern arrangements are discussed.
54. Charles Lock Mowat, *Britain Between the Wars, 1918-1940* (4th ed.; Chicago: The University of Chicago Press, 1963),p. 115. The British Cabinet specifically hoped that a limitation on aerial warfare would result. Great Britain, Public Record Office, Cab. 83 (21) in CAB 23/27.
55. According to a statement by Secretary Hughes to the press. *The New York Times*, 8 September 1921.
56. As quoted in John C. Cooper, 'Notes on Air Power in Time of Peace', *Air Affairs*, I (September, 1946), 86-87.
57. As quoted in Arnold J. Toynbee, *Survey of International Affairs, 1920-1923* (London: Oxford University Press, 1925), p. 498.
58. As quoted in Andrew Boyle, *Trenchard* (London: Collins, 1962), pp. 532-3. To a considerable degree, Trenchard was pushing at an open door. Neither Prime Minister Baldwin nor the Secretary of the Cabinet committee on disarmament, Ivone Kirkpatrick, believed in the utility of disarmament at that time. Ivone Kirkpatrick, *The Inner Circle* (London: Macmillan & Co., Ltd., 1959), pp. 38-9.
59. A.J.P. Taylor, *English History, 1914-1945* (New York: Oxford University

Press, 1965), p. 84.
60. J.M. Kenworthy, *Peace or War?* (New York: Boni & Liveright, 1927), pp. 292-3 and Ashmore, *op. cit.*, p. 154.
61. Air Commodore R.H. Clark-Hall, 'The Value of Civil Aviation as a Reserve to the Royal Air Force in the Time of War', *Journal of the Royal United Service Institution*, LXIX (August, 1924), 418.
62. Major C.C. Turner, 'British and Foreign Air Exercises of 1931', *Journal of the Royal United Service Institution*, LXXVI (November, 1931), 739.
63. Templewood, *op. cit.*, p. 96.
64. Edward P. Warner, 'Air Forces', *Proceedings of the Academy of Political Science*, XII (May, 1926), 66-68.
65. John Killen, *A History of the Luftwaffe* (Garden City, New York: Doubleday & Company, Inc., 1968), p. 46 and Higham, *The Military Intellectuals in Britain*, p. 171 and n.
66. This point is explicitly made in Eugene M. Emme, 'The Renaissance of German Air Power, 1919-1932', *Air Power Historian*, II (July, 1958), 146.
67. General Werner Baumbach, *The Life and Death of the Luftwaffe*, trans. Frederick Holt (New York: Ballantine Books, 1967), p. 2.
68. Woodward and Butler, *op. cit.*, p. 822 and also Vol. II (First Series): *1919* (London: HMSO, 1948), pp. 389, 393-4. The Germans were not the only ones to find fault. Jan Christian Smuts was strongly against the aerial provisions of the German treaty and he took pains to let Lloyd George know his feelings in that regard. W.K. Hancock and Jean van der Poel (eds.), *Selections from the Smuts Papers* Vol. IV: *November 1918-August 1919* (Cambridge: The University Press, 1966), pp. 148, 187.
69. Derek Wood and Derek Dempster, *The Narrow Margin: The Battle of Britain and the Rise of Air Power, 1930-40* 2d ed. (London: Hutchinson, 1963), pp. 36-7 and Killen, *op. cit.*, p. 38.
70. Wesley Phillips Newton, 'The Role of the Army Air Arm in Latin America, 1922-1931', *Air University Review*, XVIII (September-October, 1967), pp. 77, 79 and 81-83.
71. As quoted in Pound and Harmsworth, *op. cit.*, pp. 767-8.
72. Lt.-Col. Charles à Court Repington, *After the War: London-Paris-Rome-Athens-Prague-Vienna-Budapest-Bucharest-Berlin-Sofia-Coblenz-New York-Washington; A Diary* (Boston: Houghton Mifflin Company, 1922), p. 83.
73. In the customary section entitled 'Royal Air Force Notes', *Journal of the Royal United Service Institution*, LXVII (May, 1922), 392-4.
74. Wood and Dempster, *op. cit.*, p. 37.
75. Baumbach, *op. cit.*, p. 3 and George H. Quester, *Deterrence before Hiroshima: The Airpower Background of Modern Strategy* (New York: John Wiley & Sons, Inc., 1966), p. 76.
76. As quoted in Hans W. Gatzke, *Stresemann and the Rearmament of Germany* (New York: The Norton Library, 1969), p. 45 n. 115.
77. Killen, *op. cit.*, p. 44. In 1926 the RAF air staff were certainly worried that the unleashed German civil aircraft industry would soon represent a serious military threat. Great Britain, Public Record Office, AIR 9/20, Folio 13.
78. Rear-Admiral Murray F. Sueter, *Airmen or Noahs: Fair Play for our Airmen; the Great 'Neon' Air Myth Exposed* (London: Sir Isaac Pitman & Sons, Ltd., 1928), pp. 318-20, 421. One more example to illustrate this pattern: the Labour MP Kenworthy wrote in 1927 that the Germans had already achieved the best commercial air service in Europe and that 'this

German air service and the German chemical industry are the nucleus on which a modern fighting force could be rapidly built up'. Kenworthy, *op. cit.*, p. 155.
79. Emme, 'The Renaissance of German Air Power', p. 145 and Roger Manvell and Heinrich Fraenkel, *Hermann Göring* (London: Heinemann, 1962), p. 45.
80. Emme, 'The Impact of Air Power upon History', p. 9.
81. Some important references in this regard are: Emme, 'The Renaissance of German Air Power', pp. 140-51, Gatzke, *op. cit., passim*, F.L. Carsten, 'The Reichswehr and the Red Army, 1920-1933', *Survey*, No. 44/45 (October, 1962), pp. 114-32 and, especially for the later years in this history, Georges Castellan, *Le Réarmement Clandestin du Reich, 1930-1935: Vu par le 2e Bureau de l'État-Major Francias* (Paris: Librairie Plon, 1954), *passim.*
82. Woodward and Butler, *op. cit.*, 11, pp. 44-47.
83. Carsten, *op. cit.*, pp. 118-9. These negotiations formed part of a much broader series of arrangements sought by Germany and Russia from 1920 on. The variety of motives is well discussed in Max Beloff, 'L'U.R.S.S. et L'Europe', in Max Beloff *et al.* (eds.) *L'Europe du XIXe et du XXe Siècle-Problèmes et Interprétations Historiques*, Vol. VI: *1914-Aujourd-hui* (Paris: Librairie Fischbacher Marzorati, 1964), pp. 883-7.
84. Killen, *op. cit.*, pp. 40-41.
85. B.H. Liddell Hart in his *The German Generals Talk* (New York: William Morrow & Co., 1948), pp. 13-14, makes this sensible historical judgement and the expert opinion of the *Luftwaffe* leader, Adolf Galland, supports that view in that he claims that German military air development during the Weimar period was generally retarded. Generalleutnant Adolf Galland, 'Defeat of the Luftwaffe: Fundamental Causes', *Air University Quarterly Review*, VI (Spring, 1953), 19-20.
86. Walter Laqueur, *Russia and Germany: A Century of Conflict* (Boston: Little, Brown and Company Encounter Book, 1965), p. 131.
87. The *Manchester Guardian*, 3 and 6 December 1926, Kenworthy, *op. cit.*, p. 54 and Sueter, *op. cit.*, p. xxi.
88. The 'official' history of World War II civil defence planning and execution is the essential source for this information. The present description is from this source unless otherwise cited. Terence H. O'Brien, *Civil Defence* (London: HMSO, 1955), pp. 13-39.
89. Lord Salter, *Memoirs of a Public Servant* (London: Faber and Faber, 1961), p. 59. The excellent Anderson has been graced with an excellent biography; John W. Wheeler-Bennett, *John Anderson, Viscount Waverley* (London: Macmillan & Co., Ltd., 1962).
90. Sir Harold Scott, *Your Obedient Servant* (London: Andre Deutsch, 1959), pp. 63-4.
91. Wheeler-Bennett, *op. cit.*, pp. 85, 89, 315-8 in which Churchill's letter to his sovereign on this matter is reprinted. The author has also reproduced this letter in another, more readily accessible work. John W. Wheeler-Bennett, *King George VI: His Life and Reign* (New York: St. Martin's Press, 1958), pp. 544-6.
92. Wheeler-Bennett, *John Anderson*, p. 96.
93. As quoted in O'Brien, *op. cit.*, pp. 14-15.
94. *Ibid.*, p. 15.
95. E.J. Kingston-McCloughry, *Defence Policy and Strategy* (New York: Frederick A. Praeger, 1960), p. 213.

96. O'Brien, *loc. cit.*
97. Richard M. Titmuss, *Problems of Social Policy* (London: HMSO, 1950), p. 5 gives these 1924 estimates as 100 tons (first day), 75 tons (second day) and 50 tons (thereafter), which is very misleading as this accepts the most optimistic defensive reductions as fact. Titmuss' work is one volume in the *History of the Second World War: United Kingdom Civil Series*, edited by W.K. Hancock.
98. *Ibid.*, pp. 12-13. In 1938, based upon Spanish Civil War reports, British ARP experts raised the expected casualty multiplier to 72 per ton. Fred Charles Iklé, *The Social Impact of Bomb Destruction* (Norman, Oklahoma: University of Oklahoma Press, 1958), p. 17.
99. Titmuss, *loc. cit.*, n. 1.
100. 'The very limited evidence of the First World War was expanded out of all value to back up these [the RAF's] theories.' Robin Higham, *Armed Forces in Peacetime: Britain 1918-1940; A Case Study* (London: G.T. Foulis & Co., Ltd., 1962), p. 156.
101. Iklé, *loc. cit.*
102. Sir Charles Webster and Noble Frankland, *The Strategic Air Offensive Against Germany, 1939-1945*, Vol. I: *Preparation* (London: HMSO, 1961), p. 46.
103. Wheeler-Bennett, *John Anderson*, p. 97. The RAF air staff were certainly pleased with the report. See their comments in Great Britain, Public Record Office, AIR 9/69, Folio #34.
104. O'Brien, *op. cit.*, p. 18.
105. Titmuss, *op. cit.*, p. 23.
106. *Ibid.*
107. Wells, *The War in the Air*, p. 353.
108. The watchdog committee set up by Lloyd George to appease the 'Anti-waste' mood of the taxpayers. D.C. Somervell, *Modern Britain, 1870-1939* (London: Methuen & Co., Ltd., 1941), pp. 144-5.
109. The military recommendations within the early reports of the Geddes Committee are discussed in the 'Military Notes' section of the *Journal of the Royal United Service Institution*, LXVII (May, 1922), pp. 378-85.
110. As quoted in Webster and Frankland, *op. cit.*, p. 45.
111. Kingston-McCloughry, *Defense Policy and Strategy*, p. 90.
112. Raymond J. Sontag, 'Between the Wars', *Pacific Historical Review*, XXIX (February, 1960), pp. 5-6.
113. James Joll, 'Shapers of the Twentieth Century', a review in the *Manchester Guardian Weekly*, March 14, 1968.
114. Maurice Baumont, *La Faillitte de la Paix (1918-1939)*, Vol. I: *De Rethondes à Stresa (1918-1935)* 3d ed. rev. (Paris: Presses Universitaires de France, 1951), pp. 28-43. This two volume study by Baumont is the concluding work in the prestigious *Peoples et Civilisations* series edited by Louis Halpern and Philippe Sagnac.
115. Beloff, *op. cit.*, p. 870.
116. Dennis Bardens, *Churchill in Parliament* (London: Robert Hale, 1967), p. 140.
117. Higham, *Armed Forces in Peacetime*, p. 93.
118. Arthur Marwick, *Britain in the Century of Total War: War, Peace and Social Change, 1900-1967* (Boston: Little, Brown and Company, 1968), pp. 158-9.
119. Nesta Webster, 'Bolshevism and Secret Societies', *Journal of the Royal United Service Institution*, LXVII (February, 1922), 1-15.
120. Lieut.-Col. D.M.F. Hoysted, 'A Plea for a Strong Anti-Aircraft Defence', *The Royal Engineers Journal*, XXXVIII (September, 1924), 421. The RAF

leaders were naturally ready to mention the potential connection between air attack and revolution to bolster their arguments concerning defence priorities. See, for example, Great Britain, Public Record Office, AIR 9/8, Folio #10 and AIR 9/69, Folio #54.
121. As quoted in Wheeler-Bennett, *John Anderson*, p. 96.
122. As quoted in O'Brien, *op. cit.*, p. 19.
123. Some striking examples are A.J.P. Taylor's review in the *Observer*, 30 May 1965, Michael Howard's review in the *Sunday Times*, 30 May 1965 and Lord Chalfont's review in *The Times*, 27 May 1965. The completed work concerned is Liddell Hart, *Memoirs* (2 vols.; New York: G.P. Putnam's Sons, 1965-1966). Somewhat less flattering is the present writer's review in *The Journal of Modern History*, XL (December, 1968), 630-631.
124. President Kennedy's and General Chassin's opinions are both cited in George A. Panichas (ed.), *Promise of Greatness: The War of 1914-1918* (New York: The John Day Company, 1968), p. 97.
125. Major Ray L. Bowers, 'The Peril of Misplaced Loyalties', *Air University Review*, XVII (May-June, 1966), 97.
126. Liddell Hart, *Memoirs*, 1, pp. 15-16, 25-6, 63-5, Panichas, *loc. cit.*, and Higham, *The Military Intellectuals in Britain*, pp. 46-9.
127. *Ibid.*, p. 48.
128. B.H. Liddell Hart, *Paris: or the Future of War* (New York: E. P. Dutton & Company, 1925). The analysis here of this book is taken from pp.4-56. Thereafter, the author turns to other military theory and makes suggestions on more typical military matters. Nevertheless, the book's basic theme is already stated in its early pages as is evidenced by the title.
129. Theodore Ropp is normally a very reliable military historian and thus his paradoxical comment that Liddell Hart 'had always opposed gas and terror bombing as inhumane and counterproductive . . . ' is difficult to understand. Theodore Ropp, 'A Theorist in Power', *Air University Review*, XVIII (March-April, 1967), p. 107. That comment is especially puzzling in that Liddell Hart maintained his support of gas warfare in his post-World War II writings. See, e.g., B.H. Liddell Hart, 'Is Gas A Better Defence than Atomic Weapons?', *Survival* I (September-October, 1959), pp. 139-141.
130. Liddell Hart, *Memoirs*, I, p. 143.
131. *Ibid.*, pp. 98-9, 142-3. Just how strong a believer Hoare became in the current air power theories is revealed in his memoirs of his years as Air Minister. Templewood, *op. cit., passim.*
132. Higham, *The Military Intellectuals in Britain*, pp. 163, 166, 171. As stated, Foch was a convert. In 1909, he saw no military value at all in aircraft which, in fact, he considered useful only for 'sport'. Captain Philip M. Flammer, 'Image of the Aces, A Writer's Bonanza', *Air University Review*, XVII (January-February, 1966), 92.
133. Wing Commander C.H.K. Edmonds, 'Air Strategy', *Journal of the Royal United Service Institution*, LXIX (May, 1924), 195. Major E.R. MacPherson, 'The Development of Chemical Warfare', *ibid.*, LXX (May, 1925), p. 318. Group-Captain W.F. MacNeece, 'Certain Aspects of Air Defence', *ibid.*, LXXI (February, 1926), 98. The piece by Edmonds is the reprinting of an address given before the RUSI at a meeting in which Colonel Lord Gorell was chairman. After Edmonds' speech, Gorell presented some concluding remarks including this on the Foch commentary: 'It was interesting to me to hear again the quotation from Marshal Foch. It has played its part in a number of discussions upon the air, and I am not aware that anyone of equal authority has ever attempted

to deny that opinion.' *Ibid.*, LXIX (May, 1924), p. 209.
134. Sueter, *op. cit.*, pp. 99, 267.
135. B.H. Liddell Hart, 'Colonel Bond's Criticisms: A Reply', *The Royal Engineers Journal*, XXXVI (November, 1922), p. 308.
136. The article 'The Napoleonic Fallacy: The Moral Objective in War', originally published in the *Empire Review*, is quoted at length in Liddell Hart, *Memoirs*, I, pp. 138-42. Many passages from this article are restated verbatim in *Paris, or the Future of War*.
137. B.H. Liddell Hart, 'Armament and its Future Use', *The Yale Review*, XIX (June, 1930), 666-667.
138. V.W. Germains, 'The Army in War', *National Review*, CXII (June, 1939), p. 761.
139. Higham, *The Military Intellectuals in Britain*, p. 170, Sykes, *From Many Angles*, pp. 218-9 and Emme (ed.), *The Impact of Air Power*, p. 176.
140. Among those who note the importance of these articles are Boyle, *op. cit.*, p. 419, Higham, *Armed Forces in Peacetime*, p. 156 and E.M. Eddy, 'Britain and the Fear of Aerial Bombardment, 1935-1939', *Aerospace Historian*, XIII (Winter, 1966), pp. 177-8.
141. P.R.C. Groves, 'Our Future in the Air', in Emme (ed.) *The Impact of Air Power*, pp. 176-81 is a reprinting, in great part, of these articles, The description of these articles here is taken from this source.
142. Groves' terminology with respect to the future war being one of 'areas' has an interesting echo in the Second World War wherein the RAF referred to generalised urban raids (as compared to attacks upon specific target systems) as 'Area Bombing'. A discussion of the RAF changeover to the 'area bombing' method during the war is in Webster and Frankland, *op. cit.*, pp. 127-306.
143. The result of this was a real, if short-lived, 'flap' over the French air force disparity, a subject which will be noted in Chapter 6.
144. P.R.C. Groves, 'For France to Answer', *The Atlantic Monthly*, CXXXIII (February, 1924), pp. 145-153.
145. Hilton P. Goss, *Civilian Morale under Aerial Bombardment, 1914-1939* (Maxwell Air Force Base, Alabama: Documentary Research Division, Air University, 1948), p. 27. This is United States Air University Documentary Research Study, Number 14. Also see Higham, *The Military Intellectuals in Britain*, p. 230.
146. Spaight, *op. cit.*, p. 87.
147. This book by Spaight, *Aircraft in War*, is described in both Quester, *op. cit.*, p. 15 and Higham, *The Military Intellectuals in Britain*. pp. 230-31.
148. It is indicative of the basic theme of this chapter that one of the more temperate spokesmen on aerial concerns during the interwar period had no doubt in the ability of air power to wholly destroy a city.
149. Spaight's *Air Power and the Cities* (1930) as described and quoted in Goss, *op. cit.*, pp. 49-50.
150. J.M. Spaight, *Pseudo-Security* (London: Longmans, Green and Co., 1928), pp. 10-13, 102.
151. *Ibid.*, p. 15, 79, 100-128, 158, 160. The quoted passage concerning the fate of Geneva is on p. 79.
152. The argument for a League police force is the basic theme in David Davies, *The Problem of the Twentieth Century: A Study in International Relationships* (London: Ernest Benn Limited, 1930), Davies, later Baron Davies in 1932, was an important business and philanthropic leader in the interwar years. He was, for example, a director of the Midland Bank and the Great Western Railway Company, Chairman of the Council of the

University College of Wales and President of the National Library of Wales. He was also a Member of Parliament for the Liberal Party from 1906 to 1929. Davies is thus another example of how men of accepted judgement shared the doomsday views about air power during the interwar years. (The biographical material is taken from *The Dictionary of National Biography, 1941-1950.)*

153. Davies, *op. cit.,* pp. 1-3, 35, 39-42, 55, 149-50, 315. The quoted passage is from p. 40.
154. Carlton, *op, cit.,* p. 313.
155. Cecil has written of his party doubts. Viscount Cecil, *A Great Experiment: An Autobiography* (London: Jonathan Cape, 1941), pp. 115, 189-90.
156. Taylor, *op. cit.,* p. 116. Lord Hugh, the brother of Lord Robert, was later created first Baron Quickswood.
157. 152 H.C. Deb. 5s., Cols. 332-339.
158. J. Holland Rose, *The Indecisiveness of Modern War, and Other Essays* (Port Washington, New York: Kennikat Press, Inc., 1968), p.48. (This is a reproduction of the 1927 work.)
159. For a representative sampling of this interpretation, see Feuchter, *op. cit.,* pp. 33-4, Norman H. Gibbs, 'Le rôle de la guerre totale dans les transformations subies par l'Europe', in Beloff *et al., op. cit.,* Vol. V: *1914-Aujourd'hui* (Paris: Librairie Fischbacher Marzorati, 1964), pp. 35-6, and Webster and Frankland, *op. cit.,* pp. 10, 49-50.
160. Marwick, *op. cit.,* p. 54.
161. The article 'Shall We All Commit Suicide?' is reprinted in Winston S. Churchill, *Amid These Storms: Thoughts and Adventures* (New York: Charles Scribner's Sons, 1932), pp. 245-52.
162. Winston S. Churchill, *The Aftermath: The World Crisis, 1918-1928* (New York: Charles Scribner's Sons, 1929), pp. 11, 479, 482-3.
163. As quoted in Catherine Gavin, *Britain and France: A Study of Twentieth Century Relations; the Entente Cordiale* (London: Jonathan Cape, 1941), p. 268.
164. Templewood, *op. cit.,* pp. 39-42.
165. Spaight, *Pseudo-Security,* p. 116 gives quoted passages from these speeches.
166. Lord Thomson, *Air Facts and Problems* (New York: George H. Doran Company, 1927), pp. 5, 20-27.
167. Sykes, *Aviation in Peace and War,* pp. 96-101.
168. R.J. Minney, *No. 10 Downing Street: A House in History* (Boston: Little, Brown and Company, 1963), p. 366.
169. Macmillan, *op. cit.,* pp. 211, 237, 242, 247, 250, 381.
170. Baldwin's major biographers wrote that by the spring of 1932, Baldwin had 'a horror of air warfare and bombing, which became something of an obsession'. Keith Middlemas and John Barnes, *Baldwin: A Biography* (New York: The Macmillan Co., 1970), p. 732.
171. As quoted in O'Brien, *op. cit.,* p. 31.
172. Ian Hay, *Arms and the Men* (London: HMSO, 1950), p. 63.
173. Sir F. Maurice, *Governments and War: A Study of the Conduct of War* (London: William Heinemann Ltd., 1926), p. 153.
174. Viscount Grey of Fallodon, *Twenty-Five Years, 1892-1916,* II (New York: Frederick A.Stokes Company, 1925), p. 285.
175. Cf. Chapter 2, p. 47.
176. Boyle, *op. cit.,* pp. 398-400, Titmuss, *op. cit.,* pp. 4-5 and Basil Collier, *The Defence of the United Kingdom* (London: HMSO, 1957), p. 11. This is one of the *United Kingdom Military Series* in the official *History of the Second World War* edited by J.R.M. Butler.

177. Major-General Sir Frederick Maurice, *The Life of Viscount Haldane of Cloan, K.T., O.M.*, Vol. II: *1915-1928* (London: Faber and Faber Ltd., 1939), pp. 154, 159 and Dudley Sommer, *Haldane of Cloan: His Life and Times, 1856-1928* (London: George Allen & Unwin Ltd., 1960), p. 389.
178. Templewood, *op. cit.*, pp. 61-2. Hankey's importance is confirmed by many knowledgeable writers. See e.g. Earl of Swinton, *Sixty Years of Power: Some Memories of the Men Who Wielded It* (New York: James H. Heineman, Inc., 1966), p. 45.
179. A.J.P. Taylor, in his typically caustic manner, describes Joynson-Hicks as 'preposterous'. Taylor, *op. cit.*, p. 242.
180. As quoted in Major R. Chenevix-Trench, 'Gold Medal (Military) Prize Essay for 1922', *Journal of the Royal United Service Institution*, LXVIII (May, 1923), p. 205.
181. Mowat, *op. cit.*, pp. 22-3.
182. 123 H.C. Deb. 5s. (December 15, 1919), Cols. 87-155.
183. *Ibid.*, Cols. 91, 103-104 and 107.
184. *Ibid.*, Cols. 94, 113-115 and 126.
185. The debates are printed in 126 H.C. Deb. 5s. (11 March 1920), Cols. 1579-1674.
186. A good, if not overly sympathetic, account is in Lord Ismay, *Memoirs* (London: Heinemann, 1960), pp. 29-32.
187. Falls, *op. cit.*, p. 270.
188. Templewood, *op. cit.*, p. 284.
189. 126 H.C. Deb. 5s., Cols. 1599, 1602, 1610-1611, 1647-1652, 1657 and 1661.
190. *Ibid.*, Cols. 1591-1592, 1629, 1643 and 1656. On 23 February 1920 Churchill had made a similar statement. 125 H.C. Deb. 5s., Cols. 1353-1354.
191. 126 H.C. Deb. 5s., Cols. 1638 and 1654.
192. 138 H.C. Deb. 5s. (1 March 1921), Cols. 1655-1721 and 140 H.C. Deb. 5s. (21 April 1921), Cols. 2132-2143.
193. One observer notes such World War I pilots as Mosley, Göring and Lindberg all shared this characteristic: 'between the wars, the pioneers of aviation enjoyed great popularity everywhere, both for being brave soldiers and because of a certain halo of technical-utopian perfection'. Brigitte Granzow, *A Mirror of Nazism: British Opinion and the Emergence of Hitler, 1929-1933* (London: Victor Gollancz Ltd., 1964), p. 104.
194. 138 H.C. Deb. 5s., Cols. 1673, 1675-1676, 1686-1687, and 1692-1701 and 140 H.C. Deb. 5s., Cols. 2141-2143. Other members who spoke about civil air needs (including the military ties) were Colonel J.R.P. Newman and Mr Bartley Denniss. Denniss also remarked that England erred in not pursuing a strategic bombing policy right from the beginning of the Great War. 138 H.C. Deb. 5s., Col. 1698.
195. Raper thus reversed his position in the previous year's debate; Cf. *supra*, p. 140.
196. 138 H.C. Deb. 5s., Cols. 1666 and 1669-1670 and 140 H.C. Deb. 5s., Cols. 2132-2133.
197. 152 H.C. Deb. 5s. (21 March 1922), Cols. 285-422.
198. See *supra*, pp. 130-1.
199. This event will also be looked at in the next chapter. One useful reference for these tests is Samuel Taylor Moore, *U.S. Airpower: Story of American Fighting Planes and Missiles from Hydrogen Bags to Hydrogen War-heads* (New York: Greenburg, 1958), pp. 77-80.
200. 152 H.C. Deb. 5s., Cols. 285-310.
201. *Ibid.*, Cols. 316, 324 and 339.
202. *Ibid.*, Cols. 310-317. Relevant to prior remarks, Seely used Foch as one of his supporting authorities for his grim forecasts.

Notes

203. *Ibid.*, Cols. 309-310, 320, 326, 339-340, 352, 359-360, 369-370, 376, 384 and 404.
204. *Ibid.*, Cols. 322-327 and 332-339.
205. *Ibid.*, Cols. 329-331, 350-351 and 383.
206. 161 H.C. Deb. 5s. (14 March 1923), Cols. 1605-1667.
207. *Ibid.*, Cols. 1609-1615, 1643-1644, 1650, 1653 and 1659-1660.
208. *Ibid.*, Cols. 1605, 1636-1637, 1652 and 1659.
209. *Ibid.*, Cols. 1610 and 1627.
210. 171 H.C. Deb. 5s. (20 and 24 March 1924), Cols. 675-743 and 1065-1086.
211. *Ibid.*, Cols. 681-686 and 692.
212. *Ibid.*, Cols. 1072-1074.
213. Viscount Simon, *Retrospect: Memoirs* (London: Hutchinson, 1952), pp. 112-15.
214. See Chapter IV, p. 76, for a discussion of the 'frightfulness' issue during World War I.
215. 171 H.C. Deb. 5s., Cols. 686 and 712-723.
216. *Ibid.*, Col. 728.
217. *Ibid.*, Cols. 686-693, 701-702 and 705-709.
218. *Ibid.*, Cols. 688, 711, 736, 1079, 1081 and 1084.
219. 169 H.C. Deb. 5s., Cols. 1678-1679. Eden wrote of that maiden speech in his memoirs as did his biographer, Lewis Broad. The Earl of Avon, *The Eden Memoirs*, Vol. II: *Facing the Dictators* (London: Cassell, 1962), p. 15 and Lewis Broad, *Sir Anthony Eden: The Chronicles of a Career* (London: Hutchinson, 1955), pp. 26-7.
220. 180 H.C. Deb. 5s. (26 February 1925), Cols. 2195-2331 and 181 H.C. Deb. 5s. (12 March 1925), Cols. 1567-1698.
221. Gibbs, *op. cit.*, pp. 30-31, discusses interwar pacifism and its World War I ties, and states that 'en Grande-Bretagne ce sentiment était particulièrement bien enraciné'. In Col. E.G. Keogh, 'The Study of Military History', *Australian Army Journal*, No. 188 (January 1965), p. 15, there is a good listing of 'profoundly' influential English works of pacifist bent and based on the past war's experience. A clear linkage between the pacifist movement and the doomsday visions of the air war expected next is presented in Gerald Fay, 'Unrest', in John Raymond (ed.), *The Baldwin Age* (London: Eyre and Spottiswoode, 1960), pp. 132-6. Andre Maurois in his *A History of England*, trans. Hamish Miles (rev. ed.; New York: Grove Press, Inc., 1960), p. 486, adds the interesting suggestion that the women's vote after the war added to the political impact of pacifism.
222. Cf. *supra*, pp. 175-7.
223. A sympathetic and valuable account of Lansbury is in Ronald Blythe, *The Age of Illusion: England in the Twenties and Thirties, 1919-1940* (London: Penguin Books, 1964), pp. 270-93.
224. 180 H.C. Deb. 5s., Cols. 2261-2282, 2286-2288 and 2295-2308.
225. *Ibid.*, Cols. 2211-2224 and 181 H.C. Deb. 5s. Cols. 1568-1570.
226. 180 H.C. Deb. 5s., Cols. 2256-2260 and 2295 and 181 H.C. Deb. 5s., Cols. 1573-1574 and 1647-1651. A Colonel C. de W. Crookshank allowed even less time than did Reid for urban disaster. He claimed a city's destruction might be accomplished 'in a few minutes by bombs . . . '. *Ibid.*, Col. 1654.
227. 180 H.C. Deb. 5s., 2195-2211 and 181 H.C. Deb. 5s., Cols. 1573-1574, 1631 and 1679-1680.
228. 192 H.C. Deb. 5s. (25 February and 8 March 1926), Cols. 765-876 and 1961-2068.
229. *Ibid.*, Cols. 765-780 and 2021.
230. There were actually seven interrelated treaties concluded at Locarno. They are usefully summarised by John Connell in his *The 'Office': The Story of the British Foreign Office, 1919-1951* (New York: St Martin's Press, 1958), p. 84 n. 1. Connell describes the results of Locarno as being, in

part, 'a tragic temptation to think that the basic causes of international difference, and therefore of war, had been removed . . . ' (p. 80). Such sanguine thinking is commonly depicted as the 'spirit of Locarno'. An excellent summary of the events leading to Locarno and its immediate results is in Baumont, *op. cit.*, pp. 321-32.
231. A striking example of this new mood of confidence is King George V's diary note of 1 December 1925 (when the final signatures on the Pact were affixed): 'This morning the Locarno Pact was signed at the Foreign Office. I pray this may mean peace for many years. Why not forever?' Quoted in Harold Nicolson, *King George the Fifth: His Life and Reign* (Garden City, New York: Doubleday, 1953), p. 409. Other 'typical utterances' of 'the Locarno spirit' are given in Stephen King-Hall, *Our Own Times, 1913-1938: A Political and Economic Survey* 2d. rev. ed. (London: Nicholson and Watson, 1938), pp. 275-6.
232. 192 H.C. Deb. 5s., Cols. 772, 783, 800, 807, 849, 856, 1976-1977, 1985, 1988-1989 and 2001-2002.
233. *Ibid.*, Cols. 780-790 and 1962-1968.
234. *Ibid.*, Cols. 849-852, 860 and 1990.
235. *Ibid.*, Cols. 794 and 859.
236. 203 H.C. Deb. 5s. (10 and 17 March 1927), Cols. 1395-1517 and 2227-2312.
237. The Preparatory Commission had been appointed in December 1925 by the League Council under instructions from the League Assembly. Its first meeting was in May 1926 – after the 1926 air estimates debate – and was primarily involved in procedural matters. The Spring 1927 meeting was to be its third session, one of two held that year. Cecil, *op. cit.*, pp. 170-71 and Frank P. Chambers, *This Age of Conflict: The Western World – 1914 to the Present* 3d. rev. ed. (New York: Harcourt, Brace & World, Inc., 1962), pp. 127-8.
238. 203 H.C. Deb. 5s., Cols. 1411-1413, 1427, 1451, 1457, 1472, 1488-1493, 2228-2232, 2235-2242 and 2266.
239. *Ibid.*, Cols. 1412-1413, 1454-1461 and 1473.
240. Sassoon was a true air enthusiast and he became a commander of one of the RAF auxiliary fighter squadrons (Higham, *Armed Forces in Peacetime*, p. 258). He was also something of an exotic, at least according to the memory of Earl Winterton who described one of Sassoon's homes as 'very harem and brothel-like . . . '. As quoted in Alan Houghton Broderick, *Near to Greatness: A Life of the Sixth Earl Winterton* (London: Hutchinson, 1965), p. 187.
241. 203 H.C. Deb. 5s., Cols. 1495-1501.
242. *Ibid.*, Cols. 1426, 1488-1493 and 2266-2269.
243. 214 H.C. Deb. 5s. (12 March 1928), Cols. 1535-1664 and 215 H.C. Deb. 5s. (20 March 1928), Cols. 285-349.
244. 214 H.C. Deb. 5s., Cols. 1578, 1585, 1608-1615, 1618-1620, 1633-1635, and 1644-1646 and 215 H.C. Deb. 5s., Cols. 314-316.
245. 214 H.C. Deb. 5s., Cols. 1611 and 1619.
246. 226 H.C. Deb. 5s. (7 and 12 March 1929), Cols. 595-716 and 1005-1071.
247. *Ibid.*, Cols. 612, 623-625, 632, 637-640, 657 and 671.
248. *Ibid.*, Cols. 633-636, 675-690 and 705-708.
249. *Ibid.*, Cols. 632, 678, 688-689, 698, 705-706 and 712-714.
250. 236 H.C. Deb. 5s. (18 March 1930), Cols. 1925-2056 and 237 H.C. Deb. 5s. (25 March 1930), Cols. 307-347.
251. 236 H.C. Deb. 5s., Cols. 1925-1927, 1940-1941, 1986-1992, 2031-2034, 2039-2044 and 2047-2052.
252. *Ibid.*, Cols. 1977-1980.
253. 249 H.C. Deb. 5s. (17 March 1931), Cols. 1887-1996 and 250 H.C. Deb. 5s. (24 March 1931), Cols. 231-280.
254. 249 H.C. Deb. 5s., Cols. 1887-1901, 1903, 1906-1907, 1933, 1975-1979,

1984-1986, and 1989-1990.
255. *Ibid.*, Cols. 1901-1910 and 250 H.C. Deb. 5s., Col. 266.
256. John Lukacs, *Historical Consciousness; or the Remembered Past* (New York: Harper & Row, Publ., 1968), pp. 79-80.
257. The Inter-Parliamentary Union, *What Would Be the Character of a New War?* (London: P.S. King & Son, Ltd., 1931).
258. *Ibid.*, pp. vii-ix and Goss, *op. cit.*, p. 53.

CHAPTER 6

1. Cf. Chapter 4, pp. 146-151.
2. David Lloyd George, *War Memoirs*, Vol. IV: *1917* (Boston: Little, Brown and Company, 1934), p. 131, Winston S. Churchill, *The Aftermath: The World Crisis, 1918-1928* (New York: Charles Scribner's Sons, 1929), p. 482, and George H. Quester, *Deterrence before Hiroshima: The Airpower Background of Modern Strategy* (New York: John Wiley & Sons, Inc., 1966), pp. 44-5.
3. Peter Lewis, *The British Bomber Since 1914: Fifty Years of Design and Development* (London: Putnam, 1967), pp. 96, 112-13, and Major-General Sir F.H. Sykes, *Aviation in Peace and War* (London: Edward Arnold & Co., 1922), p. 86.
4. Ralph Michaelis, *From Bird Cage to Battle Plane: The History of the R.A.F.* (New York: Thomas Y.Crowell Compamy, 1943), p. 120 and E. Colston Shepherd, *The Air Force of To-day* (rev. ed.; London: Blackie & Son Limited, 1942), p. 65.
5. Winston S. Churchill, *The World Crisis*, IV (New York: Charles Scribner's Sons, 1927), pp. 24-31 and 252.
6. *Parliamentary Papers (Accounts and Papers)*, 1919), Vol. XXXIII, Cmd. 100, p. 16, Sykes, *op. cit.*, pp. 83, 115, Major-General E.B. Ashmore, *Air Defence* (London: Longmans, Green & Co., 1929), p. 78, Lloyd George, *loc. cit.*, John Ehrman, *Cabinet Government and War, 1890-1940* (Hamden, Conn.: Archon Books, 1969), p. 103 (this is a reproduction of the original 1958 edition published by the Cambridge University Press), Air Commodore J.A. Chamier, *The Birth of the Royal Air Force: The Early History and Experiences of the Flying Services* (London: Sir Isaac Pitman & Sons, Ltd., 1943), p. 174, David Divine, *The Broken Wing: A Study in the British Exercise of Air Power* (London: Hutchinson, 1966), p. 143 and Hans Daalder, *Cabinet Reform in Britain, 1914-1963* (Stanford, Calif.: Stanford University Press, 1964), p. 161.
7. Air Marshal Sir Robert Saundby, *Air Bombardment: The Story of its Development* (London: Chatto & Windus, 1961), p. 21.
8. Divine, *loc. cit.*
9. Major-General Sir Frederick Sykes, *From Many Angles: An Autobiography* (London: George G. Harrap & Company Ltd., 1942), p. 243.
10. Andrew Boyle, *Trenchard* (London: Collins, 1962), pp. 314-15.
11. Sir Charles Webster and Noble Frankland, *The Strategic Air Offensive Against Germany, 1939-1945*, Vol. I: *Preparation* (London: HMSO, 1961), p. 46.
12. Mark Kerr, *Land, Sea, and Air: Reminiscences* (London: Longmans, Green and Co., 1927), p. 287.
13. Cmd. 100, *op. cit.*, p. 5.
14. The airpower enthusiast, Major-General J.F.C. Fuller, did not even mention the air effort when he analysed the reasons for the German collapse in his *The Conduct of War, 1789-1961: A Study of the Impact of the French, Industrial and Russian Revolutions on War and its Conduct* (New Brunswick, New Jersey: Rutgers University Press, 1962), pp. 174-82. A sensible (and deprecatory) estimate of the impact of air power on the

course of World War I is given in Field-Marshal Viscount Montgomery of Alamein, *A History of Warfare* (New York: The World Publishing Company, 1968), p. 482.
15. Robert Blake (ed.), *The Private Papers of Douglas Haig, 1914-1919: Being Selections from the Private Diary and Correspondence of Field-Marshal the Earl Haig of Bemersyde, K.T., G.C.B., O.M. etc.* (London: Eyre & Spottiswoode, 1952), p. 280. It is very possible that this prospect was not yet a grim one to Trenchard, whose loyalty had been to the army for almost his entire career; it is certain that Haig was pleased with that expectation.
16. C.G. Grey, *A History of the Air Ministry* (London: George Allen & Unwin Ltd., 1940), p. 88.
17. This document is reprinted in full in Sykes, *From Many Angles*, pp. 544-54.
18. Sykes' 'Review' is well analysed in Robin Higham, *The Military Intellectuals in Britain: 1918-1939* (New Brunswick, New Jersey: Rutgers University Press, 1966), pp. 158-9.
19. This document is reprinted in full in Sykes, *From Many Angles*, pp. 558-74.
20. *Ibid.*, p. 265.
21. *Ibid.*, pp. 265-6.
22. As quoted in Grey, loc. cit.
23. Keynes has expressed an acid version of this common charge: 'Lloyd George is rooted in nothing; he is void and without content . . . '. John Maynard Keynes, *Essays in Biography*, Geoffrey Keynes (ed.) (rev. ed.; New York: Horizon Press Inc., 1951), p. 36.
24. Frank Owen, *Tempestuous Journey: Lloyd George, His Life and Times* (New York: McGraw-Hill Book Company, Inc., 1955), p. 419 and also see *supra*, Chapter IV, p. 98.
25. As quoted in Churchill, *The Aftermath*, p. 40.
26. Lewis Broad, *Winston Churchill*, Vol. I: *The Years of Preparation* (London: Sidgwick and Jackson, 1963), p. 255.
27. 123 H.C. Deb. 5s., Cols. 87-90, 101-106, 111, 114 and 132.
28. Lord Beaverbrook, *Men and Power, 1916-1918* (New York: Duell, Sloan and Pearce, 1956), p. 361.
29. Norman Macmillan, *Sir Sefton Brancker* (London: William Heinemann Ltd., 1935), p. 240 and J.M. Spaight, *The Beginnings of Organized Air Power: A Historical Study* (London: Longmans, Green and Co., Ltd., 1927), p. 205.
30. Divine, *op. cit.*, pp. 147-8.
31. Franklyn Arthur Johnson, *Defence by Committee: The British Committee of Imperial Defence, 1885-1959* (London: Oxford University Press, 1960), p. 173, Spaight, *op. cit.*, p. 206 and Daalder, *op. cit.*, pp. 161, 164.
32. The Marchioness of Londonderry, *Retrospect* (London: Frederick Muller Ltd., 1938), p. 170.
33. Sykes, *From Many Angles*, pp. 266-7.
34. Boyle, *op. cit.*, pp. 327-32, Higham, *op. cit.*, pp. 124-5, 165, and Divine, *op. cit.*, pp. 150-51.
35. Higham, *op. cit.*, pp. 134, 157 and Captain Liddell Hart, *Memoirs*, I (London: Cassell, 1965), p. 318.
36. This document was approved as a Government White Paper and is reprinted in *Parliamentary Papers (Accounts and Papers), 1919)*, Vol. XXXIII, as Cmd. 467. An entire chapter (entitled 'The Master Plan') in Viscount Templewood, *Empire of the Air: The Advent of the Air Age, 1922-1929* (London: Collins, 1957), is devoted to an analysis of this White Paper (pp. 72-84). A less sympathetic analysis is found in Divine, *op. cit.*,

pp. 154-8.
37. Templewood, *op. cit.*, p. 72.
38. Cf. Chapter IV, p. 100.
39. Divine, *op. cit.*, p. 156.
40. See Chapter IV, pp. 85-6.
41. Boyle, *op. cit.*, pp. 723-7.
42. Higham, *op. cit.*, pp. 165-7. One RAF authority, Air Chief Marshal Joubert de la Ferté, contends that Trenchard's conversion had at least not occurred by the armistice as is indicated by his immediate return of the Independent Air Force to the overall BEF command. Sir Philip Joubert de la Ferté, *The Third Service: The Story behind the Royal Air Force* (London: Thames and Hudson, 1955), p. 59. The air historian, Raymond H. Fredette, takes the same view in his *The Sky on Fire: The First Battle of Britain, 1917-1918, and the Birth of the Royal Air Force* (New York: Holt, Rinehart and Winston, 1966), p. 226. On the other hand, Trenchard's biographer states that Trenchard's conversion came in 1918 while he was still in command of the IAF. Boyle, *op. cit.*, p. 308.
43. An eminent military historian has discovered a similar pattern in the interwar United States aerial development; first came the desire for independence for the air service, then came the development of the strategic bombing doctrine to justify this independence, and then finally came the push to produce the bombers which (presumably) made the doctrine feasible. Walter Millis, *Arms and Men: A Study in American Military History* (New York: Mentor, 1956), p. 231. The English story is not, of course, so linear; the Great War made for much overlapping of these steps. But its postwar development does reflect that ordering of cause and effect: the desire for independence fostered the stress on the independent function of strategic bombing and then, thirdly and certainly of lesser importance for years, the wish grew for the equipment to implement that function.
44. As quoted in Boyle, *op. cit.*, p. 343.
45. Cmd. 467, *op. cit.*, p. 5.
46. Higham, *op. cit.*, pp. 195-6 and Noble Frankland, *The Bombing Offensive Against Germany: Outlines and Perspectives* (London: Faber and Faber, 1965), pp. 38-40.
47. Webster and Frankland, *op. cit.*, p. 38.
48. Ralph Barker, *The Thousand Plan: The Story of the First Thousand Bomber Raid on Cologne* (London: Chatto & Windus, 1965), p. 7.
49. William P. Snyder, *The Politics of British Defense Policy, 1945-1962* (Columbus, Ohio: Ohio State University Press, 1964), pp. 25, 137.
50. Economic prudence was not a trait common to the RAF officers. One leading airman, Sir Godfrey Paine, complained, for example, to the military correspondent, Colonel Repington, about potential budget cuts. As Repington's diary reports (on 7 November 1918); 'I think that he would like fifty millions. I told him that the only chance of retaining the RAF was to commercialise it'. Lieut-Col. C. à Court Repington, *The First World War, 1914-1918: Personal Experiences* (London: Constable and Company Ltd., 1920), II, p. 478. Repington was right that economy was essential. Lloyd George was moved in part to abolish the air service because he had seen how costly it had become and he was searching for ways to cut government expenses. Boyle, *op. cit.*, p. 327.
51. *Ibid.*, pp. 329-30.
52. Thomas Jones, *Whitehall Diary*, Vol. I: *1916-1925*, Keith Middlemas (ed.) (London: Oxford University Press, 1969), p. 84.
53. Derek Wood and Derek Dempster, *The Narrow Margin: The Battle of*

Britain and the Rise of Air Power, 1930-40 (2d. ed.; London: Hutchinson, 1963), p. 65. Shepherd, *op. cit.*, p. 78 and Spaight, *op. cit.*, p. 211.
54. 'Bomber' Harris, for example, refers to it as 'this absurd procedure'. Marshal of the RAF Sir Arthur Harris, *Bomber Offensive* (New York: The Macmillan Company, 1947), p. 13.
55. Stephen Roskill, *Naval Policy Between the Wars*, Vol. I: *The Period of Anglo-American Antagonism, 1919-1929* (New York: Walker and Company, 1969), pp. 214-15.
56. Boyle, *op. cit.*, p. 340.
57. D.C. Watt, *Personalities and Policies: Studies in the Formulation of British Foreign Policy in the Twentieth Century* (London: Longmans, 1965), p. 180 and Max Nicholson, *The System: The Misgovernment of Modern Britain* (New York: McGraw-Hill Book Company, 1967), p. 265. The informed reaction to this change has been condemnatory. See, for example, Harold Nicholson, *Curzon; The Last Phase, 1919-1925: A Study in Post-War Diplomacy* (2d. ed.; London: Constable & Co. Ltd., 1937), pp. 185-7 and John Connell, *The 'Office': The Story of the British Foreign Office, 1919-1951* (New York: St Martin's Press, 1958), p. 9, who asserts 'the demagogic politicians on their rostrums were far clumsier negotiators than the old diplomatists in their quiet libraries'.
58. Roskill, *op. cit.*, p. 215.
59. As quoted in R.J. Minney, *The Private Papers of Hore-Belisha* (Garden City, New York: Doubleday & Company, Inc., 1961), p. 32.
60. Terence H. O'Brien, *Civil Defence* (London: HMSO, 1955), pp. 32-3. O'Brien also describes the 'optimistic atmosphere' of the summer of 1928 wherein this decision can be understood as part of the general mood. For other details on the Ten Year Rule, see Denis Richards, *Royal Air Force, 1939-1945*, Vol. I: *The Fight at Odds* (7th ed.; London: HMSO, 1961), p. 18 and fn.
61. Boyle, *loc. cit.*
62. Michael Howard (ed.), *Soldiers and Governments* (Bloomington, Indiana: Indiana University Press, 1959), p.13.
63. Stefan T. Possony, *Strategic Air Power: The Pattern of Dynamic Security* (Washington: Infantry Journal Press, 1949), p. 199.
64. That the budget division lay at the heart of the early interservice fight over the future of the RAF is attested to, for example, by Air Chief Marshal Sir Basil E. Embry, 'Tallyho: To the Aid of the RAF,' *Air University Review*, XVIII (July-August, 1967), 69 and by the aerial historian Peter Lewis, *op. cit.*, p. 100.
65. As quoted in Spaight, *op. cit.*, p. 207.
66. Grey, *op. cit.*, p. 104 and Spaight, *loc. cit.*
67. See Chapter 5, pp. 114-18.
68. Sykes, *From Many Angles*, pp. 272-3, 291 and Macmillan, *op. cit.*, p. 234.
69. Spaight, *op. cit.*, pp. 208-211.
70. Ashmore *op. cit.*, pp. 113-18 and Basil Collier, *The Defence of the United Kingdom* (London: HMSO, 1957), pp. 5-6.
71. Cyril Falls, *A Hundred Years of War, 1850-1950* (3d ed.; New York: Collier Books, 1967), pp. 270-3.
72. L.S. Amery, *My Political Life*, Vol. II: *War and Peace, 1914-1929* (London: Hutchinson, 1953), pp. 201-202. Amery claims the credit for the idea to use the RAF (he was Parliamentary Under-Secretary for the Colonies at the time) but one can assume Lord Ismay is correct in his assertion that the air staff were pressing their own cause too; they were looking for useful assignments. Ismay is also a nice balance to the tendency to

overstate the role of air in this campaign; it was hardly 'the Air Ministry's first private war'. Grey, *op. cit.*, p. 173 and General the Lord Ismay, *Memoirs* (London: Heinemann, 1960), pp. 20-35.

73. Churchill, *The Aftermath*, p. 491, Air Chief Marshal Sir Arthur Longmore, *From Sea to Sky, 1910-1945* (London: Geoffrey Bles, 1946), p. 103, Saundby, *op. cit.*, p. 37 and Marshal of the Royal Air Force Sir John Slessor, *The Central Blue: Recollections and Reflections* (London: Cassell and Company, Limited, 1956), p. 52.

74. Aaron S. Klieman, *Foundations of British Policy in the Arab World: The Cairo Conference of 1921* (Baltimore: The Johns Hopkins Press, 1970), pp. 87, 110-11.

75. *Ibid.*, pp. 111-12, Templewood, *op. cit.*, pp. 47-8 and Robert Graves and Liddell Hart, *T.E. Lawrence to his Biographers* (London: Cassell, 1963), part II, p. 143. The Cabinet discussion and approval is in Great Britain, Public Record Office, Cab. 70 (21) in CAB 23/26.

76. Slessor, *op. cit.*, p. 66, Saundby, *op. cit.*, p. 41 and Boyle, *op. cit.*, pp. 508-9. Contrary to statements by some of the Labour rank and file, the party leadership endorsed the RAF approach to imperial policing as being a humane one. Lord Thomson, *Air Facts and Problems* (New York: George H. Doran Company, 1927), pp. 80-81 and David Carlton, 'The Problem of Civil Aviation in British Disarmament Policy, 1919-1934', *Journal of the Royal United Service Institution*, CXI (November, 1966), 309 and 313.

77. *Ibid.*, pp. 510-511 and Basil Collier, *Leader of the Few: The Authorised Biography of Air Chief Marshal the Lord Dowding of Bentley Priory, G.C.B., G.C.V.O., C.M.G.* (London: Jarrolds, 1957), p. 125.

78. As quoted in Slessor, *op. cit.*, p. 52. See also pp. 62-3 with respect to the advanced warning policy.

79. Air Chief Marshal Sir Basil Embry, *Mission Completed* (London: Methuen & Co. Ltd., 1957), p. 29 and J.M. Spaight, *Pseudo-Security* (London: Longmans, Green and Co., 1928), p. 113. Illogical or not, the charge kept being repeated. Lord Lloyd, for example, spoke in the House of Lords (in 1930) against air control policies in the Empire and reasoned that 'a bomb dropped from the air could not distinguish between the innocent and the guilty . . . '. Colin Forbes Adam, *Life of Lord Lloyd* (London: Macmillan & Co. Ltd., 1948), p. 233. Even as recently as December 1953, like arguments were being made about air action against the Mau Mau. Slessor, *op. cit.*, p. 51. fn.

80. 'Air Control became the accepted term for a system under which these small tribal wars were dealt with primarily by air action, in which the RAF was the predominant arm, the responsible commander was an airman and aircraft played a prominent part in the ordinary routine administration of the country'. *Ibid.*, p. 56.

81. Saundby, *op. cit.*, p. 42 and Georg W. Feuchter, *Geschichte des Luftkriegs: Entwicklung und Zukunft* (Bonn: Athenäum-Verlag, 1954), p. 41.

82. Lieutenant-General E.L.M. Burns, *Megamurder* (New York: Pantheon Books, 1966), pp. 32-33.

83. Slessor, *op. cit.*, p. 65. The relationship to strategic bombing also includes that dependency upon accurate intelligence. Possony, *op. cit.*, p. xi.

84. Churchill, *The Aftermath*, p. 495.

85. Slessor, *op. cit.*, pp. 51-2. When the Cabinet authorised this RAF command, the Secretary of State for War, Sir L. Worthington-Evans, made the very unusual gesture of requesting that his dissent be recorded. Great Britain, Public Record Office, Cab. 70 (21) in CAB 23/26.

86. Saundby, *op. cit.*, p. 34, Lewis, *op. cit.*, p. 100 and Thomson, *op. cit.*, p. 43.
87. This charge has been given recent endorsement as holding true for the entire interwar era in E.B. Potter and Chester W. Nimitz (eds.), *Sea Power: A Naval History* (Englewood Cliffs, New Jersey: Prentice-Hall, Inc., 1960), p. 477.
88. Roskill, *op. cit.*, pp. 249-55 and Grey, *op. cit.*, p. 185.
89. Roskill, *op. cit.*, pp. 190-2, 255-6. It was in June 1920 that Mustafa Kemal's forces had attacked the British on the Ismid Peninsula. G.M. Gathorne-Hardy, *A Short History of International Affairs, 1920-1939* (4th ed.; London: Oxford University Press, 1964), p. 119.
90. Liddell Hart, *op. cit.*, pp. 325-6. That battleship-cathedral was threatened not only by aircraft but also by that other major new war weapon of the Great War, the submarine. The Admiralty's hyper-sensitivity must be understood in that context of double jeopardy.
91. Samuel Taylor Moore, *U.S. Airpower: Story of American Fighting Planes and Missiles from Hydrogen Bags to Hydrogen Warheads* (New York: Greenberg, 1958), pp. 33-4 and Claude Grahame-White, *The Story of the Aeroplane* (Boston: Small, Maynard and Company, 1911), p. 268.
92. As quoted in David Davies, *The Problem of the Twentieth Century: A Study in International Relationships* (London: Ernest Benn Limited, 1930), pp. 440-41, fn. 2. For another example of Lord Fisher's rhetoric concerning air power, see Isaac Don Levine, *Mitchell: Pioneer of Air Power* (New York: Duell, Sloan and Pearce, 1943), p. 205.
93. Brigadier-General P.R.C. Groves, 'For France to Answer', *The Atlantic Monthly*, CXXXIII (February, 1924), p. 147.
94. Scott's letter of 11 December 1920, is quoted in Levine, *op. cit.*, p. 206.
95. Macmillan, *op. cit.*, pp. 236-7. Brancker's suggestion shows how that basic problem of budget competition enters into this battleship controversy.
96. See, for example, Archibald Hurd, 'Is the Battleship Doomed?', *The Fortnightly Review* CVII (February 2 1920), pp. 222-35 and his 'Great Ships or . . . ?': A Footnote to the *Times* Correspondence', *ibid.*, CIX (1 February 1921), pp. 240-54.
97. Roskill, *op. cit.*, pp. 113-15.
98. Levine, *op. cit.*, p. 101 and for a sampling of Brigadier-General William Mitchell's articles (among a very wide choice) see his 'Our Army's Air Service', *The American Review of Reviews*, LXII (September, 1920), pp. 281-90, 'Aviation over the Water', *ibid.*, LXII (October, 1920), pp. 391-8, 'Air Power vs. Sea Power', *ibid.*, LXIII (March, 1921), pp. 273-7, and 'Has the Airplane made the Battleship Obsolete?', *The World's Work*, XLI (April, 1921), pp. 550-55.
99. Moore, *op. cit.*, p. 78 and Levine, *op. cit.*, p. 215.
100. There are many accounts available of Mitchell's tests. Roskill, *op. cit.*, pp. 247-9 offers a recent and good account and Bernard Brodie is always worth consulting; see his *Sea Power in the Machine Age* (2d ed.; Princeton, New Jersey: Princeton University Press, 1943), pp. 401-2.
101. Marshal of the Royal Air Force Lord Douglas of Kirtleside, *Combat and Command: The Story of the Airman in Two World Wars* (New York: Simon and Schuster, 1966), p. 307.
102. Possony, *op. cit.*, p. 214.
103. Harry H. Ransom, 'Lord Trenchard, Architect of Air Power', *Air University Quarterly Review*, VIII (Summer, 1956), p. 59.
104. As quoted in Gerald E. Wheeler, 'Mitchell, Moffett, and Air Power', *The Airpower Historian*, VIII (April, 1961), p. 83.
105. The *New York Times*, 30 July 1921 and Levine, *op. cit.*, p. 267.

106. Lieutenant Colonel William H. Tomlinson, 'The Father of Airpower Doctrine', *Military Review*, XLVI (September 1966), pp. 27-31.
107. Bernard Brodie, *Strategy in the Missile Age* (Princeton, New Jersey: Princeton University Press, 1959), p. 22.
108. Theodore Ropp, *War in the Modern World* (rev. ed.; New York: Collier Books, 1965), p. 292.
109. See, for example, Raymond Aron, *The Century of Total War* (Boston: The Beacon Press, 1955), p. 39, Burns, *op. cit.*, p. 23. Millis, *op. cit.*, p. 227, Laurence Thompson, *The Greatest Treason: The Untold Story of Munich* (New York: William Morrow & Company, Inc., 1968), p. 6, Cy Caldwell, *Air Power and Total War* (New York: Coward-McCann, Inc., 1943), p. 98, and Hans Rumpf, *The Bombing of Germany*, trans. Edward Fitzgerald (London: Frederick Muller Limited, 1961, p. 227.
110. The most important work by Douhet, first published in Italy in 1921, is available in its best American edition as Giulio Douhet, *The Command of the Air*, trans. Dino Ferrari (rev. ed.; New York: Coward-McCann, Inc., 1942). Good analyses of his thought are available in Brodie, *Strategy in the Missile Age*, pp. 71-106 and in Raymond Richard Flugel, 'United States Air Power Doctrine: A Study of the Influence of William Mitchell and Giulio Douhet at the Air Corps Tactical School, 1921-1935' (unpublished Ph.D. dissertation, Dept. of History, University of Oklahoma, 1965), pp. 73-92. An excellent nine point condensation of Douhet's ideas is given by Lt. Col. Joseph L. Dickman, 'Douhet and the Future', *Air University Quarterly Review*, II (Summer, 1948), p. 4.
111. Robin Higham is one who postulates the self-rooted origins of British air doctrine in his downgrading of Douhet's influence. Higham, *op. cit.*, pp. 195-6. Also see pp. 257-9 wherein Higham explores 'The Place of Douhet'.
112. Slessor, *op. cit.*, p. 41 and B.H. Liddell Hart, *History of the Second World War* (New York: G.P. Putnam's Sons, 1970), p. 590 and fn.
113. *Ibid.*, and Flugel, *op. cit.*, pp. 94-5 and 201. Major Perry Smith has judged that Flugel's research is conclusive with respect to Douhet's influence, even as to the texts used at Langley Fields Tactical School in the 1920s and Langley was the school which brought forth (despite its name) the formulation of US strategic bombing doctrine. Perry M. Smith, 'Douhet and Mitchell: Some Reappraisals', *Air University Review*, XVIII (September-October, 1967), pp. 98-9.
114. Higham, *op. cit.*, p. 258.
115. Templewood, *op. cit.*, p. 258.
116. Webster and Frankland, *op. cit.*, p. 13 and Roskill, *op. cit.*, p. 243.
117. As quoted in *ibid.*, p. 257. Similar disdain towards naval priorities in the use of air power is expressed in Great Britain, Public Record Office, AIR 9/69, Folio #29.
118. As quoted in Boyle, *op. cit.*, pp. 395-6.
119. *Ibid.*, pp. 396-7 and Roskill, *op. cit.*, p. 264.
120. *Ibid.*, pp. 264-5 and Boyle, *op. cit.*, 398.
121. Bloomsbury, typically, could have a dissenting view. Lytton Strachey wrote of Balfour (in 1928): 'I like watching him — the perfection of his manners — the curious dimness — the wickedness one catches glimpses of underneath. But of course any communication of ideas is totally out of the question. One might as well talk to the man in the moon'. As quoted in Michael Holroyd, *Lytton Strachey: A Critical Biography*, Vol. II: *The Years of Achievement (1910-1932)* (New York: Holt, Rinehart and Winston, 1968), p. 573.

122. Johnson, *op. cit.*, pp. 170-72, Collier, *The Defence of the United Kingdom*, pp. 8-9, and Lord Chandos, *Memoirs: An Unexpected View from the Summit* (New York: New American Library, 1963), pp. 66-7.
123. Owen, *op. cit.*, p. 408.
124. E.L. Woodward and Rohan Butler (eds.), *Documents on British Foreign Policy, 1919-1939*, Vol. I: *1919* (London: HMSO, 1947), pp. 672-3.
125. As quoted in Longmore, *op. cit.*, p. 101.
126. As quoted in Boyle, *op. cit.*, pp. 399-400.
127. *Ibid.*, p. 398.
128. *Ibid.*, p. 400 and Roskill, *op. cit.*, pp. 265-6. For representative examples of air staff memos connected with this Balfour inquiry, see Great Britain, Public Record Office, AIR 8/2, Part III, Folios 1 and 2.
129. Roskill, *op. cit.*, p. 267 and Boyle, *op. cit.*, p. 403.
130. D.C. Somervell, *Modern Britain, 1870-1939* (London: Methuen & Co., Ltd., 1941), pp. 144-5, Charles Loch Mowat, *Britain Between the Wars, 1918-1940* (Chicago, University of Chicago Press, 1963), pp. 129-131 and Andrew Boyle, *Montagu Norman: A Biography* (London: Cassell, 1967), pp. 130-31.
131. Macmillan, *op. cit.*, p. 156 and Roskill, *op. cit.*, p. 20.
132. The Geddes Committee's first and second interim reports are analysed and quoted in the 'Military Notes' section of the *Journal of the Royal United Service Institution*, LXVII (May, 1922), pp. 378-86. Also see Boyle, *Trenchard*, pp. 401-8 and Roskill, *op. cit.*, p. 268.
133. *Ibid.*, pp. 356-9 and Boyle, *Trenchard*, p. 409. Beatty, in fact, was censured by the Cabinet for his antagonistic dealings with the Geddes Committee. Great Britain, Public Record Office, Cab. 9 (22) and Cab. 11 (22) in CAB 23/29.
134. Boyle, *Trenchard*, pp. 430-31.
135. Lord William Strang, *Britain in World Affairs: The Fluctuation in Power and Influence from Henry VIII to Elizabeth II* (New York: Frederick A. Praeger, 1961), pp. 297-8. Another authority has stated that, except for the talks starting in 1924, that led to Locarno, the two governments 'were never in step throughout the whole of the interwar period'. Dorothy Pickles, *The Uneasy Entente: French Foreign Policy and Franco-British Misunderstandings* (London: Oxford University Press, 1966), p. 1.
136. Watt, *op. cit.*, p. 145.
137. Reginald Pound and Geoffrey Harmsworth, *Northcliffe* (London: Cassell, 1959), pp. 852-3 and the Earl of Swinton, *Sixty Years of Power: Some Memories of the Men Who Wielded It* (New York: James H. Heineman, Inc., 1966), p. 50.
138. The caustic exchange of comments over this programme are quoted in I.I. Minz, 'Die Washingtoner Konferenz und der Neunmächtepakt', *Geschichte der Diplomatie*, Vol. III, part I: *Die Diplomatie in der Periode der Vorbereitung des zweiten Weltkrieges (1919-1939)*, ed. W.P. Potjomkin (Berlin: SWA-Verlag, 1948), p. 172. England had already attempted to persuade France to give up her submarine programme in 1920. Georges Bonnet, *Quai d'Orsay* (Isle of Man: Times Press and Anthony Gibbs & Phillips, 1965), p. 49.
139. Roskill, *op. cit.*, pp. 306, 323, fn. 1 and Jones, *op. cit.*, p. 178. RAF archival evidence abounds of warnings over the French danger. For representative examples, see Great Britain, Public Record Office, AIR 9/8, Folios, 9, 10 and 11.
140. Collier, *The Defence of the United Kingdom*, pp. 11-14, O'Brien, *op. cit.*,

Notes

p. 12, Boyle, *Trenchard*, pp. 431-2 and Higham, *op. cit.*, p. 177.
141. See Chapter 5, pp. 129-31.
142. J.R.M. Butler, *Lord Lothian (Philip Kerr), 1882-1940* (London: Macmillan & Co. Ltd., 1960), p. 78.
143. Lord Salter, *Memoirs of a Public Servant* (London: Faber and Faber, 1961), p. 160, Connell, *op. cit.*, p. 41 and W.N. Medlicott, *Contemporary Europe, 1914-1964* (New York: David McKay Company, Inc., 1967), p. 169.
144. F.S. Northedge, *The Troubled Giant: Britain Among the Great Powers, 1916-1939* (New York: Frederick A. Praeger, 1966), pp. 150-51 and Ivone Kirkpatrick, *The Inner Circle* (London: Macmillan & Co., Ltd., 1959), p. 35.
145. Alfred Fabre-Luce, *Histoire de la Révolution Européenne* (Paris: Librairie Plon, 1960), p. 28.
146. Lt.-Col. Charles à Court Repington, *After the War: London-Paris-Rome-Athens-Prague-Vienna-Budapest-Bucharest-Berlin-Sofia-Coblenz-New York-Washington: A Diary* (Boston: Houghton Mifflin Company, 1922), pp. 63, 70, 194.
147. Herbert Tint, *The Decline of French Patriotism, 1870-1940* (London: Weidenfeld and Nicolson, 1964), pp. 172-3. I have extensively analysed this French Ruhr policy and the British reaction to it in a paper, 'Locarno: The Easement of the *Mésentente*', delivered at the Mars Hill College symposium on 'European Security in the Locarno Era' held 16-18 October 1975.
148. Tint, *op. cit.*, p. 173, Amery, *op. cit.*, p. 247, Medlicott, *op. cit.*, p. 179 and Keith Middlemas and John Barnes, *Baldwin, a Biography* (London: Macmillan, 1969), p. 136.
149. *Ibid.*, pp. 150-51, 191, 193 fn. and Amery, *op. cit.*, p. 268.
150. Sir Charles Petrie, *Diplomatic History, 1713-1933* (London: Hollis and Carter Ltd., 1948), pp. 337, 339, Catherine Gavin, *Britain and France: The Entente Cordiale: a Study of Twentieth Century Relations* (London: Jonathan Cape, 1941), pp. 266-7 and Alan Campbell Johnson, *Viscount Halifax: A Biography* (London: Robert Hale Limited, 1941), p. 109.
151. Douglas, *op. cit.*, p. 307.
152. Oswald Spengler, *Spengler Letters, 1913-1936*, trans. and ed. Arthur Helps (London: George Allen & Unwin Ltd,, 1966), pp. 118, 121, 124-5, Fabre-Luce, *op. cit.*, p. 60 and Nicolson *op.cit.*, p. 365.
153. The Committee's terms of reference are quoted in Roskill, *op. cit.*, pp. 371-2. Also see Templewood, *op. cit.*, pp. 36 and 60 and Boyle, *Trenchard*, pp. 458-464.
154. Franklyn Arthur Johnson, *op. cit.*, p. 196.
155. Robin Higham, *Armed Forces in Peacetime: Britain, 1918-1940: a case study* (London: G.T. Foulis & Co. Ltd., 1962), p. 161 and Templewood, *op. cit.*, pp. 61-2. As to the importance of Lord Weir, see Boyle, *Trenchard*, p. 464 and Roskill, *op. cit.*, p. 376.
156. *Ibid.*, pp. 373-5 and Templewood, *op. cit.*, pp. 66-7. Relevant RAF documents concerning the Salisbury Committee are found in Great Britain, Public Record Office, AIR 8/67.
157. Templewood, *op. cit.*, p. 65, Boyle, *Trenchard*, pp. 487-90 and Amery, *op. cit.*, pp. 264-5.
158. The French menace was, in fact, specifically referred to in the Salisbury Committee's interim report. Middlemas and Barnes, *op. cit.*, p. 320. Also see Boyle, *Trenchard*, p. 468 and Roskill, *op. cit.*, pp. 364, 373 fn. 6. The Cabinet approved the Salisbury Committee's recommendations with

a spirit of 'melancholy necessity'. Great Britain, Public Record Office, Cab. 32 (23) in CAB 23/46.
159. Jones, *op. cit.*, p. 239, Major-General Sir Frederick Maurice, *The Life of Viscount Haldane of Cloan, K.T., O.M.*, Vol. II: *Haldane, 1915-1928* (London: Faber and Faber Ltd., 1939), pp. 126-7 and Dudley Sommer, *Haldane of Cloan: His Life and Times, 1856-1928* (London: George Allen & Unwin Ltd., 1960), p. 389.
160. Collier, *The Defence of the United Kingdom*, p. 15 and Joubert de la Ferté, *op. cit.*, p. 105.
161. As quoted (including emphasis) in Boyle, *Trenchard*, p. 520.
162. Saundby, *op. cit.*, p. 50 and Higham, *The Military Intellectuals*, pp. 177-8.
163. The minutes of this conference are reprinted in Sir Charles Webster and Noble Frankland, *The Strategic Air Offensive against Germany, 1939-1945* Vol. IV: *Annexes and Appendices* (London: HMSO, 1961), pp. 62-70.
164. Saundby, *op. cit.*, p. 51 and Embry, 'Tallyho: To the Aid of the RAF,' pp. 72-3.
165. Ashmore, *op. cit.*, pp. 148-52.
166. Cf. Chapter 5, p. 149.
167. Viscount Templewood, *Nine Troubled Years* (London: Collins, 1954), pp. 112, 116.
168. Richards, *op. cit.*, p. 8. Middlemas and Barnes, *op. cit.*, p. 320 and Collier, *The Defence of the United Kingdom*, pp. 18-19. The delays in the air expansion programme are recorded in Great Britain, Public Record Office, AIR 8/73.
169. The Steel-Bartholomew Plan is analysed and diagrammed in Collier, *The Defence of the United Kingdom*, pp. 12-16. Also see General Sir Frederick Pile, *Ack-Ack: Britain's Defence Against Air Attack during the Second World War* (London: George G. Harrap & Co. Ltd., 1949), pp. 52-53, Wood and Dempster, *op. cit.*, pp. 69-71 and Webster and Frankland, *op. cit.*, pp. 61-62.
170. Ashmore, *op. cit.*, pp. 108-10, 127, 131-43, Collier, *The Defence of the United Kingdom*, pp. 16-20, Pile, *op. cit.*, pp. 53-6 and Major-General E. B. Ashmore, 'Anti-Aircraft Defence', *Journal of the Royal United Service Institution*, LXXII (February, 1927), pp. 3-14.
171. The Air Raid Precautions Committee was analysed in Chapter 5 with respect to its purview, as well as the concepts held by its political chiefs, especially those of Sir John Anderson, its chairman. See pp. 120-24.
172. The official history on this subject is O'Brien, *op. cit.*, cf. pp. 13-42 for the period covered in this chapter. Also see John W. Wheeler-Bennett, *John Anderson, Viscount Waverley* (London: Macmillan & Co. Ltd., 1962), pp. 96-8, Richard M. Titmuss, *Problems of Social Policy* (London: HMSO, 1950), pp. 8-21, Higham, *The Military Intellectuals*, pp. 179-85, Webster and Frankland, *op. cit.*, I, pp. 62-3, E.J. Kingston-McCloughry, *Defense Policy and Strategy* (New York: Frederick A. Praeger, 1960), pp. 208, 213 and E.M. Eddy, 'Britain and the Fear of Aerial Bombardment, 1935-1939', *Aerospace Historian*, XIII (Winter, 1966), p. 178.
173. O'Brien, *loc. cit.*
174. Higham, *The Military Intellectuals*, p.35.
175. Liddell Hart claimed that the Institution's official ties tended to create aridity — but he also described it as being one of the three important military journals in Britain and it is also to the point that his first published work was in this journal. Liddell Hart, *Memoirs*, I, pp. 35 and

11 (New York: G.P. Putnam's Sons, 1965), p. 77. *The Journal of the Royal United Service Institution* will henceforth be denoted as *JRUSI.*
176. Air Commodore H.R. Brooke-Popham, 'The Air Force', *JRUSI,* LXV (February, 1920), pp. 43-70.
177. Major-General Sir Louis C. Jackson, 'Possibilities of the Next War', *ibid.,* pp. 78-89.
178. Group Captain J.A. Chamier, 'Strategy and Air Strategy', *ibid.,* LXVI (November 1921), pp. 641-61.
179. Flight-Lieut. C.J. Mackay, 'The Influence in the Future of Aircraft upon Problems of Imperial Defence', *ibid.,* LXVII (May, 1922), pp. 274-310.
180. Squadron Leader B.E. Sutton, 'Some Aspects of the Work of the Royal Air Force with the BEF in 1918', *ibid.,* pp. 336-48.
181. Major R. Chenevix-Trench, 'Gold Medal (Military) Prize Essay for 1922', *ibid.,* LXVIII (May, 1923), pp. 199-227.
182. Major-General W.D. Bird, 'Thoughts on our Requirements in Relation to an Air Force', *ibid.,* pp. 291-4.
183. Wing Commander C.H.K. Edmonds, 'Air Strategy', *ibid.,* LXIX (May, 1924), pp. 191-210.
184. Major E.R. Macpherson, 'The Development of Chemical Warfare', *ibid.,* LXXI (February, 1926), pp. 94-106.
185. Group Captain W.F. MacNeece, 'Certain Aspects of Air Defence,' *ibid.,* LXXI (February, 1926), pp. 94-106.
186. Cf. *supra,* p. 191.
187. Colonel J.P. Villiers-Stuart, 'The Nation in Relation to its Armed Forces', *JRUSI,* LXXI (August, 1926), pp. 500-13.
188. Flight Lieutenant W.T.S. Williams, 'Air Exercises, 1927', *ibid.,* LXXII (November, 1927), pp. 739-45.
189. For an early example, see Rear-Admiral Murray F. Sueter, *Airmen or Noahs: Fair Play for our Airmen; the Great 'Neon' Air Myth Exposed* (London: Sir Isaac Pitman & Sons, Ltd., 1928), p. 176.
190. Higham, *Armed Forces in Peacetime,* p. 165. Of course it is easier for the defence to detect raids when it is expecting them, as in these exercises.
191. Group-Captain W.F. MacNeece Foster, 'Air Power and Its Application', *JRUSI,* LXXII (May, 1928), pp. 247-61.
192. Major Victor Lefebure, 'Chemical Warfare', *ibid.,* LXXIII (August, 1928), pp. 492-507.
193. Major C.C. Turner, 'The Aerial Defence of Cities: Some Lessons of the Air Exercises, 1928', *ibid.,* LXXIII (November, 1928), pp. 692-9. Major Turner later expanded his aerial fears into a book. Major C.C. Turner, *Britain's Air Peril: The Danger of Neglect, Together with Considerations on the Role of an Air Force* (London: Sir Isaac Pitman & Sons, Ltd., 1933).
194. Colonel H.W. Hill, 'Air Defence', *JRUSI,* LXXV (February, 1930) pp. 103-16.
195. Flight Lieutenant W.M. Yool, 'Air Exercises 1930', *ibid.,* LXXV (November, 1930), pp. 755-62.
196. Squadron Leader J.C. Slessor, 'The Development of the Royal Air Force', *ibid.,* LXXVI (May, 1931), pp. 324-34.
197. Cf. *supra,* p. 198.
198. Major C.C. Turner, 'British and Foreign Air Exercises of 1931', *JRUSI,* LXXVI (November, 1931), pp. 731-9.
199. Squadron Leader J.O. Andrews, 'The Strategic Role of Air Forces', *ibid.,* pp. 740-3.
200. Air Marshal Sir H.M. Trenchard, 'Aspects of Service Aviation', *The Army Quarterly,* II (April, 1921), pp. 10-21.
201. See *supra,* pp. 193-4.
202. Group Captain J.A. Chamier, 'Co-operation of Aircraft with Artillery', *The*

Army Quarterly, IV (April, 1922), pp. 46-62.
203. Major-General W.D. Bird, 'Some Speculations on Aerial Strategy', *ibid.*, IV (July, 1922), pp. 248-52.
204. Squadron Leader A.A. Walser, 'The Influence of Aircraft on Problems of Imperial Defence', *ibid.*, V (October, 1922), pp. 38-49.
205. Major-General W.D. Bird, 'One Air Force or Three?', *ibid.*, V (January, 1923), pp. 352-5.
206. Major C.F. Stoehr, 'The Bertrand Stewart Prize Essay, 1923', *ibid.*, VI (July, 1923), pp. 237-73. Stoehr was another who used Foch's statement to bolster his argument. See Chapter 5, pp. 127 and 129.
207. Captain K.M. Loch, 'Anti-Aircraft Ground Defences', *The Army Quarterly*, VIII (April, 1924), pp. 131-42.
208. Captain McA. Hogg, 'Aeroplanes in Future Warfare', *ibid.*, IX (October, 1924), pp. 98-107.
209. Captain Rupert de la Bere, 'The Bertrand Stewart Prize Essay, 1925', *ibid.*, X (July, 1925), pp. 236-57.
210. 'OTAC', 'The Offensive Side of Chemical Warfare', *ibid.*, XI (October, 1925), pp. 111-21.
211. Brigadier-General H. Hartley, 'Chemical Warfare', *ibid.*, XIII (January, 1927), pp. 240-51. This article was a reprinting of a lecture given by General Hartley at the University of London on November 25, 1926.
212. Major Oliver Stewart, 'The Air Exercises', *ibid.*, XV (October, 1927), pp. 114-25.
213. Major Oliver Stewart, 'The Air Exercises', *ibid.*, XVII (January, 1929), pp. 262-75.
214. Captain C.B. Thorne, 'The Defence of the Civil Population', *ibid.*, XXI (October, 1930), pp. 70-80.
215. Major Oliver Stewart, 'The Air Exercises, 1930: Balance of Air Forces', *ibid.*, pp. 87-98 and also his 'The Royal Air Force Exercises of 1931', *ibid.*, XXIII (October, 1931), pp. 109-15, from which the quotation is taken.
216. Wing-Commander A.G.B. Garrod, 'Air Strategy', *The Royal Air Force Quarterly*, I (January, 1930), pp. 28-36.
217. Liddell Hart, *The Defence of Britain* (London: Faber and Faber Limited, 1939), p. 342.
218. A good account of Fuller's military theories is in Higham, *The Military Intellectuals in Britain*, pp. 67-81. At least by 1923, Fuller was an outspoken claimant for air power, at one time even foreseeing a London insane with panic and giving in after, perhaps, only two days of air raids. Spaight, *Pseudo-Security*, pp. 116, 125. Also see Edward Mead Earle (ed.), *Makers of Modern Strategy: Military Thought from Machiavelli to Hitler* (Princeton, New Jersey: Princeton University Press, 1943), p. 376 fn. and Quester, *op. cit.*, pp. 56-7.
219. J.F.C. Fuller, 'The Supremacy of Air Power', *The Royal Air Force Quarterly*, I (April, 1930), pp. 240-6.
220. Lieut.-Colonel P.T. Etherton, *Adventures in Five Continents* (London: Hutchinson & Co. Ltd., 1928), pp. 278-84.
221. Major General Sir F. Maurice, *Governments and War: A Study of the Conduct of War* (London: William Heinemann Ltd., 1926), pp. 156-7.
222. Field-Marshal Sir William Robertson, *From Private to Field-Marshal* (Boston: Houghton Mifflin Company, 1921), pp. 350-51.
223. See *supra*, p. 175.
224. Lord Fisher, *Memories and Records*, Vol. I: *Memories* (New York: George H. Doran Company, 1920), p. 131.
225. See *supra*, p. 188.

226. Air Vice-Marshal H.R.M. Brooke-Popham, 'Air Warfare', in Major-General Sir George Aston (ed.), *The Study of War for Statesmen and Citizens: Lectures Delivered in the University of London during the years 1925-26* (London: Longmans, Green and Co. Ltd., 1927), pp. 151-71.
227. Thomas Jones, *A Diary with Letters, 1931-1950* (London: Oxford University Press, 1954), pp. 159-60.
228. Colonel Roderick Macleod and Denis Kelly (eds.), *The Ironside Diaries, 1937-1940* (London: Constable, 1962), p. 61. The diary note is dated 19 September 1938.
229. Marshal of the RAF Sir John Slessor, 'Air Power and the Future of War', *JRUSI*, XCIX (August, 1954), p. 345.

EPILOGUE

1. This story has received the attention of 'official' historians. M.M. Postan, D. Hay and J.D. Scott, *Design and Development of Weapons: Studies in Government and Industrial Organisation* (London: HMSO, 1964), pp. 373-430.
2. Great Britain, Public Record Office, AIR 9/114. I am grateful to Mr E.B. Haslam, Head, Air Historical Branch (RAF) for bringing this document to my attention.
3. P.M.S. Blackett, *Studies of War: Nuclear and Conventional* (New York: Hill and Wang, 1962), pp. 169-234. The quoted phrase is from p. 173.
4. Vincent Massey, *What's Past is Prologue: Memoirs* (New York: St Martin's Press, 1964), pp. 268-9.
5. For just a few titles as an indication: the 'official' history is Basil Collier's *The Defence of the United Kingdom* (London: HMSO, 1957), Basil Collier has also written an excellent biography of Lord Dowding of the famed Fighter Command, *Leader of the Few: The Authorised Biography of Air Chief Marshal the Lord Dowding of Bentley Priory, G.C.B., G.C.V.O., C.M.G.* (London: Jarrolds, 1957). A companion volume to Collier's official account of Britain's air defence is Terence H. O'Brien, *Civil Defence* (London: HMSO, 1955). A good, popular introduction to the Battle of Britain and its prelude is Peter Townsend's *Duel of Eagles* (New York: Pocket Books, 1972) as is Derek Wood and Derek Dempster's *The Narrow Margin: The Battle of Britain and the Rise of Air Power, 1930-40*, 2nd ed. (London: Hutchinson, 1963).

BIBLIOGRAPHY

Books

Adam, Colin Forbes. *Life of Lord Lloyd.* London: Macmillan & Co., Ltd., 1948.

Addison, Christopher. *Politics from Within, 1911-1918: Including Some Records of a Great National Effort.* Vol. II. London: Herbert Jenkins Limited, 1924.

Amery, L.S. *My Political Life.* Vol. II: *War and Peace, 1914-1929.* London: Hutchinson, 1953.

Andler, Charles. *'Frightfulness' in Theory and Practise as Compared with Franco-British War Usages.* Translated by Bernard Miall. London: T. Fisher Unwin Ltd., 1916.

Aron, Raymond. *The Century of Total War.* Boston: The Beacon Press, 1965.

Arthur, Sir George. *Life of Lord Kitchener.* Vol. III. New York: The Macmillan Company, 1920.

Ashmore, E.B. *Air Defence.* London: Longmans, Green & Co., 1929.

Asquith, Lady Cynthia. *Diaries, 1915-1918.* London: Hutchinson, 1968.

Asquith, Herbert Henry. *Letters of the Earl of Oxford and Asquith to a Friend: First Series, 1915-1922.* Edited by Desmond MacCarthy. London: Geoffrey Bles, 1933.

——. *Memories and Reflections, 1852-1927.* Vol. II. London: Cassell and Company Limited, 1928.

Avon, Earl of. *The Eden Memoirs.* Vol. II: *Facing the Dictators,* London: Cassell, 1962.

Azeau, Henri, Caseneuve, Gilbert, and Saurel, Louis. *Histoire Vivante du xxe Siècle.* Vol. II: *L'Enfance du Siècle (1900-1912).* Paris: Robert Laffont, 1966.

Baldwin, Hansen W. *World War I: An Outline History.* New York: Harper & Row, 1962.

Bardens, Dennis. *Churchill in Parliament.* London: Robert Hale, 1967.

Baring, Maurice. *Flying Corps Headquarters, 1914-1918.* Revised edition. London: William Heinemann Ltd., 1930.

Barker, Ralph. *The Thousand Plan: The Story of the First Thousand Bomber Raid on Cologne.* London: Chatto & Windus, 1965.

Batchelder, Robert C. *The Irreversible Decision, 1939-1950.* Boston: Houghton Mifflin Company, 1962.

Baumbach, Werner. *The Life and Death of the Luftwaffe.* Translated by Frederick Holt. New York: Ballantine Books, 1967.
Baumont, Maurice. *La Faillite de la Paix (1918-1939).* Vol. I: *De Rethondes à Stresa (1918-1935).* 3d ed. revised. Paris: Presses Universitaires de France, 1951.
Beaver, Daniel R. *Newton D. Baker and the American War Effort, 1917-1919.* Lincoln, Nebraska: University of Nebraska Press, 1966.
Beaverbrook. Lord. *Men and Power, 1916-1918.* New York: Duell, Sloan and Pearce, 1956.
Bergonzi, Bernard. *Heroes' Twilight: A Study of the Literature of the Great War.* New York: Coward-McCann, Inc., 1966.
Bertie of Thame, Lord. *Diary, 1914-1918.* Vol. II. Edited by Lady Algernon Gordon Lennox. London: Hodder and Stoughton, Ltd., 1924.
Blackett, P.M.S. *Studies of War: Nuclear and Conventional.* New York: Hill and Wang, 1962.
Blake, Robert (ed.). *The Private Papers of Douglas Haig, 1914-1919: Being Selections from the Private Diary and Correspondence of Field-Marshal the Earl Haig of Bemersyde, K.T., G.C.B., O.M., etc.* London: Eyre & Spottiswoode, 1952.
Blumenfeld, R.D. *All in a Lifetime.* London: Ernest Benn Limited, 1931.
Blythe, Ronald. *The Age of Illusion: England in the Twenties and Thirties, 1919-1940.* London: Penguin Books, 1964.
Bonham-Carter, Victor. *The Strategy of Victory, 1914-1918: The Life and Times of the Master Strategist of World War I: Field-Marshal Sir William Robertson.* New York: Holt, Rinehart and Winston, 1963.
Bonham Carter, Violet. *Winston Churchill As I Knew Him.* London: Eyre & Spottiswoode and Collins, 1965.
Bonnet, Georges, *Quai d'Orsay.* Isle of Man: Times Press and Anthony Gibbs & Phillips, 1965.
Boyle, Andrew, *Montagu Norman: A Biography.* London: Cassell, 1967.
——. *Trenchard.* London: Collins, 1962.
Broad, Lewis, *Sir Anthony Eden: The Chronicles of a Career.* London: Hutchinson, 1955.
——. *Winston Churchill.* Vol. I: *The Years of Preparation.* London: Sidgwick and Jackson, 1963.
Broderick, Alan Houghton. *Near to Greatness: A Life of the Sixth Earl Winterton.* London: Hutchinson, 1965.
Brodie, Bernard. *Sea Power in the Machine Age.* 2d edition. Princeton, New Jersey: Princeton University Press, 1943.

——. *Strategy in the Missile Age.* Princeton, New Jersey: Princeton University Press, 1959.
Bruun, Geoffrey and Mamatey, Victor S. *The World in the Twentieth Century.* 5th ed. revised. Boston: D.C. Heath and Company, 1967.
Burns, E.L.M. *Megamurder.* New York: Pantheon Books, 1966.
Butler, J.R.M. *Lord Lothian (Philip Kerr), 1882-1940.* London: Macmillan & Co., Ltd., 1960.
Butler, Rohan and Bury, J.P.T. (eds.). *Documents on British Foreign Policy, 1919-1939.* Vol. VII (First Series). *1920.* London: HMSO, 1958.
Caldwell, Cy. *Air Power and Total War.* New York: Coward-McCann, Inc., 1943.
Carr, Edward Hallett. *The Twenty Years' Crisis, 1919-1939: An Introduction to the Study of International Relations.* 2d edition. New York: Harper Torchbooks, 1964.
Castellan, Georges. *Le Réarmement Clandestin du Reich, 1930-1935: Vu par le 2e Bureau de l'État-Major Français.* Paris: Librairie Plon, 1954.
Cecil, Viscount. *A Great Experiment: An Autobiography.* London: Jonathan Cape, 1941.
Chambers, Frank P. *This Age of Conflict: The Western World – 1914 to the Present.* 3d ed. revised. New York: Harcourt, Brace & World, Inc., 1962.
——. *The War behind the War, 1914-1918: A History of the Political and Civilian Fronts.* London: Faber and Faber Ltd., 1939.
Chamier, J.A. *The Birth of the Royal Air Force: The Early History and Experiences of the Flying Services.* London: Sir Isaac Pitman & Sons, Ltd., 1943.
Chandos, Lord. *Memoirs: An Unexpected View from the Summit.* New York: New American Library, 1963.
Charlton, L.E.O. *War Over England.* London: Longmans, Green and Co., 1936.
Churchill, Randolph S. *Lord Derby, 'King of Lancashire': The Official Life of Edward, Seventeenth Earl of Derby; 1865-1948.* London: Heinemann, 1959.
Churchill, Winston S. *The Aftermath: The World Crisis, 1918-1928.* New York: Charles Scribner's Sons, 1929.
——. *Amid these Storms: Thoughts and Adventures.* New York: Charles Scribner's Sons, 1932.
——. *The World Crisis.* Vols. I, II & IV. New York: Charles Scribner's Sons, 1923 and 1927.

Clark, Ronald W. *The Rise of the Boffins*. London: Phoenix House Ltd., 1962.
Clarke, I.F. *Voices Prophesying War, 1763-1984*. London: Oxford University Press, 1966.
Collier, Basil. *The Defence of the United Kingdom*. London: HMSO, 1957.
———. *Leader of the Few: The Authorised Biography of Air Chief Marshal the Lord Dowding of Bentley Priory, G.C.B., G.C.V.O., C.M.G.* London: Jarrolds, 1957.
Colvin, Ian. *The Life of Lord Carson*. Vol. III. New York: The Macmillan Company, 1937.
Connell, John. *The 'Office': The Story of the British Foreign Office, 1919-1951*. New York: St Martin's Press, 1958.
Crafford, F.S. *Jan Smuts: A Biography*. 2d edition. London: George Allen & Unwin Ltd., 1946.
Craig, Gordon A. *War, Politics and Diplomacy: Selected Essays*. New York: Frederick A. Praeger, Publishers, 1966.
Daalder, Hans. *Cabinet Reform in Britain, 1914-1963*. Stanford, Calif.: Stanford University Press, 1964.
Davies, David. *The Problem of the Twentieth Century: A Study in International Relationships*. London: Ernest Benn Limited, 1930.
Dickson, Lovat. *H.G. Wells: His Turbulent Life and Times*. New York: Atheneum, 1969.
Divine, David. *The Broken Wing: A Study in the British Exercise of Air Power*. London: Hutchinson, 1966.
Douglas of Kirtleside, Lord. *Combat and Command: The Story of the Airman in Two World Wars*. New York: Simon and Schuster, 1966.
Douhet, Giulio. *The Command of the Air*. Translated by Dino Ferrari. Rev. edition. New York: Coward-McCann, Inc., 1942.
Dugdale, Blanche E.C. *Arthur James Balfour: First Earl of Balfour: K.G., O.M. F.R.S., etc.*, Vol. II: *1906-1930*. New York: G.P. Putnam's Sons, 1937.
Earle, Edward Mead (ed.). *Makers of Modern Strategy: Military Thought from Machiavelli to Hitler*. Princeton, New Jersey: Princeton University Press, 1943.
Eckener, Hugh. *Count Zeppelin: The Man and his Work*. Translated by Leigh Farnell. London: Massie Publishing Company, Ltd., 1938.
Ehrman, John. *Cabinet Government and War, 1890-1940*. Hamden, Conn.: Archon Books, 1969.
Embry, Sir Basil E. *Mission Completed*. London: Methuen & Co., Ltd., 1957.

Emme, Eugene M. (ed.). *The Impact of Air Power: National Security and World Politics.* Princeton, New Jersey: D. Van Nostrand Co., Inc., 1959.
Etherton, P. T. *Adventures in Five Continents.* London: Hutchinson & Co., Ltd., 1928.
Fabre-Luce, Alfred. *Histoire de la Révolution Européenne.* Paris: Librairie Plon, 1960.
Falls. Cyril. *The Great War.* New York: Capricorn Books, 1961.
—. *A Hundred Years of War, 1850-1950.* New York: Collier Books, 1962.
Feuchter, Georg N. *Geschichte des Luftkriegs: Entwicklung und Zukunft.* Bonn: Athenäum-Verlag, 1954.
Fischer, Fritz. *Germany's Aims in the First World War.* New York: W.W. Norton & Co., Inc., 1967.
Fisher, Lord. *Memories and Records.* Vol. I: *Memories:* New York George H. Doran Company, 1920.
Fitzroy, Sir Almeric. *Memoirs.* Vol. II. 5th edition. London: Hutchinson & Co., n.d.
Foulois, Benjamin D. and Glines, C.V. *From the Wright Brothers to the Astronauts: The Memoirs of Major General Benjamin D. Foulois.* New York: McGraw-Hill Book Company, 1968.
Frankland, Noble. *The Bombing Offensive Against Germany: Outlines and Perspectives.* London: Faber and Faber, 1965.
Fredette, Raymond H. *The Sky on Fire: The First Battle of Britain 1917-1918 and the Birth of the Royal Air Force.* New York: Holt, Rinehart and Winston, 1966.
Fuller, J.F.C. *The Conduct of War, 1789-1961: A Study of the Impact of the French, Industrial and Russian Revolutions on War and its Conduct.* New Brunswick, New Jersey: Rutgers University Press, 1962.
Fyfe, Hamilton. *Northcliffe: An Intimate Biography.* New York: The Macmillan Company, 1930.
Gathorne-Hardy, G.M. *A Short History of International Affairs, 1920-1939.* 4th edition. London: Oxford University Press, 1964.
Gatzke, Hans W. *Stresemann and the Rearmament of Germany.* New York: The Norton Library, 1969.
Gavin, Catherine. *Britain and France: The Entente Cordiale; A Study of Twentieth Century Relations.* London: Jonathan Cape, 1941.
Glover, C.W. *Civil Defence: A Practical Manual Presenting with Working Drawings the Methods Required for Adequate Protection Against Aerial Attack.* London: Chapman & Hall Ltd., 1938.

Görlitz, Walter (ed.). *The Kaiser and His Court: The Diaries, Note Books and Letters of Admiral Georg Alexander von Müller, Chief of the Naval Cabinet, 1914-1918.* New York: Harcourt, Brace & World, Inc., 1959.
Goss, Hilton P. *Civilian Morale under Aerial Bombardment, 1914-1939.* (United States Air University Documentary Research Study Number 14). Montgomery, Alabama: Air University, 1948.
Grahame-White, Claude. *The Story of the Aeroplane.* Boston: Small, Maynard and Company, 1911.
—. and Harper, Harry. *The Aeroplane in War.* Philadelphia: J.B. Lippincott Company, 1912.
—. *Air Power: Naval, Military, Commercial.* London: Chapman & Hall, Ltd., 1917.
Granzow, Brigitte. *A Mirror of Nazism: British Opinion and the Emergence of Hitler, 1929-1933.* London: Victor Gollancz, Ltd., 1964.
Graves, Robert and Liddell Hart. *T.E. Lawrence to his Biographers.* London: Cassell, 1963.
Grey, C.G. *A History of the Air Ministry.* London: George Allen & Unwin Ltd., 1940.
Grey of Fallodon, Viscount. *Twenty-Five Years, 1892-1916.* Vol. II. New York: Frederick A. Stokes Company, 1925.
Guinn, Paul. *British Strategy and Politics: 1914 to 1918.* Oxford: The Clarendon Press, 1965.
Haldane, Richard Burdon. *An Autobiography.* Garden City, New York: Doubleday, Doran & Company, Inc., 1929.
Hancock, W.K. *Four Studies of War and Peace in this Century.* Cambridge: The University Press, 1961.
—. *Smuts.* Vol. I: *The Sanguine Years, 1870-1919.* Cambridge: The University Press, 1962.
—. and Poel, Jean van der (eds.). *Selections from the Smuts Papers.* Vol. IV: *November 1918-August 1919.* Cambridge: The University Press, 1966.
Harlech, Lord. *Must the West Decline?* New York: Columbia University Press, 1966.
Harris, Sir Arthur. *Bomber Offensive.* New York: The Macmillan Company, 1947.
Havighurst, Alfred F. *Twentieth-Century Britain.* Evanston, Illinois: Row, Peterson and Company, 1962.
Hay, Ian. *Arms and the Men.* London: HMSO, 1950.
Hersh, Samuel M. *Chemical and Biological Warfare: America's Hidden*

Arsenal. Indianopolis: The Bobbs-Merrill Company, 1968.
Higham, Robin. *Armed Forces in Peacetime: Britain 1918-1940, a Case Study.* London: G.T. Foulis & Co., Ltd., 1962.
—. *The British Rigid Airship, 1908-1931: A Study in Weapons Policy.* London: G.T. Foulis & Co., Ltd., 1961.
—. *The Military Intellectuals in Britain: 1918-1939.* New Brunswick, New Jersey: Rutgers University Press, 1966.
Hill, A.V. *The Ethical Dilemma of Science and Other Writings.* London: The Scientific Book Guild, 1962.
Hill, Sir Norman, *et al. War and Insurance.* London: Oxford University Press, 1927.
Hoeppner, Ernest Wilhelm von. *Germany's War in the Air: A Retrospect on the Development and the Work of our Military Aviation Forces in the World War.* Translated by J. Hawley Larnev. Leipsig: A.F. Koehler, 1921.
Holley, I.B., Jr. *Ideas and Weapons: Exploitation of the Aerial Weapon by the United States during World War I; A Study in the Relationships of Technological Advance, Military Doctrine, and the Development of Weapons.* New Haven, Conn.: Yale University Press, 1953.
Holroyd, Michael. *Lytton Strachey: A Critical Biography,* Vol. II; *The Years of Achievement (1910-1932).* New York: Holt, Rinehart and Winston, 1968.
Howard, Michael (ed.), *Soldiers and Governments.* Bloomington, Indiana: Indiana University Press, 1959.
Hudson, James J. *Hostile Skies: A Combat History of the American Air Service in World War I.* Syracuse, New York: Syracuse University Press, 1968.
Hull, Cordell. *Memoirs.* Vol. I. New York: The Macmillan Company, 1948.
Iklé, Fred Charles. *The Social Impact of Bomb Destruction.* Norman, Okla.: University of Oklahoma Press, 1958.
Inter-Parliamentary Union. *What Would Be the Character of a New War?* London: P.S. King & Son, Ltd., 1931.
Ismay, Lord. *Memoirs.* London: Heinemann, 1960.
Jasper, Ronald C.D. *George Bell, Bishop of Chichester.* London: Oxford University Press, 1967.
Jenkins, Roy. *Asquith.* London: Collins, 1964.
Johnson, Alan Campbell. *Viscount Halifax: A Biography.* London: Robert Hale Limited, 1941.
Johnson, Franklyn Arthur. *Defence by Committee: The British*

Committee of Imperial Defence, 1885-1959. London: Oxford University Press, 1960.

Jones, Thomas. *A Diary with Letters, 1931-1950.* London: Oxford University Press, 1954.

———. *Whitehall Diary.* Vol. I: *1916-1925.* Edited by Keith Middlemas. London: Oxford University Press, 1969.

Joubert de la Ferté, Sir Philip. *The Third Service: The Story behind the Royal Air Force.* London: Thames and Hudson, 1955.

Joynson-Hicks, W. *The Command of the Air or Prophecies Fulfilled: Being Speeches Delivered in the House of Commons.* London: Nisbet & Co., Ltd., 1916.

Kelly, Charles J., Jr. *The Sky's the Limit: The History of the Airlines.* New York: Coward-McCann, Inc., 1963.

Kenworthy, J.M. *Peace or War?* New York: Boni & Liveright, 1927.

Kerr, Mark. *Land, Sea, and Air: Reminiscences.* London: Longmans, Green and Co., 1927.

Kertesz, G.A. (ed.). *Documents in the Political History of the European Continent, 1815-1939.* Oxford: The Clarendon Press, 1968.

Keynes, John Maynard. *Essays In Biography.* Edited by Geoffrey Keynes. Rev. edition. New York: Horizon Press Inc., 1951.

Killen, John. *A History of the Luftwaffe.* Garden City, New York: Doubleday & Company, Inc., 1968.

King-Hall, Stephen. *Our Own Times, 1913-1938: A Political and Economic Survey.* 2d ed. revised. London: Nicholson and Watson, 1938.

Kingston-McCloughry, E.J. *Defense Policy and Strategy.* New York: Frederick A. Praeger, 1960.

———. *The Direction of War: A Critique of the Political Direction and High Command in War.* New York: Frederick A. Praeger, 1955.

Kirkpatrick, Ivone. *The Inner Circle.* London: Macmillan & Co., Ltd., 1959.

Klieman, Aaron S. *Foundations of British Policy in the Arab World: The Cairo Conference of 1921.* Baltimore: The Johns Hopkins Press, 1970.

Laqueur, Walter. *Russia and Germany: A Century of Conflict.* Boston: Little, Brown and Company Encounter Book, 1965.

Leasor, James. *The Clock with Four Hands: Based on the Experiences of General Sir Leslie Hollis.* New York: Reynal & Co., 1959.

Leonard, Roger Ashley (ed.). *A Short Guide to Clausewitz on War.* London: Weidenfeld and Nicolson, 1967.

Levine, Isaac Don. *Mitchell: Pioneer of Air Power.* New York: Duell,

Sloan and Pearce, 1943.
Lewis, Cecil. *Sagittarius Rising.* New York: Harcourt, Brace and Company, 1936.
Lewis, Peter. *The British Bomber Since 1914: Fifty Years of Design and Development.* London: Putnam, 1967.
Liddell Hart, B.H. *The Defence of Britain.* London: Faber and Faber Limited, 1939.
—. *The German Generals Talk.* New York: William Morrow & Co., 1948.
—. *History of the Second World War.* New York: G.P. Putnam's Sons, 1970.
—. *Memoirs,* 2 vols. New York: G.P. Putnam's Sons, 1965-1966.
—. *Paris: or the Future of War.* New York: E.P. Dutton & Company, 1925.
—. *The Real War, 1914-1918.* Boston: Little, Brown and Company, 1930.
Littell, Philip. *Books and Things.* New York: Harcourt, Brace and Howe, 1919.
Lloyd George, David. *War Memoirs.* Vol. II: *1915-1916* and Vol. IV: *1917.* Boston: Little, Brown and Company, 1933-1934.
Londonderry, Marchioness of. *Retrospect.* London: Frederick Muller Ltd., 1938.
Longmore, Sir Arthur. *From Sea to Sky, 1910-1945.* London: Geoffrey Bles, 1946.
Ludendorff, General Erich. *The General Staff and its Problems: The History of the Relations between the High Command and the German Imperial Government as Revealed by Official Documents. Vol. I.* Translated by F.A. Holt. New York: E.P. Dutton and Company, n.d.
Lukacs, John. *Historical Consciousness: or the Remembered Past.* New York: Harper & Row, Publ., 1968.
Luvaas, Jay. *The Education of an Army: British Military Thought. 1815-1940.* Chicago: The University of Chicago Press, 1964.
Macleod, Roderick and Kelly, Denis (eds.). *The Ironside Diaries, 1937-1940.* London: Constable, 1962.
Macmillan, Norman. *Sir Sefton Brancker.* London: William Heinemann Ltd., 1935.
—. *Tales of Two Air Wars.* London: G. Bell and Sons, Ltd., 1963.
Manvell, Roger and Fraenkel, Heinrich. *Herman Göring.* London: Heinemann, 1962.
Marder, Arthur J. (ed.). *Fear God and Dread Nought: The*

Correspondence of Admiral of the Fleet Lord Fisher of Kilverstone.
Vol. III: *Restoration, Abdication, and Last Years, 1914-1920.*
London: Jonathan Cape, 1959.

Marwick, Arthur, *Britain in the Century of Total War: War, Peace and Social Change, 1900-1967.* Boston: Little, Brown and Company, 1968.

——. *The Deluge: British Society and the First World War.* Boston: Little, Brown and Company, 1965.

Massey, Vincent. *What's Past is Prologue: Memoirs.* New York: St. Martin's Press, 1964.

Maurice, Sir Frederick. *Governments and War: A Study of the Conduct of War.* London: William Heinemann Ltd., 1926.

——. *The Life of Viscount Haldane of Cloan, K.T., O.M.* Vol.II: *Haldane, 1915-1928.* London: Faber and Faber Ltd., 1939.

Maurois, André. *A History of England.* Translated by Hamish Miles. Rev. edition. New York: Grove Press, Inc., 1960.

May, Arthur James. *Europe and Two World Wars.* New York: Charles Schribner's Sons, 1947.

McKee, Alexander. *The Friendless Sky: The Story of Air Combat in World War I.* New York: William Morrow and Company, 1964.

Medlicott, W.N. *Contemporary Europe, 1914-1964.* New York: David McKay Company, Inc., 1967.

Michaelis, Ralph. *From Bird Cage to Battle Plane: The History of the R.A.F.* New York: Thomas Y. Crowell Company, 1943.

Middlemas, Keith and Barnes, John. *Baldwin: A Biography.* New York: The Macmillan Company, 1970.

Millis, Walter. *Arms and Men: A Study in American Military History.* New York: Mentor, 1956.

Minney, R.J. *No. 10 Downing Street: A House in History.* Boston: Little, Brown and Company, 1963.

——. *The Private Papers of Hore-Belisha.* Garden City, New York: Doubleday & Company, Inc., 1961.

Mitchell, David. *Monstrous Regiment: The Story of the Women of the First World War.* New York: The Macmillan Company, 1965.

Montgomery of Alamein, Viscount. *A History of Warfare.* New York: The World Publishing Company, 1968.

Moore, Samuel Taylor. *U.S.Airpower: Story of American Fighting Planes and Missiles from Hydrogen Bags to Hydrogen War-heads.* New York: Greenberg, 1958.

Mowat, Charles Loch. *Britain Between the Wars, 1918-1940.* 4th edition. Chicago: University of Chicago Press, 1963.

Nicholson, Max. *The System: The Misgovernment of Modern England.* New York: McGraw-Hill Book Company, 1967.

Nicholson, Harold. *Curzon: The Last Phase, 1919-1925; A Study in Post-War Diplomacy.* 2d edition. London: Constable & Co., Ltd., 1937.

—. *King George the Fifth: His Life and Reign.* Garden City, New York: Doubleday, 1953.

Northedge, F.S. *The Troubled Giant: Britain among the Great Powers, 1916-1939.* New York: Frederick A. Praeger, 1966.

Norwich, Viscount. *Old Men Forget: The Autobiography of Duff Cooper.* London: Rupert Hart-Davis, 1953.

O'Brien, Terence H. *Civil Defence.* London: HMSO, 1955.

Owen, Frank. *Tempestuous Journey: Lloyd George, His Life and Times.* London: Hutchinson, 1954.

Panichas, George A. (ed.). *Promise of Greatness: The War of 1914-1918.* New York: The John Day Company, 1968.

Pemberton-Billing, N. *Air War: How to Wage It.* London: Gale & Polden, Ltd., 1916.

Petrie, Sir Charles. *Diplomatic History, 1713-1933.* London: Hollis and Carter Ltd., 1948.

Pickles, Dorothy. *The Uneasy Entente: French Foreign Policy and Franco-British Misunderstandings.* London: Oxford University Press, 1966.

Pile, Sir Frederick. *Ack-Ack: Britain's Defence against Air Attack during the Second World War.* London: George G. Harrap & Co. Ltd., 1949.

Possony, Stefan T. *Strategic Air Power: The Pattern of Dynamic Security.* Washington: Infantry Journal Press, 1949.

Postan, M. M., Hay, D. and Scott, J.D. *Design and Development of Weapons: Studies in Government and Industrial Organisation.* London: HMSO., 1964.

Postgate, Raymond and Vallance, Aylmer. *England Goes to Press: The English People's Opinion on Foreign Affairs as Reflected in their Newspapers Since Waterloo (1815-1937).* New York: The Bobbs-Merrill Company, 1937.

Potjomkin, W.P. (ed.). *Geschichte der Diplomatie.* Vol. I: *Die Diplomatie in der Periode der Vorbereitung des Zweiten Weltkrieges (1919-1939).* Berlin: SWA-Verlag, 1948.

Potter, E.B. and Nimitz, Chester W. (eds.). *Sea Power: A Naval History.* Englewood Cliffs, New Jersey: Prentice-Hall, Inc., 1960.

Pound, Reginald and Harmsworth, Geoffrey. *Northcliffe.* London: Cassell, 1959.

Prentiss, Augustin M. *Chemicals in War: A Treatise on Chemical Warfare.* New York: McGraw-Hill Book Company, Inc., 1937.

Quester, George H. *Deterrence before Hiroshima: The Airpower Background of Modern Strategy.* New York: John Wiley & Sons, Inc., 1966.

Raleigh, Sir Walter Alexander and Jones, H.A. *The War in the Air: Being the Story of the Part Played in the Great War by the Royal Air Force.* 6 vols. Oxford: The Clarendon Press, 1922-1937.

Rawlinson, Alfred. *The Defence of London: 1915-1918.* 2d edition. London: Andrew Melrose, Ltd., 1923.

Raymond, John (ed.). *The Baldwin Age.* London: Eyre & Spottiswoode, 1960.

Repington, C. à Court. *After the War: London-Paris-Rome-Athens-Prague-Vienna-Budapest-Bucharest-Berlin-Sofia-Coblenz-New York-Washington; A Diary.* Boston: Houghton Mifflin Company, 1922.

——. *The First World War, 1914-1918: Personal Experiences of Lieut.-Col. C. à Court Repington.* Vol. II. London: Constable and Co. Ltd., 1920.

Richards, Denis. *Royal Air Force, 1939-1945.* Vol. I: *The Fight at Odds.* 7th edition. London: HMSO, 1961.

Rickards, Maurice. *Posters of the First World War.* New York: Walker and Company, 1968.

Riddell, Lord. *War Diary, 1914-1918.* London: Ivor Nicholson & Watson Ltd., 1933.

Robertson, Sir William. *From Private to Field-Marshal.* Boston: Houghton Mifflin Company, 1921.

——. *Soldiers and Statesmen, 1914-1918.* Vol. II. New York: Charles Scribner's Sons, 1926.

Ropp, Theodore. *War in the Modern World.* Rev. edition. New York: Collier Books, 1965.

Rose, J. Holland. *The Indecisiveness of Modern War, and other Essays.* Port Washington, New York: Kennikat Press, Inc., 1968.

Roskill, Captain S.W. (ed.). *Documents Relating to the Naval Air Service.* Vol. I: *1908-1918.* London: The Navy Records Society, 1969.

——. *Naval Policy Between the Wars.* Vol. I: *The Period of Anglo-American Antagonism, 1919-1929.* New York: Walker and Company, 1969.

Rumpf, Hans. *The Bombing of Germany.* Translated by Edward Fitzgerald. London. Frederick Muller Limited, 1961.

Salter, Lord. *Memoirs of a Public Servant.* London: Faber and Faber,

1961.
Saundby, Sir Robert. *Air Bombardment: The Story of Its Development.* London: Chatto & Windus, 1961.
Scott, Sir Harold. *Your Obedient Servant.* London: Andre Deutsch, 1959.
Scott, Sir Percy. *Fifty Years in the Royal Navy.* New York: George H. Doran Company, 1919.
Sellier, Henri, Bruggeman, A., and Poëte, Marcel. *Paris pendant la Guerre.* Paris: Les Presses Universitaires, 1926.
Seversky, Alexander P. de. *Victory through Air Power.* New York: Simon and Schuster, 1942.
Shepherd, E. Colston. *The Air Force of To-day.* Revised by H.T. Winter. Rev. edition. London: Blackie & Son Limited, 1942.
Shirer, William L. *The Collapse of the Third Republic: An Inquiry into the Fall of France in 1940.* New York: Simon and Schuster, 1969.
Simon, Viscount. *Retrospect: Memoirs.* London: Hutchinson, 1952.
Slessor, Sir John. *The Central Blue: Recollections and Reflections.* London: Cassell and Company Limited, 1956.
Snyder, William P. *The Politics of British Defense Policy, 1945-1962.* Columbus, Ohio: Ohio State University Press, 1964.
Somervell, D.C. *Modern Britain, 1870-1939.* London: Methuen & Co., Ltd., 1941.
Sommer, Dudley. *Haldane of Cloan: His Life and Times, 1856-1928.* London: George Allen & Unwin Ltd., 1960.
Spaight, J.M. *The Beginnings of Organised Air Power: A Historical Study.* London: Longmans, Green and Co., Ltd., 1927.
—. *Pseudo-Security.* London: Longmans, Green and Co., Ltd., 1928.
Spengler, Oswald. *Spengler Letters, 1913-1936.* Translated and edited by Arthur Helps. London: George Allen & Unwin Ltd., 1966.
Strang, Lord William. *Britain in World Affairs: The Fluctuation in Power and Influence from Henry VIII to Elizabeth II.* New York: Frederick A. Praeger. 1961.
Sueter, Murray F. *Airmen or Noahs: Fair Play for our Airmen; the Great 'Neon' Air Myth Exposed.* London: Sir Isaac Pitman & Sons, Ltd., 1928.
Swinton, Earl of. *Sixty Years of Power: Some Memories of the Men Who Wielded it.* New York: James H. Heineman, Inc., 1966.
Sykes, Sir F.H. *Aviation in Peace and War.* London: Edward Arnold & Co., 1922.
—. *From Many Angles: An Autobiography.* London: George G. Harrap & Company Ltd., 1942.

Taylor, A.J.P. *English History, 1914-1945.* Oxford: Oxford University Press, 1965.
Taylor, H.A. *Jix, Viscount Brentford: Being the Authoritative and Official Biography of the Rt. Hon. William Joynson-Hicks, First Viscount Brentford of Newick.* London: Stanley Paul & Co., Ltd., 1933.
Tedder, Lord. *Air Power in War.* London: Hodder and Stoughton, 1948.
Templewood, Viscount. *Empire of the Air: The Advent of the Air Age, 1922-1929.* London: Collins, 1957.
—. *Nine Troubled Years.* London: Collins, 1954.
Tennyson, Alfred, Lord. *Works: With Notes by the Author.* Edited by Hallam, Lord Tennyson. New York: The Macmillan Company, 1935.
Thompson, Laurence. *The Greatest Treason: The Untold Story of Munich.* New York: William Morrow & Company, Inc., 1968.
Thomson, Sir Basil. *The Scene Changes.* Garden City, New York: Doubleday, Doran & Company, Inc., 1937.
Thomson, Lord. *Air Facts and Problems.* New York: George H. Doran Company, 1927.
Tint, Herbert. *The Decline of French Patriotism, 1870-1940.* London: Weidenfeld and Nicolson, 1964.
Tirpitz, Alfred von. *My Memoirs.* 2 vols. New York: Dodd, Mead and Company, 1919.
Titmuss, Richard M. *Problems of Social Policy.* London: HMSO, 1950.
Townsend, Peter. *Duel of Eagles.* New York: Pocket Books, 1972.
Toynbee, Arnold J. *Survey of International Affairs, 1920-1923.* London: Oxford University Press, 1925.
Trenchard, Lord. *Air Power: Three Papers.* London: Air Ministry Pamphlet Number 229, 1946.
Trevelyan, George Macaulay. *Grey of Fallodon: The Life and Letters of Sir Edward Grey, afterwards Viscount Grey of Fallodon.* Boston: Houghton Mifflin Company, 1937.
Turner, C.C. *Britain's Air Peril: The Danger of Neglect, Together with Considerations on the Role of an Air Force.* London: Sir Isaac Pitman & Sons, Ltd., 1933.
Watt, D.C. *Personalities and Policies: Studies in the Formulation of British Foreign Policy in the Twentieth Century.* London: Longmans, 1965.
Webster, Sir Charles and Frankland, Noble. *The Strategic Air Offensive Against Germany, 1939-1945.* Vol. I: *Preparation* and Vol. IV: *Annexes and Appendices.* London: HMSO, 1961.
Weller, Jac. *Weapons and Tactics: Hastings to Berlin.* London: Nicholas

Vane, 1966.
Wells, H.G. *Anticipations of the Reaction of Mechanical and Scientific Progress upon Human Life and Thought.* New York: Harper & Brothers, 1902.
——. *First and Last Things: A Confession of Faith and a Rule of Life.* New York: G.P. Putnam's Sons, 1908.
——. *Journalism and Prophecy, 1893-1946.* Edited by W. Warren Wager. Boston: Houghton Mifflin, 1964.
——. *The War in the Air: And Particularly how Mr. Bert Smallways Fared While it Lasted.* London: George Bell and Sons, 1908.
——. *Washington and the Riddle of Peace.* New York: The Macmillan Company, 1922.
Wheeler-Bennett, John W. *John Anderson, Viscount Waverley.* London: Macmillan & Co., Ltd., 1962.
——. *King George VI: His Life and Reign.* New York: St Martin's Press, 1958.
Whitehouse, Arch. *Decisive Air Battles of the First World War.* New York: Duell, Sloan and Pearce, 1963.
——. *The Years of the Sky Kings.* Garden City, New York: Doubleday & Company Inc., 1959.
Wolfers, Arnold. *Britain and France between Two Wars: Conflicting Strategies of Peace from Versailles to World War II.* New York: The Norton Library, 1966.
Wood, Derek and Dempster, Derek. *The Narrow Margin: The Battle of Britain and the Rise of Air Power, 1930-40.* 2d edition. London: Hutchinson, 1963.
Wood, James Playsted. *I Told You So!: A Life of H.G. Wells.* New York: Pantheon Books, 1969.
Woodward, Sir Llewellyn. *Great Britain and the War of 1914-1918.* London: Methuen & Co., Ltd., 1967.
Woodward, E.L. and Butler, Rohan (eds.). *Documents on British Foreign Policy, 1919-1939.* Vols. I & II (First Series): *1919.* London: HMSO, 1947-1948.
Wrench, John Evelyn. *Struggle: 1914-1920.* London: Ivor Nicholson & Watson Limited, 1935.

Articles, Periodicals, Archives and other sources

Andrews, J.O. 'The Strategic Role of Air Forces'. *Journal of the Royal United Service Institution,* LXXVI (November, 1931).
Anon. 'It's a Bird, It's a Plane, It's the Riesenflugzeug of WWI', *Esquire,*

LXVI (October, 1966).

———. 'Les forces de l'air en Angleterre', *Bulletin Belge des Sciences Militaires,* II (September 1921).

Ashmore, E.B. 'Anti-Aircraft Defence', *Journal of the Royal United Service Institution,* LXXII (February, 1927).

Barclay, Sir Thomas. 'Aircraft Bombs and International Law', *The Nineteenth Century and After,* LXXVI (November, 1914).

Beloff, Max. 'LURSS et L'Europe', *L'Europe du XIXe et du XXe Siècle: Problèmes et Interprétations Historiques.* Vol. VI: *1914 Aujourd-hui.* Edited by Max Beloff et. al., Paris: Librairie Fischbacher Marzorati, 1964.

Bere, Rupert de la. 'The Bertrand Stewart Prize Essay, 1925', *The Army Quarterly,* X (July, 1925).

Bird, W.D. 'One Air Force or Three?' *The Army Quarterly,* V (January, 1923).

———. 'Some Speculations on Aerial Strategy', *The Army Quarterly,* IV (July, 1922).

———. 'Thoughts on our Requirements in Relation to our Air Force', *Journal of the Royal United Service Institution.* LXVIII (May, 1923).

Bowers, Ray L. 'The Peril of Misplaced Loyalties', *Air University Review,* XVII (May-June, 1966).

Boyd, J.H. 'A.A. Development in the Royal Engineers', *The Royal Engineers Journal,* XLIX (September, 1935).

Brooke-Popham, H.R.M. 'Air Warfare' in Sir George Aston (ed.), *The Study of War for Statesmen and Citizens: Lectures Delivered in the University of London during the Years 1925-26.* London: Longmans, Green and Co., Ltd., 1927.

Carlton, David. 'The Problem of Civil Aviation in British Disarmament Policy, 1919-1934', *Journal of the Royal United Service Institution,* CXI (November, 1966).

Carranza, Alberto Salinas. 'The Historic Flight Across the English Channel — July 25 1909', *The Air Power Historian,* VI (April, 1959).

Carsten, F.L. 'The Reichswehr and the Red Army, 1920-1933', *Survey,* No. 44/45 (October, 1962).

Chamier, J.A. 'Co-operation of Aircraft with Artillery', *The Army Quarterly,* IV (April, 1922).

———. 'Strategy and Air Strategy', *Journal of the Royal United Service Institution,* LXVI (November, 1921).

Charlton, L.E.O. 'The Great Unprepared', *The Fortnightly Review,* CXLIV (October, 1938).

Chenevix-Trench, R. 'Gold Medal (Military) Prize Essay for 1922',

Journal of the Royal United Service Institution, LXVIII (May, 1923).

Clark-Hall, R.H. 'The Value of Civil Aviation as a Reserve to the Royal Air Force in the Time of War', *Journal of the Royal United Service Institution,* LXIX (August, 1924).

Cooper, John C. 'Notes on Air Power in Time of Peace', *Air Affairs,* I (September, 1946).

The *Courier Journal* (Louisville, Kentucky).

The *Daily Mail* (London).

Desmet, Cap-Commandant aviateur. 'Rôle des aviations belge et française: sur le front occidental pendant la grande guerre', *Bulletin Belge des Sciences Militaires,* III (January and March, 1922).

Dickman, Joseph. L. 'Douhet and the Future', *Air University Quarterly Review,* II (Summer, 1948).

The Dictionary of National Biography.

Earle, Edward Mead. 'H.G. Wells, British Patriot in Search of a World State', *Nationalism and Internationalism: Essays Inscribed to Carlton J.H. Hayes.* Edited by Edward Mead Earle. New York: Columbia University Press, 1950.

——. 'The Influence of Air Power upon History', *The Yale Review,* XXXV (June, 1946).

Eddy, E.M. 'Britain and the Fear of Aerial Bombardment, 1935-1939', *Aerospace Historian,* XIII (Winter, 1966).

Edmonds, C.H.K. 'Air Strategy', *Journal of the Royal United Service Institution.* LXIX (May, 1924).

Embry, Sir Basil E. 'Tallyho: To the Aid of the RAF', *Air University Review,* XVIII (July-August, 1967).

Emme, Eugene M. 'The Impact of Air Power upon History', *Air University Quarterly Review,* II (Winter, 1948).

——. 'The Renaissance of German Air Power, 1919-1932', *Air Power Historian,* II (July, 1958).

Farman, T.F. 'Aeroplanes in War', *Blackwood's Magazine, CC (December, 1916).*

——. 'The American Aviation Program', *The Contemporary Review,* CXII (September, 1917).

Flammer, Philip M. 'Image of the Aces, A Writer's Bonanza', *Air University Review,* XVII (January-February, 1966).

Flugel, Raymond Richard. 'United States Air Power Doctrine: A Study of the Influence of William Mitchell and Giulio Douhet at the Air Corps Tactical School, 1921-1935'. Unpublished Ph.D. dissertation, University of Oklahoma, 1965.

Fredette, Raymond H. 'Bombers of the Black Cross: German

Bombardment Aviation in World War I', *The Airpower Historian*, VII (July, 1960).
——. 'First Gothas over London: The Story of a Daring Raid and Its Aftermath', *The Airpower Historian*, VIII (October, 1961).
Fuller, J.F.C. 'The Supremacy of Air Power', *The Royal Air Force Quarterly*, I (April, 1930).
Galland, Adolph. 'Defeat of the Luftwaffe: Fundamental Causes', *Air University Quarterly Review*, VI (Spring, 1953).
Garrod, A.G.B. 'Air Strategy', *The Royal Air Force Quarterly*, I (January, 1930).
Germains, V.W. 'The Army in War', *National Review*, CXII (June, 1939).
Gibbs, Norman H. 'Le rôle de la<<guerre totale>>dans les transformations subies par l'Europe', *L'Europe du XIXe et du XXe Siècle: Problèmes et Interprétations Historiques.* Vol. V: *1914-Aujourd'hui.* Edited by Max Beloff *et al.* Paris: Librairie Fischbacher Marzorati, 1964.
Grahame-White, Claude. 'Aircraft in the War', *The Fortnightly Review*, XCVI (October 1, 1914).
——. and Harper, Harry. 'The Dawn of the Air Age', *The Contemporary Review*, CXII (July, 1917).
——. 'Two Years of Aerial War', *The Fortnightly Review*, C (August 1, 1916).
——. 'Zeppelin Airships': Their Record in the War', *The Fortnightly Review*, XCVIII (1 September 1915).
Great Britain. Public Record Office. Cabinet records.
——. Public Record Office, Royal Air Force records.
Groves, P.R.C. 'For France to Answer', *The Atlantic Monthly*, CXXXIII (February, 1924).
Hartley, H. 'Chemical Warfare', *The Army Quarterly*, XIII (January, 1927).
Hearson, T.G. 'Balloon Barrages', *Journal of the Royal United Service Institution*, LXXXIII (February, 1938).
Hill, H.W. 'Air Defence', *Journal of the Royal United Service Institution*, LXXV (February, 1930).
Hogg, McA. 'Aeroplanes in Future Warfare', *The Army Quarterly*, IX (October, 1924).
House of Commons Debates.
House of Lords Debates.
Hoysted, D.M.F. 'A Plea for a Strong Anti-Aircraft Defence', *The Royal Engineers Journal*, XXXVIII (September 1924).
Hurd, Archibald. 'Great Ships or . . . ?': A Footnote to the *Times*

Correspondence', *The Fortnightly Review,* CIX (1 February 1921).
——. 'Is the Battleship Doomed?' *The Fortnightly Review,* CVII (2 February 1920).
Hurley, Alfred Francis. 'The Aeronautical Ideas of General William Mitchell'. Unpublished Ph.D. dissertation, Princeton University, 1961.
Jackson, Sir Louis C. 'Possibilities of the Next War', *Journal of the Royal United Service Institution,* LXV (February, 1920).
Jenkins, Rees. 'Civil Aspects of Air Defence', *Journal of the Royal United Service Institution,* LXXII (August, 1927).
Joll, James. 'Shapers of the Twentieth Century', *Manchester Guardian Weekly,* 14 March 1968.
Journal of the Royal United Service Institution.
Keate, R.H. 'Searchlights', *Journal of the Royal United Service Institution,* LXI (November, 1916).
Keogh, E.G. 'The Study of Military History', *Australian Army Journal,* No. 188 (January, 1965).
Lefebure, Victor, 'Chemical Warfare', *Journal of the Royal United Service Institution,* LXXIII (August, 1928).
Liddell Hart, B.H. 'Armament and its Future Use', *The Yale Review, XIX (June, 1930).*
——.' 'Colonel Bond's Criticisms: A Reply', *The Royal Engineers Journal,* XXXVI (November, 1922).
——. 'Is Gas a Better Defence than Atomic Weapons?' *Survival,* I (September-October, 1959).
Liversedge, A.J. 'Possibilities of the Large Airship', *The Fortnightly Review,* XCVIII (1 December 1915).
Loch, K.M. 'Anti-Aircraft Defence', *Journal of the Royal United Service Institution,* LXXXII (May, 1937).
——. 'Anti-Aircraft Ground Defences', *The Army Quarterly,* VIII (April, 1924).
Mackay, C.J. 'The Influence in the Future of Aircraft upon Problems of Imperial Defence', *Journal of the Royal United Service Institution,* LXVII (May, 1922).
MacNeece Foster, W.F. 'Air Power and its Application', *Journal of the Royal United Service Institution,* LXXIII (May, 1928).
——. 'Certain Aspects of Air Defence', *Journal of the Royal United Service Institution,* LXXI (February, 1926).
MacPherson, E.R. 'The Development of Chemical Warfare', *Journal of the Royal United Service Institution,* LXX (May, 1925).
The *Manchester Guardian.*
Mills, Sir George H. 'Bomber Command of the Royal Air Force', *Air*

University Quarterly Review, VII (Spring, 1955).
Mitchell, William. 'Air Power vs. Sea Power', *The American Review of Reviews*, LXIII (March, 1921).
—. 'Aviation over the Water', *The American Review of Reviews*, LXII (October, 1920).
—. 'Has the Airplane made the Battleship Obsolete?' *The World's Work*, XLI (April, 1921).
—. 'Our Army's Air Service', *The American Review of Reviews*, LXII (September, 1920).
The *New York Times*.
Newton, Wesley Phillips. 'The Role of the Army Air Arm in Latin America, 1922-1931', *Air University Review*, XVIII (September-October, 1967).
The *Observer* (London).
'OTAC'. 'The Offensive Side of Chemical Warfare', *The Army Quarterly*, XI (October, 1925).
Parliamentary Papers.
Powers, Barry D. 'Locarno: The Easement of the *Mésentente*'. Paper presented at the European Security in the Locarno Era Symposium. Mars Hill, N.C.: Mars Hill College, 1975.
Ranson, Harry H. 'Lord Trenchard, Architect of Air Power', *Air University Quarterly Review*, VIII (Summer, 1956).
Robinson, Douglas H. 'The Zeppelin Bomber: High Policy Guided by Wishful Thinking', *The Airpower Historian*, VIII (July, 1961).
Ropp, Theodore. 'A Theorist in Power', *Air University Review*, XVIII (March-April, 1967).
Saundby, Sir Robert. 'The Air Battle', *Journal of the Royal United Service Institution*, XCVIII (February, 1953).
Slessor, Sir John. 'Air Power and the Future of War', *Journal of the Royal United Service Institution*, XCIX (August, 1954).
—. 'The Development of the Royal Air Force', *Journal of the Royal United Service Institution*, LXXVI (May, 1931).
Smith, Dale O. 'The Morality of Retaliation', *Air University Quarterly Review*, VII (Winter, 1954-1955).
Smith, Frederick H., Jr. 'Current Practise in Air Defence: Principles and Problems', *Air University Quarterly Review*, VI (Spring, 1953).
Smith, Perry M. 'Douhet and Mitchell: Some Reappraisals', *Air University Review*, XVIII (September-October, 1967).
Sontag, Raymond J. 'Between the Wars', *Pacific Historical Review*, XXIX (February, 1960).
Stewart, Oliver. 'The Air Exercises', *The Army Quarterly*, XV (October,

1927). —, 'the Air Exercises', *The Army Quarterly,* XV (October, 1927).

—. 'The Air Exercises', *The Army Quarterly,* XVII (January, 1929).

—. 'The Air Exercises, 1930: Balance of Air Forces', *The Army Quarterly,* XXI (October, 1930).

—. 'The Doctrine of Strategical Bombing', *Journal of the Royal United Service Institution,* LXXXI (February, 1936).

—. 'The Royal Air Force Exercises of 1931', *The Army Quarterly,* XXIII (October, 1931).

Stoehr, C.F. 'The Bertrand Stewart Prize Essay, 1923', *The Army Quarterly,* VI (July, 1923).

Stone, F.G. 'Aeroplanes', *The Nineteenth Century and After,* LXXXVIII (July, 1920).

The *Sunday Times* (London).

Sutton, B.E. 'Some Aspects of the Work of the Royal Air Force with the BEF in 1918', *Journal of the Royal United Service Institution,* LXVII (May, 1922).

Thorne, C.B. 'The Defence of the Civil Population', *The Army Quarterly, XXI* (October, 1930).

The Times (London).

Tomlinson, William H. 'The Father of Airpower Doctrine', *Military Review,* XLVI (September, 1966).

Trenchard, Sir H.M. 'Aspects of Service Aviation', *The Army Quarterly,* II (April, 1921).

Turner, C.C. 'The Aerial Defence of Cities: Some Lessons of the Air Exercises, 1928', *Journal of the Royal United Service Institution,* LXXIII (November, 1928).

—. 'British and Foreign Air Exercises of 1931', *Journal of the Royal United Service Institution,* LXXVI (November, 1931).

Villiers-Stuart, J.P. 'The Nation in Relation to its Armed Forces', *Journal of the Royal United Service Institution,* LXXI (August, 1926).

Walser, A.A. 'The Influence of Aircraft on Problems of Imperial Defence', *The Army Quarterly,* V (October, 1922).

Warner, Edward P. 'Air Forces', *Proceedings of the Academy of Political Science,* XII (May, 1926).

Webster, Nesta. 'Bolshevism and Secret Societies', *Journal of the Royal United Service Institution,* LXVII (February, 1922).

Wells, W.T. 'Air Raid Precautions and the Public', *The Fortnightly Review,* CXLIII (February, 1938).

Wheeler, Gerald E. 'Mitchell, Moffett, and Air Power', *The Airpower Historian,* VIII (April, 1961).

Williams, W.T.S. 'Air Exercises, 1927', *Journal of the Royal United Service Institution*, LXXII (November, 1927).

Worthington, P. 'Work of the Kite Balloon on Land and Sea', *Journal of the Royal United Service Institution*, LXV (August, 1920).

Wyatt, H.F. 'Air Raids and the New War', *The Nineteenth Century and After*, LXXXII (July, 1917).

—. 'The Air War and the Bishops' Religion', *The Nineteenth Century and After*, LXXXII (November, 1917).

—. 'Convocation versus the Church and the Bible', *The Nineteenth Century and After*, LXXXII (August, 1917).

Yool, W.M. 'Air Exercises 1930', *Journal of the Royal United Service Institution*, LXXV (November, 1930).

INDEX

Acland, F.D., 145
Admiralty, the, 15-16, 18-20, 28-9, 36, 39, 41, 43-4, 46-7, 68, 79-81, 84, 86, 90, 93, 96, 137, 153, 166, 174-5, 179-81, 184, 187
Aerocraft, 23
Afghanistan, 139-170-1
Air Defence of Great Britain Command, 16, 149, 154, 186-90, 196-7, 199
Air defence responsibility, 15-16, 28, 36, 41-9, 69, 170, 180 1, 186, 190
Air exercises, 196-9, 202-3
Air Ministry, 42-6, 90, 96-8, 101, 110, 120, 122, 123, 131, 134-6, 139, 141, 144, 146-7, 152, 160, 162-4, 166, 168-70, 172, 174, 178, 182, 184, 186-7, 190-1
Air raid casualties and damage, 50-5, 58, 122-3
Air Raid Precautions Committee, 61, 120-4, 126, 189, 191-2, 196
Air supremacy, 24, 45, 86, 91, 105, 108, 112-14, 193-4, 200-1
Alexandra, Queen, 27, 41
Amery, Leo, 187
Anderson, Sir John (Viscount Waverley), 120-1, 124, 126-7
Andler, Charles, 76
Andrews, J.O., 199-200
Anti-aircraft ammunition, 18, 31, 34-5, 42, 49-50, 66, 206
Anti aircraft gunnery, 15-19, 25, 30-2, 42, 49, 66, 69-71, 73, 145, 156, 170, 189-90, 193, 199-202
Appeasement, 145, 189, 206-7, 209
Area bombardment, 130-1, 191
Army Quarterly, 200-4
Ashley, Wilfred, 111
Ashmore, Edward B., 35, 69-73, 115, 117, 189-90
Asquith, Lady Cynthia, 58
Asquith, H.H. (Early of Oxford and Asquith), 44, 48, 61, 81
Attlee, Clement R. (Earl Attlee), 149

Baird, John L., 46, 110
Baker Newton D., 87, 105

Baldwin, Stanley (Earl Baldwin), 136-7, 186-7, 206
Balfour, Arthur James (Earl Balfour), 41, 43, 47, 87, 115, 137, 180-3, 185, 187
Balfour, Harold, 154
Balloon air defence, 66, 71, 194
Baltimore, 177
Banbury, Sir Frederick, 56-7
Barclay, Sir Thomas, 78
Barès, Commander, 80
Barnes, H., 140, 152
Barrage gunnery defence, 31, 69-71, 73
Bartholomew, W.H., 190
Bath, 60
Battle of Britain, 10, 16, 37, 120, 172, 188, 190-1, 200, 208-9
Battleships and air warfare, 26, 109, 138-42, 154, 174-7, 179
Baumont, Maurice, 125
Beatty, Earl, 168, 178-9, 182-3, 186-7
Beckett, J., 152
Behncke, Paul, 12, 14
Belfort, 79-80
Belgium, 30, 77
Bell, Alexander Graham, 114
Bellairs, Carlyon W., 142, 148, 152, 154
Bellamy, A., 153
Beloff, Max, 125
Bere, Rupert de la, 202
Beresford, Lord, 82
Berlin, 76, 82-3, 158-9
Bernard, V.H.G., 195
Bertie of Thame, Lord, 53
Bethman-Hollweg, Theobold von, 12, 54-5, 66
Birch, Sir Noel, 198
Bird, W.D., 200-1
Birkenhead, Earl of, 189
Birmingham, 43, 45
Blackett, P.M.S., 208-9
Blériot, Louis, 109-10
Blumenfeld, Ralph David, 27, 85
'Boffins', 30, 208
Bonham Carter, Lady Violet, 16
Bournemouth, 60
Boyle, Andrew, 48, 168
Brancker, Sir Sefton, 24, 45-8, 80,

289

136, 163, 175
Brandenburg, Ernst, 55
Brass, W., 149
Briand, Aristide, 185
Brighton, 60
British Expeditionary Force, 15, 63, 65 80, 104-5
British Union of Fascists, 141
Brodie, Bernard, 177
Brooke-Popham, H.R.M., 66, 190, 192-3, 205
Brookes, Warwick, 42
Buchan, John (Lord Tweedsmuir), 152
Budget considerations 110, 136, 138-55, 158-62, 164-5, 168-9, 173-5, 179, 182-3, 189, 191-2
Burgoyne A.H., 140
Burns, E.L.M., 92

Cabinet the, 20-1, 43, 45-8, 50, 56-7, 61-5, 69, 84-6, 89-90, 95-9, 103, 115-17, 134, 137, 159, 161-3, 165, 168, 171, 182-3, 185
Cairo Conference, 171-3
Calais, 11, 64, 73, 109
Carr, E.H., 13-14
Cave, Sir George, 57-9
CBW, 113-14, 134-5, 143, 151, 156-7, 193, 198, 202
Cecil, Lord Hugh, 133, 140, 143-4
Cecil, Lord Robert, 48, 133
Chamberlain, Sir Austen, 48, 183
Chamberlain, Neville, 45, 168
Chamier, J.A., 193-4, 200
Chanak crisis, 185
Charlton, L.E.O., 65, 172
Chenevix-Trench, R., 194-5
Churchill, Sir Winston 15-17, 19-21, 25, 28-30, 34, 41, 44, 50, 60-1, 79, 86, 110, 121, 125, 134, 138-43, 148, 158-9, 162-4, 166-71, 180, 189, 207
Civil defence 120-4, 156, 191-2, 196, 198
Clausewitz, Karl von, 14
Clynes, J.R., 147
Coastal Command, 178
Coffin, Howard, 88
Colonial Office, 140, 171, 180
Commercial aircraft and air war, 114 20, 130, 132, 136, 139-43, 145-6, 148-51, 161-2, 170, 180
Committee of Imperial Defence 46, 120-1, 124, 137, 169, 179-82, 184-6, 208
Communist Party (of Great Britain), 147, 153
Conservative Party, 43, 133, 137, 141, 145-6, 151, 154-5
Contemporary Review, 87
Cooper, Alfred Duff (Viscount Norwich), 27, 150-2
Cowdray, Viscount, 48-9, 81, 94, 97-8, 101
Crawford, Earl of, 95
Cuffley, 49
Curtis, Glenn, 175
Curzon of Kedleston, Earl, 45-8, 56, 80-1, 94, 184-5, 187

Daily Express, 160
Daily Mail, 21, 23, 55, 62, 75, 82, 87, 98, 101, 109, 111
Daily Telegraph, 111, 127, 138
Dalton, Hugh, 150-3
Daniel, Josephus, 176
Darwin, Erasmus, 107
Davidson, Randall Thomas, 77
Davies, A. Vernon, 152
Davies, David, 132-3
Deeds, E.A., 88
Defence of the Realm Acts, 37-8
Defence-offence ambiguity, 9-10, 16-17, 100, 103, 106, 126, 130, 135-7, 139, 142, 144, 146, 149, 154, 161, 164, 167, 177, 179, 188-9, 194-9, 202, 205-6, 208
Demobilization, 138
Derby, Earl of, 44-5, 56-7, 90, 93-4, 118
Deterrence, see defence-offence ambiguity.
Deutsche Lufthansa, 119
Devonshire, Lord, 187
Disarmament, 117, 147-53, 155
Divine, David, 92-4
Dornier aircraft, 118
Douglas, Sholto, 176, 186
Donhet, Giulio, 26, 176-8
Dover, 31, 109
Dowding, Hugh (Lord Dowding of Bentley Priory), 172
Dunkirk, 28-31
Düsseldorf, 28

Earle, Edward Mead, 81
Early warning systems, 18-19, 35-7, 39, 49, 55-8, 66, 68, 72-3, 121,

130, 190, 192, 197-8, 200, 208
Eden, Sir Anthony (Earl of Avon), 121, 146
Edison, Thomas A., 147
Edmonds, C.H K., 195
Emme, Eugene M., 119
Essen, 75, 78-9
Etherton, P.T., 204-5
Ethiopia, 206
Evacuation, 74, 121-4, 192, 198-9, 203
Evère, 30

Ferté, Joubert de la, 69
Fighter airplane defense, 19, 32-4, 49-50, 63-5, 67-72, 132, 142, 144, 146, 149-50, 156, 170, 188-90, 193-5, 197-200, 203-5, 208
Fighter Command, 172
Fisher of Kilverstone, Lord, 20-1, 85, 138, 140, 175, 205
Fitzroy, Sir Almeric, 101
Foch, Ferdinand, 104-5, 127, 129, 131, 135, 147
Focke-Wulf Company, 118
Folkestone, 53, 56, 63
Foreign Office, 115
Fortnightly Review, 79, 108, 175
"Forward air defence", 16-17, 23-5, 27-32, 64, 75, 79, 82, 99
Foster, H.F.M., 152
France, 11, 13, 15, 17, 28-9, 33, 64, 78-81, 87-8, 103-5, 109, 115, 130, 141, 143-4, 150, 154-6, 158, 183-9, 191, 194
Frankland, Noble, 105
Frederick William, Prince, 78
Freiburg, 13
Friedrichshafen, 28, 79
French, Lord, 42, 56, 63-5, 96
"Frightfulness", 15, 76-7, 145
Fuller, J.F.C., 204

Garro-Jones, G.M., 148-9
Garrod, A.G.B., 204
Gas warfare, 39-40, 77, 106, 109, 113-14, 116, 121 127-30, 132, 134-6, 139, 143, 145, 147-8, 151-3, 157-8, 177, 192, 194-8, 201-3, 205
Geddes, Sir Eric, 96, 124, 182-3, 185
Geneva, 132, 184
George V, 184
George VI, 121
Germany, 11-15, 17, 20, 29, 35, 40, 50, 52-6, 58, 63, 66-7, 71-2, 74, 78, 80, 82, 88, 91, 97-9, 103-5, 111, 115, 117-20, 130, 139-40, 142, 145, 150, 153, 155-6, 158-60, 166-7, 184-5, 189, 193, 205-6
Glyn, R., 139-40, 163
Göring, Hermann, 119
Gotha bombers, 49, 52-75, 79, 81-2, 85-6, 89-90, 93-6, 98-100, 103, 106, 177
Grahame-White, Claude, 112-14, 175
Grey of Fallodon, Viscount, 50, 137
Groves, P.R.C., 115, 129-31, 141-2, 151, 153, 185
Guest, Frederick E., 141-2, 149-50, 153, 182

Haeften, General von, 156
Hague Peace Conferences, 77-8, 135
Haig, Sir Douglas (Earl Haig), 42, 63-5, 86, 89, 96, 101-2, 160
Haldane of Cloan, Viscount, 137, 188
Halsbury, Lord, 153
Hancock, W.K., 93
Hankey, Sir Maurice (Baron Hankey), 137
Hansard, 25, 138
Hardie, G.D., 146
Hardinge, C., 53
Harmsworth, Cecil, 76
Harper, Harry, 112-14
Harris, Sir Arthur, 91-2
Hartley, H., 202
Havilland, Geoffrey de, 33
Hay, Ian, 137
Heinkel aircraft, 117
Henderson, Sir David, 46, 68
Henderson, T., 148
Hersch, Liebmann, 156
Higham, Robin, 129, 166
Hill, A.V., 30
Hill, H.W., 198
Hills, J.W., 143
Hindenburg, Paul von, 53-5, 66
Hoare, Sir Samuel (Viscount Templewood), 100, 110, 117, 129, 134, 137, 144-5, 148-9, 154-5, 165, 178, 186, 189, 206
Hobhouse, Emily, 71
Hoeppner, Ernest Wilhelm von, 32, 35, 50
Holocaust prophecies, 26, 108-9, 113-4, 123-4, 126, 131-8, 142-53, 155-57, 179, 181, 193-205, 208

Home Office, 39-40, 43, 120, 138
Horsnaill, W.O., 51
House of Commons, 24-5, 42-3, 46, 56, 59, 62, 75-6, 79, 84, 98-100, 102, 133, 136, 138-55, 163
House of Lords, 46, 63, 77, 87, 94-5
Howard, Michael, 169
Howard-Bury, C.K., 146
Hudson, James H., 147, 149, 153
Hughes, Charles Evans, 116
Hull, 21-2, 61
Hurd, Sir Archibald, 175

Imperial air control, 139, 154, 161, 165, 170-4, 194
Imperial Defence College, 187
Incendiaries, 39-41, 106, 108-9, 133-4, 130, 132, 148, 152, 157, 177, 193, 197, 205
Independence for air operations, 25, 27, 42-5, 90, 93-5, 99, 106, 138, 140-1, 158, 160, 162-4, 166-7, 174, 179-80, 183, 187, 201
Independent Air Force, 102-6, 129, 158-60, 165-6, 179
India, 170-1
Insurance for air raid damage, 41
Inter-Parliamentary Union, 155
Interservice rivalry, 43-4, 47, 93-5, 100, 137-43, 160, 166, 169, 173-4, 178-9, 182, 186-7, 200-1
Iraq, 139, 171-3
Ironside, Sir Edmund, 206
Italy, 104, 111, 154, 176, 178, 206

Jackson, Sir Louis C., 193
Japan, 117, 132
Jellicoe, Sir John, 20
Jenkins, Roy, 48
Joergersen, Joerg, 156
Joffre, Marshal Joseph, 87
Joll, James, 125
Jones, H.A., 70
Jones, Morgan, 148
Jones, Thomas, 206
Journal of the Royal United Service Institution, 117-18, 129, 192-200, 202
Joynson-Hicks, William (Viscount Brentford of Newick), 25, 42, 62-3, 75-6, 79, 82, 84, 111, 138, 141-3
Junkers aircraft, 117-19

Karlsruhe, 78, 80

Kennedy, John F., 126
Lenworthy, J.M., 117, 120, 140, 146, 148-51, 153
Kerr, Mark, 97-9, 160
Kerr, Philip (Lord Lothian), 185
Kesselring, Albert, 120
Kingston-McCloughry, E.J., 22, 39
Kitchener, Lord, 15, 78-9

Labour Party, 134, 139, 141-2, 144-7, 150-1, 154-5, 172, 185, 189
Lambert, G., 138, 163
Lana de Terzi, Count Francesco, 107
Lanchester, Frederick William, 76
Lansbury, George, 147
Law, Andrew Bonar, 43, 56, 96, 98-9, 136, 163, 185-6
Lawrence, A. Susan, 150-1
Lawrence, T.E., 172
Leach, W., 146
League of Nations, 130, 132-4, 140, 143, 149-53, 155, 171
Lee, Arthur (Lord Lee of Fareham), 111
Lees-Smith, H.B., 150-2
Lefebure, Victor, 198
Legal restrictions on air war, 76-8, 112, 114, 166-7, 126, 132, 135, 144-6, 148-9, 151, 154, 157, 193, 195, 198
Lenin, Vladimir I, 125
Lewis, Cecil, 65
Liberal Party, 147
Liddell Hart, B.H., 21, 126-9, 133, 177-8
Linnarz, Karl, 30
Liverpool, 20, 199
Locarno Conference, 119, 149, 189
Lloyd George, David (Earl Lloyd George), 21, 44, 47-8, 59-61, 68-9, 85, 94-6, 98-9, 101, 103, 106, 112, 117-8, 158-9, 162-3, 168, 180, 182, 184-5
Loch, K.M., 66
London, 12-15, 17-21, 23-4, 26, 29, 31-2, 35-6, 38, 40, 42, 49, 52-62, 64-71, 73-5, 78, 80-4, 90-1, 98-101, 109-10, 122-4, 128, 131, 137, 144, 146-8, 150-1, 153-4, 170, 180, 190-1, 193, 197, 199, 202-6, 209
London Air Defence Area, 69-70, 72-3, 190
Londonderry, Marquess of, 164

Long, Walter, 168
Lowther, Claude, 84
Ludwigshafen, 80
Luftwaffe, 119
Lukacs, John, 155
Luneville, 11
Luxeuil, 80-81
Lynch, A.A., 79

Mackay, C.J., 194
MacNeece Foster, W.F., 196-8
MacPherson, E.R., 195
Maidenhead, 60
Malone, C. L'Estrange, 142-3, 153
Manchester, 197, 199
Manchester Guardian, 120
Manual of Military Law, 75
Mannheim, 86, 89
March, Peyton, 105
Marwick, Arthur, 125
Massey, Vincent, 209
Maurice, Sir Frederick, 137, 205
Mayer, André, 156
Messerschmitt aircraft, 118
Metz, 11
Meyler, H.M., 144
Milch, Erhard, 119
Ministry of Health, 191-2
Mitchell, William ("Billy"), 26, 116, 141-2, 176-7
Montagu of Beaulieu, Lord, 44-5, 82
Montague, Frederick, 154
Moore-Brabazon, J.T.C., 139, 142-4, 146, 153, 163
Morality questionings, 13-15, 25, 75-7, 83, 89, 99, 105, 128, 131, 172
Morning Post, 59, 111
Mosley, Sir Oswald, 141, 144-6, 163
Mouillard, Louis P., 109
Munich, 206-7, 209

Nancy, 80, 103
Neuve-Chapelle, 40
New York City, 109, 177
New York Herald, 88
New York Sun, 52
New York Times, 88, 177
Newall, Cyril L.N., 103
Nineteenth Century and After, 76, 85
Norman, Montagu, 182
Northcliffe, Lord, 23, 44, 89, 101, 111, 118
Nuclear warfare, 10, 78, 92, 124

Ochey, 103
Operational research, 30-1, 40-1, 146, 148, 165, 170, 192, 202, 208-9
Ostend, 31, 52-3
"OTAC", 202

Pacifism, 142-3, 146-7, 150-1
Panic and air war, 12, 14, 21-22, 56-7, 61-2, 108, 112, 123-6, 128, 135, 156, 195-8, 202-3, 206, 208-9
Paris, 11, 74, 78, 151, 157, 185
Paris: or the Future of War, 127-9
Parliamentary Air Committee, 25
Passchendaele, 65
Passive air defence, 24, 35, 37-9, 49, 195, 197, 199-200, 203, 208
Pearce, Sir William, 57, 62
Peary, Robert, 88
Peel, Viscount, 187, 193
Pemberton Billing, Noel, 22-6, 28, 45, 59, 75-6, 84, 140
Philadelphia, 177
Pile, Sir Frederick, 35, 70
Poincaré, Raymond 185-6
Portugal, 77
Ponsonby, Arthur A., 150
Possony, Stefan T., 14, 169
Pratt, J. Davidson, 198
Prophetic visions about air power, 107-9

Radar, 31, 35, 188, 190, 198, 200, 208
Raleigh, Sir Walter, 110
Rapallo, Treaty of 119
Raper, A.P., 140-1
Rawlinson, Alfred, 18, 110
"Red Scare", 125-6
Reid, A.S.C., 148-50, 153
Religion and air war, 24, 75-6, 83, 85
Renny, F.G., 149-50
Repington, C. à Court, 59, 102, 185
Reprisal aerial policy, 13, 21, 25, 43-4, 55, 62, 74-8, 81-6, 89, 95, 98-9, 102-3, 128, 130-1, 133, 137, 142-3, 145-6, 153, 194, 198-9, 201-2
Revolution via air war, 108, 126, 156, 197, 202
Ribot, Alexandre F., 87
Richardson, Sir Philip, 146
Ridell, Lord, 48, 63

Robertson, Sir William, 62-5, 96, 205
Rolleston, Sir John, 23
Ropp, Theodore, 177
Rose, J. Holland, 133
Rothermere, Lord, 101-2
Royal Air Force, 16, 22, 44, 49, 93-5, 97-104, 106, 117, 121, 124, 126, 128, 134-5, 137-55, 158-207
Royal Air Force Quarterly, 204-5
Royal Aircraft Factory, 33
Royal Artillery, 201-2
Royal Engineers, 202
Royal Flying Corps, 15, 21, 43, 46, 49, 63, 68, 80-1, 86, 93, 96, 100, 103, 111, 129, 174
Royal Naval Air Service, 15, 17, 19, 23, 28, 34, 46, 63, 79-81, 86, 93, 100, 103, 119, 160, 174, 183
Royal United Service Institution, 37, 117, 125, 192-200, 206
Ruhr occupation, 185-6
Rumpf, Hans, 92
Russia, 78, 119-20, 125

Saarbrücken, 103
Saklatvala, S., 147
Salisbury, Marquis of, 137, 186-8
Salmond, Sir John, 190
Salter, Lord, 120
Samson, C.R., 29
Samuel, Herbert (Viscount Samuel), 43
Sassoon, Sir Philip, 151-4
Saundby, Sir Robert, 92
Schlieffen Plan, 11
Scott, Percy, 18, 41-2, 139, 175
Searchlights, 16, 19, 32, 49-50, 66, 72, 170, 190
Seely, John E.B., 30, 138-40, 142-44, 146, 163-4
Senex, Lord, 82
Sheerness, 53
Shelters, 58-61, 66, 113, 148, 192
Shepherd, A.L., 150
Siegert, Wilhelm, 11, 52
Simon, Sir John (Viscount Simon), 145
Sinclair, Sir Archibald, 148
Slessor, Sir John, 173, 177, 199, 206-7
Smith, Rennie, 149-52
Smuts, Jan Christian, 68-70, 83, 90-7, 99, 185
Snowden, Philip, 147

Somaliland, 139, 171
Sontag, Raymond J., 125
Southend, 134
Spaight, James Moloney, 131-2
Spain, 103
Spengler, Oswald, 186
Sperrle, Hugo, 120
Squier, George O., 88
Steel-Bartholomew Plan, 189-90
Stewart, Oliver, 202-4
Stoehr, C.F., 201
Strachie, Lord, 63
Strasser, Peter, 20
Strategic bombing, 11-13, 44, 46, 54-5, 65-6, 78-81, 85, 90-2, 94-5, 97-100, 102-6, 108-9, 113-14, 123, 128-36, 139-61, 164-7, 173, 176-81, 183-4, 188, 191, 193-208
Stresemann, Gustav, 119, 186
Stumpff, Hans-Juergen, 120
"Substitution" service policy, 140-3, 153-4, 173-4, 179, 183, 201
Sueter, Murray, 16, 28, 119-20, 129, 142, 149, 152-3
Supermarine Aviation Works, 23
Swinton, Earl of, 68
Sydenham, Lord, 46, 56-7, 87
Sykes, Frederick H., 104, 115-6, 129, 135-6, 144, 159-62, 164, 168

Tactical air operations, 15, 43-4, 91, 99, 102, 104-5, 144, 165-7, 182, 200-01
Tank development, 29
Taylor, A.J.P., 22, 47
Tedder, Lord, 92
Temps, Le, 185
Ten Year Rule, 168-9, 189
Tennant, H.J., 43
Tennyson, Alfred Lord, 107
Tenterden, Lord, 82
Territorial Army, 124-5, 209
Thomson, Lord, 134-5
Thomson, Sir Basil, 53
Thorne, C.B., 203
Thurtle, E., 147, 149
Times (London), 24-5, 45, 55, 59, 62-3, 82-3, 98-101, 111, 125, 127, 130, 175, 205
Tirpitz, Adm. Alfred von, 12, 14-15
Titmuss, Richard M., 122
Transjordan, 172

Trenchard, Hugh Montagu (Lord Trenchard), 47-8, 64, 80-1, 85, 92, 96, 101-5, 116-17, 128, 131, 134, 145, 158, 160, 164-9, 171-2, 174, 176-80, 182-4, 186-8, 193-4, 199-200, 204-5
Tryon, G.C., 140
Turkey, 111, 115, 174, 185
Turner, C.C., 198-9

United States of America, 26, 77, 82, 86-9, 104-6, 114, 118, 154, 175-8, 208
United Wards Club, 26-7

Venice, 71
Versailles Conference, 114-5, 118, 130, 180, 184-5
Victoria, Princess, 27
Victory through air power, 12, 14, 20, 25-6, 55, 78-80, 84, 87-8, 91-2, 95, 98, 101, 113-4, 116, 123-4, 128, 132, 135, 138, 148, 150-51, 160-61, 176-8, 193-6, 198, 200-3, 205, 207
Villiers-Stuart, J.P., 196
Viviani, René, 87
Vyvyan, A.V., 178

Walser, A.A., 201
Warneford, Reginald A.J., 34
Warner, Edward P., 117
Washington Conference, 109, 116, 144, 184
Washington D.C., 87
Watson, Sir William, 85
War Office, 15, 34, 36, 39, 42-3, 46, 49, 63, 65, 67, 70, 84, 86, 90, 96, 125, 138-9, 162-4, 166, 170, 179, 181, 190-1
Weather effects on air war, 19, 74, 105, 152, 203
Webster, Nesta, 125
Wedgwood, Josiah C., 142, 149
Wedgwood Benn, W., 139-40, 142, 144, 153, 163
Weir, Sir William (Lord Weir of Eastwood), 33, 67, 81, 94, 104, 160, 162, 164, 187
Wellock, W., 152
Wells, H.G., 107-9, 114
Westminster Gazette, 101
What Would be the Character of a New War?, 155-7

Wilhelm II, Kaiser, 12-13, 15, 78
"Will to war" (popular morale), 12, 14, 55-6, 61-2, 81, 126-31, 135, 139, 144, 156, 191, 194-8, 200-5
Wilson, Sir Henry, 143, 172-3, 182-3
Wilson, Woodrow, 88
Woker, Gertrud Johanna, 157
Women Voluntary Patrollers, 22
Woolwich, 32
Work stoppages, 22, 39, 56-7, 60 1, 193
World War, First, 9-109, 111-14, 123, 127, 130, 133-5, 137-40, 144, 147, 158-9, 171, 177, 179, 190, 193-4, 206
World War, Second, 13-14, 16-17, 20, 31, 35, 37, 40-1, 58, 60, 70-4, 91-2, 105, 117-18, 120, 123, 139, 166-7, 173, 208-9
Wrench, John Evelyn, 22, 50, 102
Wright Brothers, 109
Wyatt, Harold F., 76, 85

Yarmouth, 14
Ypres, 40

Zeppelin, Count Ferdinand von, 12
Zeppelin era, 11-53, 61-4, 68, 70, 72-3, 75-6, 78-9, 85, 112, 177

For Product Safety Concerns and Information please contact our EU
representative GPSR@taylorandfrancis.com
Taylor & Francis Verlag GmbH, Kaufingerstraße 24, 80331 München, Germany

www.ingramcontent.com/pod-product-compliance
Lightning Source LLC
Chambersburg PA
CBHW070554300426
44113CB00010B/1253